Second Edition

Automotive Suspension and Steering

A Text-Lab Manual

Sheldon L. Abbott
Former Automotive Technology Instructor
Mount San Jacinto College

Ivan D. Hinerman
Automotive Technology Instructor
Mount San Jacinto College

Glencoe Publishing Co., Inc.
Encino, California
Collier Macmillan Publishers
London

Glencoe Automotive Technology Series

Automotive Brakes: A Text-Lab Manual
Automotive Power Trains: A Text-Lab Manual
Automotive Engine Repair: A Text-Lab Manual
Automotive Transmissions: A Text-Lab Manual

Printed in the United States of America

Glencoe Publishing Co., Inc.
17337 Ventura Boulevard
Encino, California 91316
Collier Macmillan Canada, Ltd.

Library of Congress Catalog Card Number: 81-84578
1 2 3 4 5 6 7 8 9 10 86 85 84 83 82
ISBN 0-02-810350-5

CONTENTS

Preface to the Second Edition

The wide acceptance of this text in its first edition might have led the authors to believe that no major changes would be necessary. However, there have been the inevitable changes within the automotive technology field, some of which have been caused primarily by the increasing cost of energy for transportation. Among these changes are a trend toward wider use of MacPherson strut suspension systems and front-wheel-drive power trains. Also, the import vehicle market has expanded while the domestic vehicle market has declined.

The automotive technician is challenged today to be aware of these newest design changes in suspension and steering systems and to perform competent service work on the job. This second edition of *Automotive Suspension and Steering* is a response to the above-mentioned trends.

About This Text

Automotive Suspension and Steering, Second Edition, is designed to meet the training needs of the automotive suspension system specialist of today. The text-lab manual assists the automotive instructor in presenting a balance of fundamental information and current service practices at various levels of training. This text, like others in the Glencoe Automotive Technology Series, will serve equally well in a program of individualized instruction or in traditional automotive training at the secondary and postsecondary level. The material has been used successfully at Mt. San Jacinto College, San Jacinto, California, in an open-entry, open-exit, individualized instruction program. This text-lab manual is written for the vocational automotive course where the emphasis is on skill development.

This text covers the design, structure, diagnosis, and repair of suspension and steering systems. Concepts and procedures are explained in simple, everyday language, with technical vocabulary defined at first point of use. The broad use of photographs and drawings help clarify each concept and procedure to increase the student's understanding of the text material. General and specific safety hazards are discussed whenever the student has the greatest need for learning about safe procedures.

This text-lab manual is organized in unit form. Each unit begins with a short statement concerning the major topics of discussion. This is followed by a written test objective that spells out exactly what the student needs to achieve.

The objectives are labeled with a unit number and a letter indicating position in the unit; for example: 3-A. Below the objective are the text pages that the student needs to study. This is the preparation assignment. The preparation assignment provides all the general and specific information required to fully complete the objective.

A manipulative objective or objectives follow the text material. These are also labeled with the unit number, location within the unit, and a number to maintain consecutive order of completion. The manipulative objectives are stated in specific performance terms. Step-by-step procedures of a general nature follow the manipulative objective with specific directions for completing a service operation typical of those performed by today's automotive technician.

At the end of each unit or at the end of each major division within a unit is a written test that covers material and service procedures from the written and manipulative objective found earlier in each unit. There are a variety of methods and approaches in the teaching of an automotive program. Each instructor has individual preferences on assignments to be completed. This text includes the majority of assignment preferences. It is the author's experience that the most usable learning will take place when this text is used as a guide in completing the lab assignments and as a reference when completing the tests and reading assignments. Unlike many other automotive texts, *Automotive Suspension and Steering* reflects practical teaching experience at the community college level whose object is to provide the most relevant materials and the most conducive climate for students to gain the skills necessary to be competent suspension and steering technicians. It is with these same goals that this text is presented.

Sheldon L. Abbott
Ivan D. Hinerman

ACKNOWLEDGMENT

The authors wish to acknowledge with sincere appreciation the efforts and contributions of those companies whose names appear in the credit lines of each illustration.

A special thanks is offered to the authors' families, whose encouragement and patience made this book possible.

unit I

Wheel Assembly

The material in this section provides you, the student, with the basic information and techniques needed to service front wheel assemblies of American and foreign made automobiles. Emphasis is on types of bearings, what causes bearing problems, proper lubrication, tires, and wheel balancing. Commercial service procedures are stressed throughout the text.

OBJECTIVE I-A (written test)

After completing the preparation assignment, you will correctly answer a minimum of seven out of ten multiple choice questions in a written test, without reference material, on the following wheel, hub and drum assembly topics:

1. Ball bearings
2. Roller bearings
3. Front wheel grease seals
4. Spindle
5. Hub assembly
6. Wheel bearing lubricant
7. Bearing problems
8. Torque specifications

Preparation:

Study pages 1 through 24.

Front wheel assemblies have commonly been constructed in the past with two types of wheel bearings: (1) the ball bearing and (2) the roller bearing. In recent years most auto makers have switched exclusively to the use of roller bearings. Roller bearings can carry greater loads (weight) and stress (pressure) than similar sized ball bearings without becoming deformed. Modern automobiles have had tendencies toward weighing more, increasing the load, and going faster, which increases the stress. Thus, roller bearings last longer and prove more reliable than ball bearings in modern automobiles. For repair purposes, the hub, drum, and spindle assembly are the same on both ball bearing and roller bearing equipped cars.

BALL BEARING

The front wheel ball bearing has a one piece, stamped sheet metal cage that serves as a retainer and a separator (Fig. I-1). This design allows lubricating grease to be stored in chambers on each side of the ball bearing.

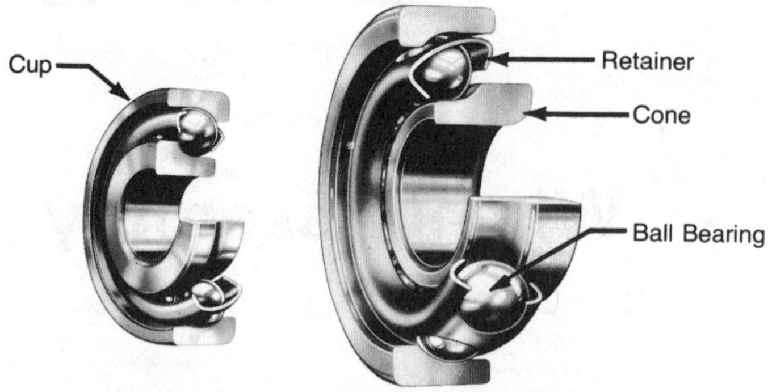

FIGURE I-1 Cutaway view of a typical front wheel ball bearing showing the cups, retainer, balls, and cones. (Courtesy of New Departure)

Both the cup and the cone are individual units. When the cup is pressed into the wheel hub, the cone is free to slip over the spindle (Fig. I-2).

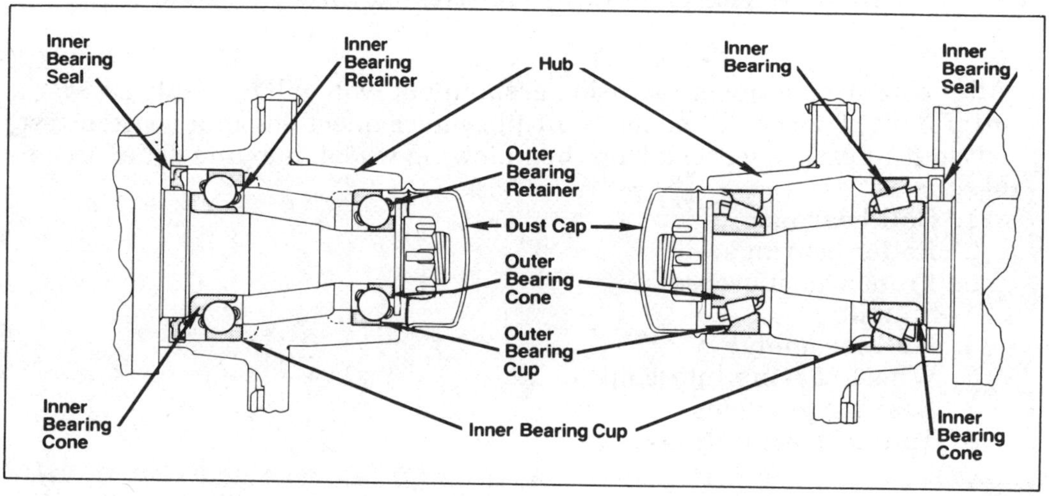

FIGURE I-2 Comparison of ball and roller bearing assemblys as they are mounted in the hub.

ROLLER BEARING

The front wheel roller bearing has a one piece stamped cage that acts both as a retainer and a separator. The cage holds the rollers against the cone and provides an area between the rollers for storage of lubricating grease.

The bearing assembly consists of two units: (1) the cup which is pressed into the wheel hub and (2) the cone and roller assembly which slides over the spindle (Fig. I-3). The roller bearing cone must be free to creep on the spindle to allow a constantly changing load contact between the cone, cup, and the roller bearings (see Fig. I-2).

FIGURE I-3 Roller Bearings are used for front wheel bearings in most modern automobiles. They require no preload as did the ball bearing. (Courtesy of New Departure)

FRONT WHEEL GREASE SEALS

The oil or grease seal is a precision part which seals out dirt and seals in the grease or oil. Only the grease seal used with front wheel bearings will be discussed in this unit. The front wheel grease seal is usually of a much simpler design than the high pressure oil seals used on engine and transmission parts (Fig. I-4).

The cutaway of the seal shown in Fig. I-4 is a typical front wheel grease seal consisting of: (1) an outer metal shell, (2) the sealing member, which is usually a synthetic material such as Buna rubber, and (3) the garter spring used to hold the sealing member against the spindle. Some seals will not have a garter spring. Other types of seals may have two sealing lips: a large one for retaining the grease and a smaller one for keeping out dust and other contaminants.

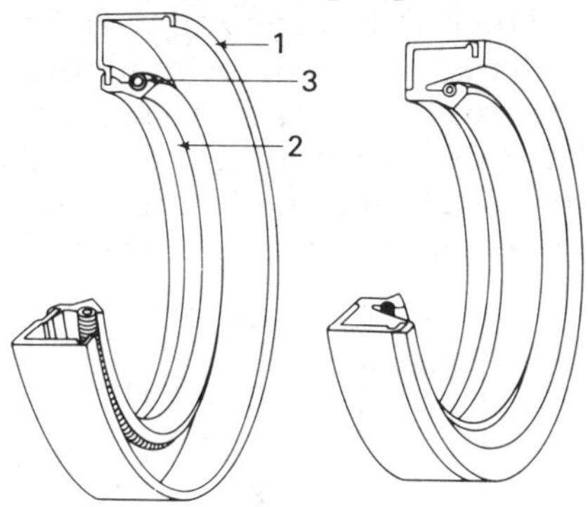

FIGURE I-4 Cutaway of front wheel grease seals. When installing seals, remember that the lip always points toward the section to be sealed. (Courtesy of Chicago Rawhide Mfg. Co.)

SPINDLE

When ball joints are used as part of the steering assembly, the spindles and steering knuckles are cast in one piece. The one piece design allows the wheel assembly to move up and down or turn, and the spindle provides the base for wheel assembly mounting and a surface for both bearing and seal placement (Fig. I-5).

FIGURE I-5 Typical front hub and drum assembly as mounted on the spindle. (Courtesy of Cadillac Motor Car Division

The cone of the bearing does not spin on the spindle, but in some cases it must be allowed to creep. Thus, the bearing placement area must be free of any burrs or roughness, and the spindle should be polished and lubricated to permit the cone to creep. The back part of the spindle is the sealing surface and also must be free of any nicks or burrs.

The end of the spindle has a slot for the washer tab and is threaded to take a spindle nut. Near the end of the spindle there is at least one hole for a cotter pin to lock the spindle nut in place.

HUB ASSEMBLY

The hub is a casting with a cup pressed into it for both the inner and outer bearing (Fig. I-6). A machined area at the back of the hub is for mounting the front-wheel grease seal. The machined area at the front of the hub holds the dust cap in place. The flange around the outside diameter of the hub is for mounting of: (1) the brake drum and (2) the wheel assembly. Lugs for wheel mounting are pressed through holes evenly spaced around this flange.

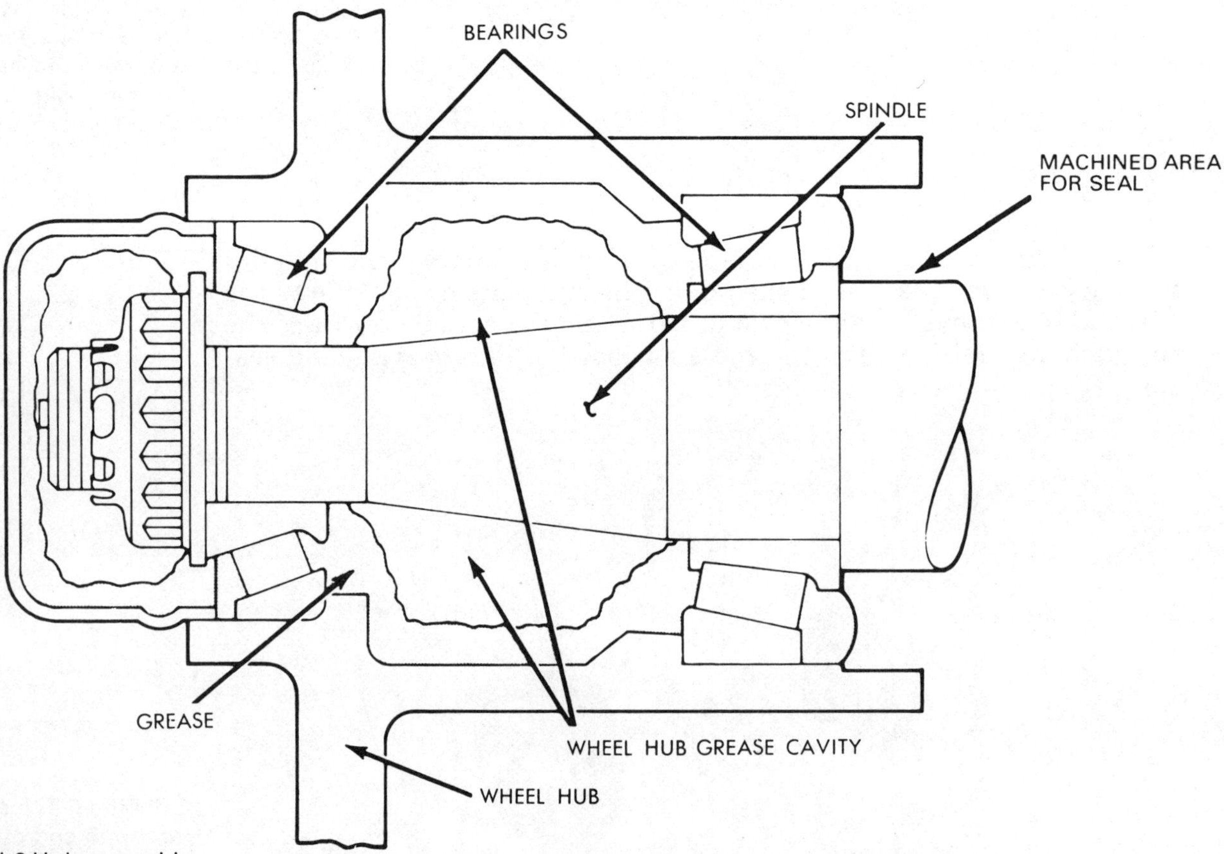

FIGURE I-6 Hub assembly with inner and outer bearing cups pressed into each end. (Courtesy of Chrysler Motors Corp.)

WHEEL BEARING LUBRICANT

A major improvement in automotive technology has been the development of better lubricating greases and oils. Not too long ago, wheel bearings required repacking with grease at 5,000 or 10,000 mile (8,050–16,100 km) intervals or after excessive wetting of the wheel assemblies from driving in flooded conditions. Modern greases are much more resistant to heat, pressure, and water. Most new car manufacturers recommend that wheels be packed with modern grease every 20,000 to 24,000 miles (32,200–38,640 km) or with each brake job.

The automobile manufacturer generally recommends that a specific type of grease be used for each model or make of car. As can be seen from Table I-1 on wheel greases, they may look the same, but the ingredients of greases vary markedly. *Use of the proper grease should extend the life of parts and improve the operation of bearing surfaces.*

TABLE I-1 WHEEL BEARING LUBRICANTS

Type of Grease	*Base Used in Grease*	*Auto Manufacturer Specifications*
Union MP	Lithium 12-hydroxystearate, containing molybdenum disulfide	American Motors, Ford, General Motors #6031-M
Valvoline DB	Aluminum complex soap	General Motors #1051195
Valvoline Special Moly EP	Lithium 12-hydroxystearate, containing molybdenum disultide	Ford Motor Co., #MIC 75B
Valvoline Val-lith No. 2 EP	Lithium 12-hydroxystearate	General Motors, #4733 and B-31-M; Chrysler Corp., #MS-3701
Valvoline Multilube Lithium	With MLGI No. 1 consistency	General Motors #6040M Ford Motor Co. #MIC57A

Obviously there is no single wheel bearing grease that meets the requirements of all manufacturers. It is, however, necessary to purchase a grease which will fit as many specifications as possible.

Each automotive manufacturer lists specification numbers for recommended wheel bearing greases in his repair and owner's manual. Likewise, each grease manufacturer lists the automotive manufacturer's specifications, which his greases will meet, in his catalog.

DIAGNOSING BEARING PROBLEMS

A. ROAD TEST

The road test is the simplest and yet most analytic tool available to the *experienced* automotive technician. Used correctly, and in connection with other diagnostic procedures, the road test is one of the speediest ways to pinpoint the car's mechanical problems.

1. Preparation for road test
   a. Check tires for irregular wear.
   b. Check tire pressure.

2. Road test procedure
   a. Drive until the engine is at normal operating temperature.
   b. Test at various speeds and conditions.
   c. At slow speeds, turn steering wheel to develop left and right hand motions (traffic permitting). Bad front wheel bearings are indicated by a noise change due to shifting loads while cornering.

B. ON THE CAR TESTS
1. Raise the car on a jack or lift, and spin the tire by hand to verify bearing roughness at the wheel. A good bearing will turn freely without a grating sound.
2. With one hand on the top of the tire and the other on the bottom, alternately pull with one hand and push with the other to check for excessive play in the bearing. A good bearing will be tight with little or no play.

C. BENCH DIAGNOSTIC PROCEDURE

The final decision to replace a bearing or other component part should be made at this point. The following factors should be considered:
1. Note general condition of all parts during teardown and examinations.
2. Classify the failure with the aid of the illustrations on pages 7–13.
3. Correct the cause of failure once you identify the problem and prevent a repeat failure.
4. Make all repairs following recommended procedures.

Table I-2 lists the do's and don't's when handling bearings.

TABLE I-2 BEFORE HANDLING BEARINGS

Things to Remember	*Things to Avoid*
1. Remove all outside dirt from housing before exposing bearing.	1. Don't work in dirty surroundings.
2. Treat a used bearing as carefully as you would a new would a new one.	2. Don't use dirty, brittle, or chipped tools.
3. Work with clean tools in clean surroundings.	3. Don't use wooden mallets or work on wooden bench tops.
4. Handle bearings with clean, dry hands or clean canvas gloves.	4. Don't handle bearings with dirty, moist hands.
5. Use clean solvents and flushing oils.	5. Don't use gasoline as a cleaner. It is a dangerous and explosive liquid.
6. Lay bearings out on clean paper.	6. Don't spin uncleaned bearings.
7. Protect disassembled bearings from rust and dirt.	7. Don't spin bearings with compressed air.
8. Use clean rags to wipe bearings.	8. Don't use cotton waste or dirty cloths to wipe bearings.
9. Keep bearings wrapped in oilproof paper when not in use.	9. Don't expose bearings to rust or dirt at any time.
10. Clean inside of housing before replacing bearing.	10. Don't scratch or nick the bearing surfaces.

The following list of defects is accompanied by a description of the proper procedure to follow in correcting the problem. Most of the defects can be detected either visually, by sound, or by touch. *With experience and practice, the automotive technician becomes an expert at spotting and correcting bearing problems.*

WEAR

Wear on the cone, the cup, and the surface the bearing works against may be checked visually (Fig. I-7). A pattern of discoloration or roughness caused by fine abrasives will be obvious when comparing the worn area against metal that does not make actual contact with the bearings (or a part of the roller bearing that does not touch the race).

When making the inspection, clean all parts and housings. Check seals and replace bearings if rough or noisy.

FIGURE I-7 Bearing wear. (Courtesy of Chevrolet Motor Division)

STEP WEAR

Wear on roller ends caused by fine abrasives (Fig. I-8).

Clean all parts and housing. Check seals and replace bearings if rough or noisy.

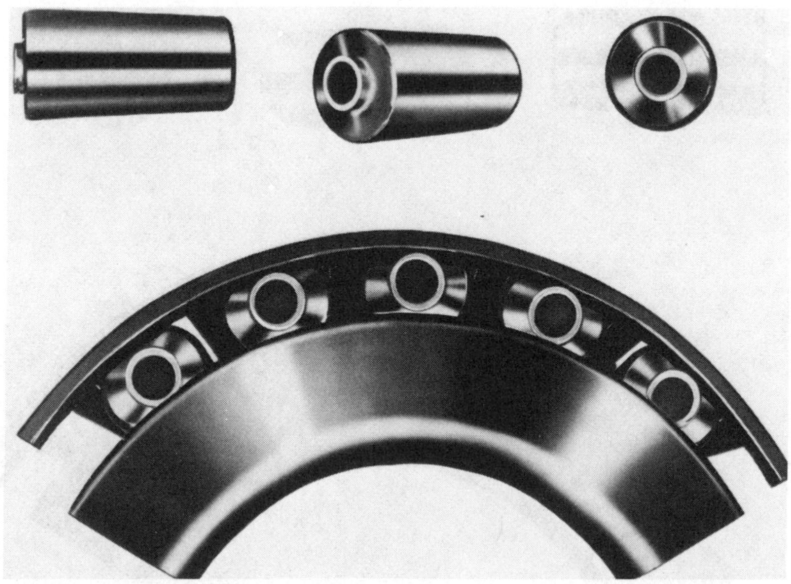

FIGURE I-8 Step wear. (Courtesy of Chevrolet Motor Division)

INDENTATIONS

Surface depressions on race and rollers are caused by hard particles of foreign materials (Fig. I-9).

Clean all parts and housings. Check seals and replace bearings if rough or noisy.

FIGURE I-9 Indentations. (Courtesy of Chevrolet Motor Division)

GALLING

Metal smears on roller ends are due to overheating caused by lubricant failure or overload (Fig. I-10).

Replace bearing—check seals and check for proper lubrication.

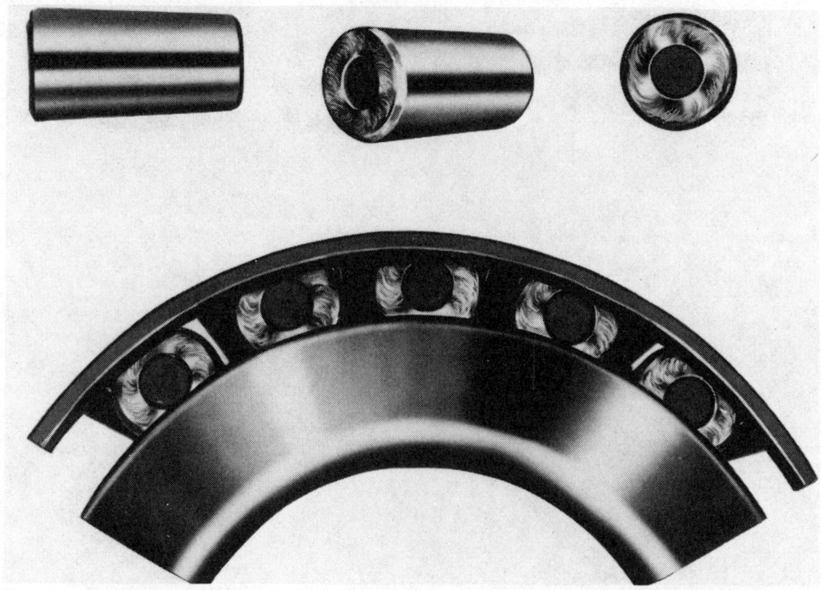

FIGURE I-10 Galling. (Courtesy of Chevrolet Motor Division)

ETCHING

Bearing surfaces appear gray or grayish black in color with related etching away of material usually at roller spacing (Fig. I-11).

Replace bearings—check seals and check bearings for proper lubrication.

FIGURE I-11 Etching. (Courtesy of Chevrolet Motor Division)

CAGE WEAR

Wear around outside diameter of cage and roller pockets is caused by abrasive material and inefficient lubrication (Fig. I-12).

Clean related parts and housing. Check seals and replace bearings.

FIGURE I-12 Cage wear. (Courtesy of Chevrolet Motor Division)

FATIGUE SPALLING

Flaking of surface metal results from fatigue (Fig. I-13).

Replace bearing—clean all related parts

FIGURE I-13 Fatigue spalling. (Courtesy of Chevrolet Motor Division)

STAIN DISCOLORATION

Discoloration ranging from light brown to black is caused by lubricant breakdown or moisture (Fig. I-14).

Reuse bearings if stains can be removed by light polishing or if no evidence of overheating is observed.

Check seals and related parts for damage.

FIGURE I-14 Stain discoloration. (Courtesy of Chevrolet Motor Division)

HEAT DISCOLORATION

Heat discoloration can range from faint yellow to dark blue caused by an overload or lubricant breakdown (Fig. I-15).

Excessive heat can cause softening of cup, cone, or rollers.

To check for loss of temper (hardness) on cup, cone, or rollers, make a simple file test. A file drawn over a damaged temper will grab and cut metal. A file drawn over a hard part will glide readily with no metal cutting.

Replace bearings if overheating damage is indicated. Check seals and other parts.

FIGURE I-15 Heat discoloration. (Courtesy of Chevrolet Motor Division)

BRINELLING

Surface indentations in raceway are caused by rollers, either under impact loading or vibration, while the bearing is not rotating (Fig. I-16).

Replace bearing if rough or noisy.

FIGURE I-16 Brinelling. (Courtesy of Chevrolet Motor Division)

BENT CAGE

Cage damage is due to improper handling (Figs. I-17 and I-18). Replace bearing.

FIGURE I-17 Bent cage. (Courtesy of Chevrolet Motor Division)

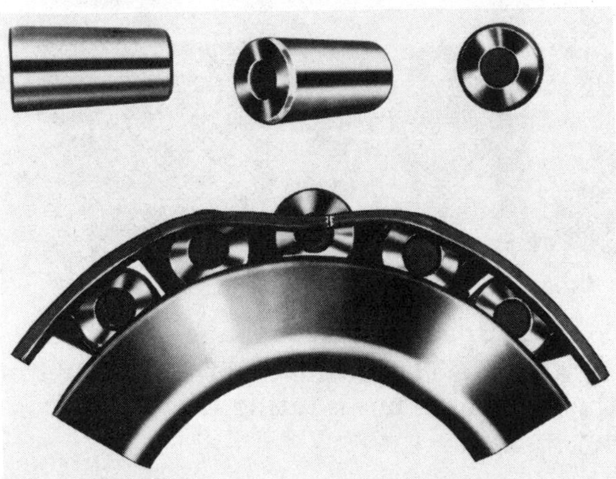

FIGURE I-18 Bent cage. (Courtesy of Chevrolet Motor Division)

MISALIGNMENT

This is generally caused by an improperly seated bearing (Fig. I-19). Replace bearing and make sure races are properly seated.

FIGURE I-19 Misalignment. (Courtesy of Chevrolet Motor Division)

CRACKED CONE RACE

Races crack because of improper fit, cocking, or poor bearing seats (Fig. I-20).

Replace bearing and correct bearing seats.

(This failure is more common in rear axle applications.)

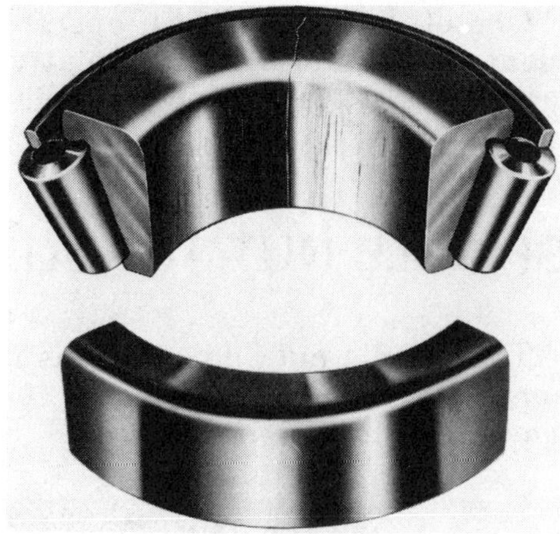

FIGURE I-20 Cracked inner race (cone). (Courtesy of Chevrolet Motor Division)

FRETTAGE

Corrosion is caused by a small relative movement of the cup and/or cone with no coating of oil to protect them (Fig. I-21).

Replace bearing. Clean related parts. Check seals and check for proper lubrication.

SMEARS

Smearing of metal is due to slippage. Slippage can be caused by poor fits, lack of lubrication, overheating, overloads, or handling damage (Fig. I-22).

Replace bearings, clean related parts, and check for proper fits and lubrication.

(This failure is more common in rear axle applications.)

FIGURE I-21 Frettage. (Courtesy of Chevrolet Motor Division)

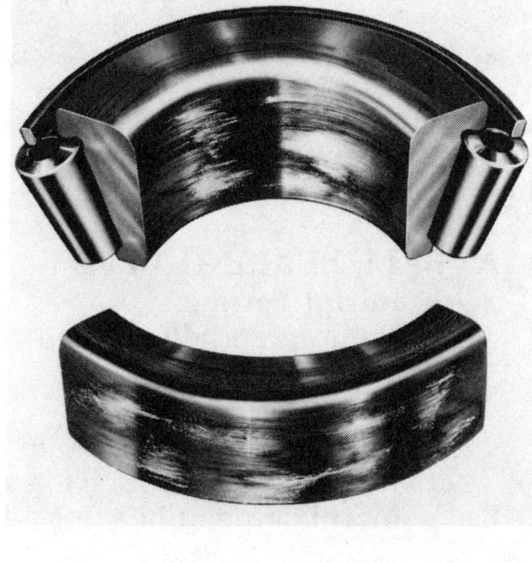

FIGURE I-22 Smears. (Courtesy of Chevrolet Motor Division)

REASSEMBLY

In the reassembly of the wheel components, great care should be taken not to create a problem that didn't exist before the repair work was undertaken. It should be noted that cage damage, misalignment, inner race cracking, and smearing can often be directly attributed to improper assembly in the first place.

One of the most critical operations in reassembly is spindle nut adjustment (or torque). Although relatively simple to perform properly, a mistake is also easy to make and could have serious consequences.

SPINDLE NUT TORQUE

The spindle nut adjustment is a very critical operation (Fig. I-23). *If not done properly, it may result in damage to the bearings or even loss of the wheel assembly at high speeds.*

FIGURE I-23 Always adjust the spindle nut with a torque wrench.

Manufacturer's torque specifications and procedures should always be followed. However, if for some reason the specifications are not available, the following are the general service procedures.

A. BALL BEARING TYPE
1. Initial Torque:
   Tighten spindle nut to 25 ft-lbs. or 33.9 N-m *while rotating the wheel by hand.*
2. Final Setting:
   Back off spindle nut one-half to one turn and retighten to 11 to 12 ft-lbs. or 16.2 N-m. This will give the approximate pre-load needed by this type of bearing.

B. ROLLER BEARING TYPE

1. Initial Torque:
   The torque required on the spindle nut depends on the bearing size. Measure the hub bore size for the outer bearing.
   a. If the hub bore is 1¾ inches (44.4 mm) or more, tighten the spindle nut to 210 inch-lbs. (17.5 ft-lbs.) or 23.7 N-m while rotating the wheel by hand.
   b. If the outer hub bore is less than 1¾ inches (44.4 mm), tighten the spindle nut to 96 inch-lbs. (8 ft-lbs.) or 10.8 N-m while rotating the wheel by hand.
2. Final Setting:
   Reverse the spindle nut 1/6 to 1/4 turn to find the cotter pin alignment. Look for .001 to .005 inch (.08 to 1.3 mm) end play.

WHEEL LUG-NUT TORQUE

Improper wheel lug nut torque will cause strain and distortion of brake drums, brakes, and the threads of the lug studs and nuts (Fig. I-24). The wheel lug nuts cannot be properly tightened using an impact gun.

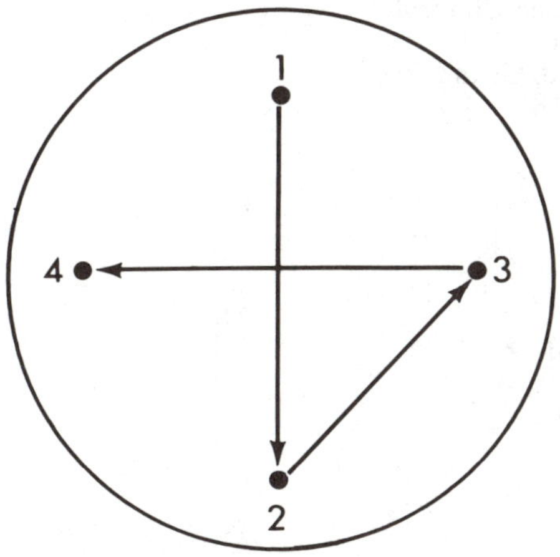

FIGURE I-24 To prevent drum or disc warpage, use the star pattern and a torque wrench when tightening lug nuts.

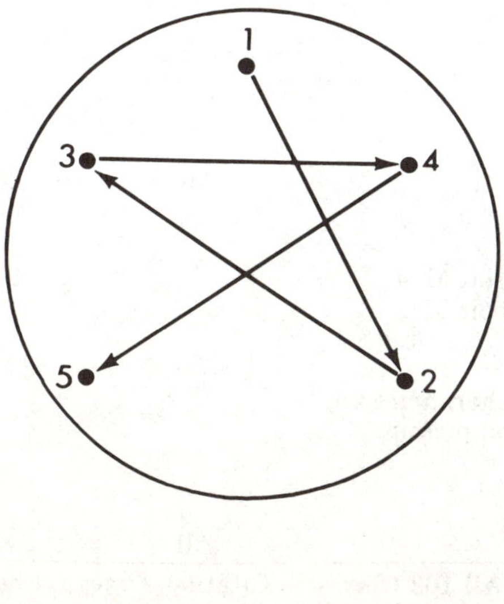

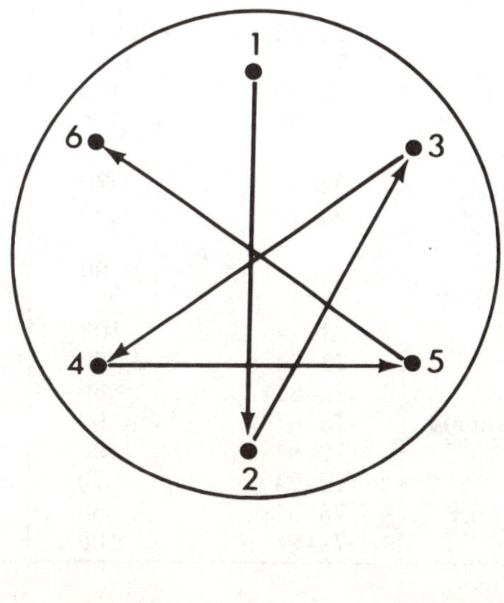

Nearly all car manufacturers now have the wheel lug nut torque requirements listed in their torque specifications. These specifications should be followed to minimize the chance of strain or distortion of the wheel and brake assembly.

All specifications are based on the use of clean, lubricated threads on both lug studs and nuts. Most manufacturers recommend the use of a star pattern while tightening the lug nuts. An example of the star pattern is shown in Figure I-24.

TABLE I-3 WHEEL LUG-NUT TORQUE SPECIFICATIONS
(ft/lbs. unless specified)

American Motors	*Year*	*Torque*
All	72–81	75
Jeep	72–81	80
Chrysler Corp.		
All 5 lug	72–75	65
5 lug	76–81	85
4 lug	72–81	55
4 lug (alum.)	72–81	65
Cricket	All	55
Omni-Horizon	All	80
Aries, Reliant	All	80
Ford Motor Co.		
All 4,5 lug (except below)		80–105
Capri, Cortina, Fiesta conical nut		75
Capri, Cortina, Fiesta flat, alum.		90
Escort, Lynx	All	80
General Motors		
Buick (all)	73–74	70
Electra, LaSabre, Wagons, Riviera, Skylark	75–81	100
All other Buicks	75–81	80
Opel	69–75	65–75
Opel	76–79	50
Riviera	80–81	140
X*, Skylark	80–81	103
Cadillac	72–76	80–100
Seville	76	60–80
El Dorado	72–80	130
Cadillac	77–81	100
Chevrolet		
Chevette, LUV	All	70
Monza, Vega (except alum.)	All	80
Monza, Vega (alum. wheels)		90
X*, Citation	All	103
Camaro, Firebird	72–74	70
Camaro, Firebird	75–81	80
Alum or special wheels	75–81	90
Corvette	72–81	90
Malibu, Chevelle	72–74	70
Monte Carlo	75–81	80
Chevrolet full-size	72–81	100

American Motors	*Year*	*Torque*
Oldsmobile		
Starfire (see Chevrolet Monza)		
88, 98, Cruiser	All	100
Cutlass	All	80
X*, Omega	80–81	103
Toronado	72–81	130–140
Pontiac		
Sunbird (see Chevrolet Monza)		
Phoenix*	80–81	103
Firebird (see Chevrolet Camaro)		
Bonneville, Grand Prix, Catalina		75
Datsun		
All	72–81	65
Honda		
600	All	32
Civic, Accord	All	58
Prelude	All	80
MG, Morris	All	42
Mazda		
Pickup	All	60
Others	All	70
Mercedes		
600	All	122
All others		72
Toyota		
All	All	66–86
VW		
Type 1,3,4 bolt M14		87–94
Type 1,3,4 bolt M12		72
Type 2 bolt		94
Type 2 nut		87–101
Rabbit, Dasher, Scirocco, Jetta, and pickup		65
Vanagon		123
Volvo		
All	All	72–94

*GMX body: All 103 (Skylark, Citation, Omega, Phoenix)

The impact wrench is used by most mechanics to speed up a job; properly used, it can be a very effective tool. While it may be used for disassembly the authors of this text and the major automotive manufacturers do not recommend its use for tightening wheel lug nuts.

OBJECTIVE I-A1, A2, (manipulative)

Working alone, you will repack the front wheel bearings on two instructor-assigned automobiles, one with drum-type brakes and one with disc-type brakes.

Follow the procedure and specifications given in the manufacturer's or general repair manual for the automobiles assigned to you. If inspection shows the bearing cups to be defective, they should be replaced.

Time recommended: Twice the flat-rate time allowed in the Motors or Chilton Flat Rate manual.

Preparation:

1. Review material for Objective I-A, pages 1 through 16.
2. Study pages 17 through 24.
3. Study manufacturer's or general repair manual procedures for the specific automobile you are assigned.
4. Carefully observe instructor's demonstration on wheel packing procedures.

Materials and Tools Needed:

1. Bearing packer with proper grease
2. Hub cap and dust cap remover
3. Lug wrench
4. Diagonal pliers
5. Ball peen hammer
6. Slide hammer puller or pry bar
7. Channel lock pliers
8. Air nozzle
9. Proper size sockets and ratchet for spindle nut (size dependent on spindle nut size and type of disc brake caliper)
10. Torque wrench
11. Hand towels or wiping rags
12. New cotter pins (select proper diameter and length)
13. Reference material for specific automobile
14. New grease seals
15. Jack and jack (safety) stands

REPACKING FRONT WHEEL BEARINGS ON DRUM-TYPE BRAKES

It is always good practice in any repair work to consult the manufacturer's repair manual and use those procedures and torque specifications. However, with most automobiles, the following basic procedure is generally sound:

1. RAISE AUTOMOBILE

Lift both front wheels off the floor and then place jack (safety) stands under the car frame. (**Caution:** *Place stands so the car will not slip or fall.*)

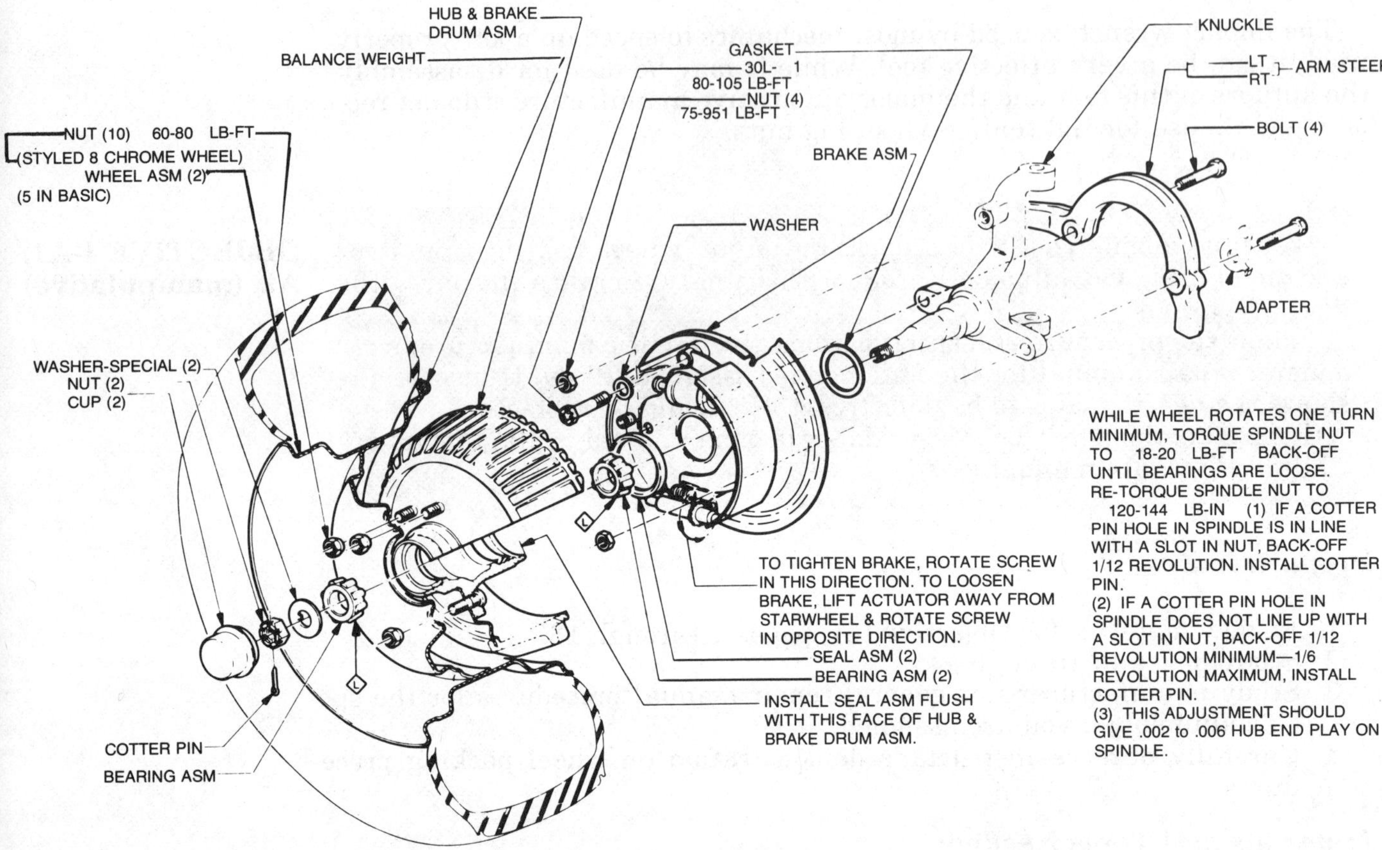

FIGURE I-25 Typical exploded view of wheel hub and drum assembly as they are mounted on the spindle. (Courtesy of Buick Motor Division)

2. REMOVE WHEELS AND HUBS

Removal of the wheel and tire assembly will make the hub and drum assembly easier to handle and reduce the chance of damage to brakes and bearing parts (Fig. I-25). Place wheel lugs or nuts in hub caps for safekeeping.

To prepare the hub and drum assembly for removal, (1) remove the wheel assembly cotter pin from the spindle and (2) remove the spindle nut and washer.

The outer wheel bearing may now be removed. Be sure to place all bearings and wheel hub and drum assembly parts on a clean shop towel inside the hub cap for safekeeping.

Remove the hub and drum assembly from the spindle by easing the drum off the spindle. If the drum does not move freely, it may be necessary to back off the brake adjusting star by using a brake spoon. In the case of self-adjusting brakes, a long, thin screwdriver will be needed to push the self-adjusting lever away from the star adjustor (Fig. I-26).

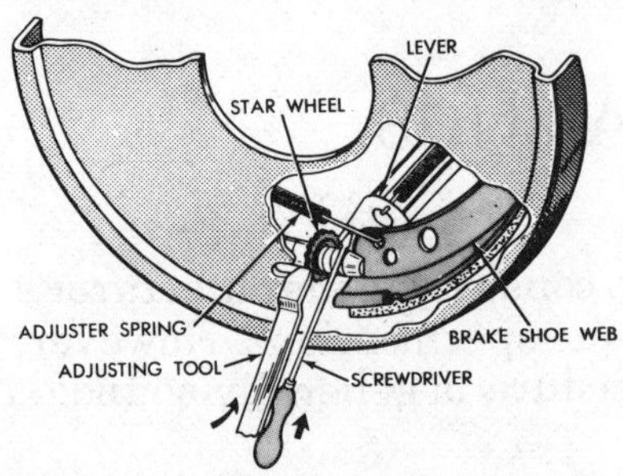

FIGURE I-26 A thin screw driver is used to push or lift the self-adjusting lever away from the star wheel. The star wheel can then be turned to loosen the brakes. (Courtesy of Buick Motor Division)

Use an approved vacuum cleaner or washer to clean the brake assembly. All brake dust and dirt must be removed. Do not blow this dust into the air, because it may result in an accumulation of harmful dust in the technician's lungs.

3. REMOVE INNER BEARING AND SEAL FROM HUB

There are many ways to remove the seal, and the amount of care you take will depend on whether you intend to replace or reuse it. It is recommended that whenever possible new seals be installed each time the bearings are repacked. The seal can be removed by using either a three-prong slide-hammer puller, or pry bar (Fig. I-27). Remove the inner bearing from that hub. Never discard the seal until you have compared it with the replacement seal.

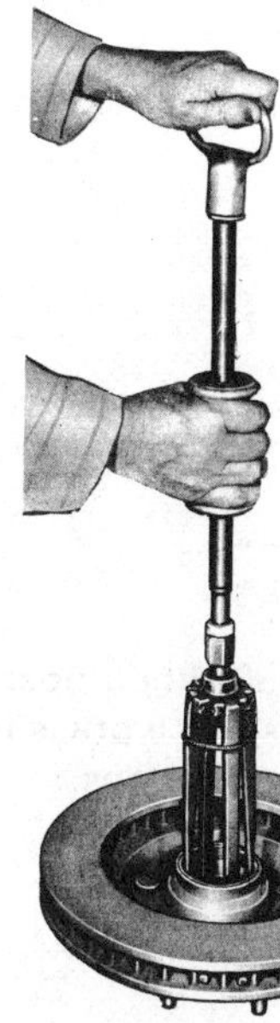

FIGURE I-27 Grease seals should be replaced when front wheel bearings are packed. Therefore, a screw driver or a pry bar or puller can be used to remove the seal. Discard all used seals. (Courtesy of Ford Customer Service Division)

4. CLEAN AND INSPECT BEARINGS AND HUB

Clean the hub by washing it in solvent or by using a steam cleaner. Make certain that all dust, dirt, and old grease is removed. Inspect the bearing and the old grease for metal particles before washing. If no metal particles or defects are present, the bearing is good and can be reused. The bearings must be cleaned before reuse, and, the best method is to wash them in cleaning solvent (Fig. I-28). *Do not use gasoline for cleaning parts.* After washing, wrap the bearing in a clean, dry towel, insert an air nozzle into an opening in the folds of the towel, and blow the bearing dry. *Do not allow the bearing to spin while drying it with compressed air.*

When the bearing is clean and dry, refer to pages 7 through 13 and compare the bearings you have removed with the descriptions and photographs of defective bearings. *Replace bearings that are defective. Always replace bearings as a complete assembly.*

FIGURE I-28 Use a soft-bristle brush when washing bearings.

5. REPLACEMENT OF BEARING CUP WHEN NECESSARY

Inside the hub there is a shoulder which locates the position of each bearing cup. This shoulder usually has at least two openings which allow a puller or punch to be placed behind the bearing cup (Fig. I-29). Remove the bearing cup from the hub with a puller or drift punch. *Inspect the hub to make certain the bearing cup has not turned inside the hub bore.*

A new bearing cup may be pressed or driven into the hub with a tubular or solid drive (Fig. I-30). Never hammer directly on a bearing cup to start it onto the seat (Fig. I-31).

Make certain the cup is tight and is seated against the inside shoulder.

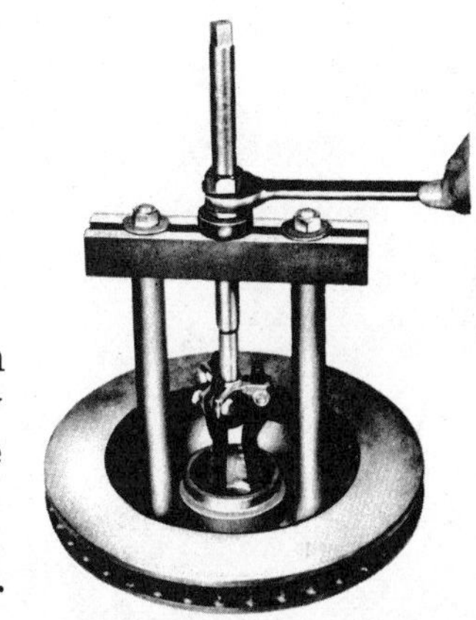

FIGURE I-29 Puller being used to remove front wheel inner bearing cup. (Courtesy of Ford Customer Service Division)

Note: *Bearings are never replaced without also replacing the cups.*

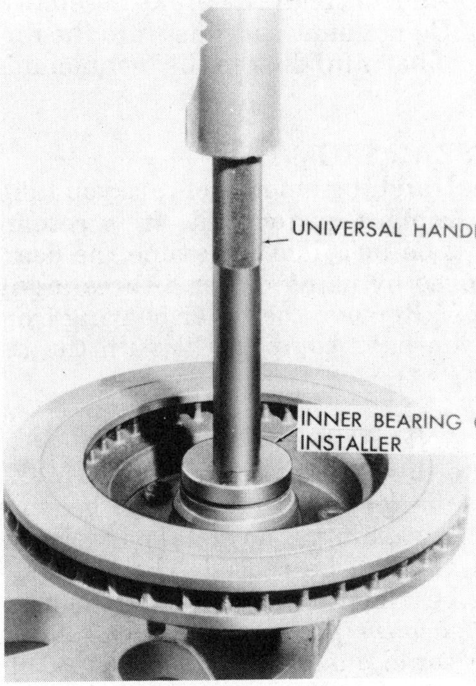

FIGURE I-30 When possible always use a press to install bearing cups. (Courtesy of Cadillac Motor Car Division)

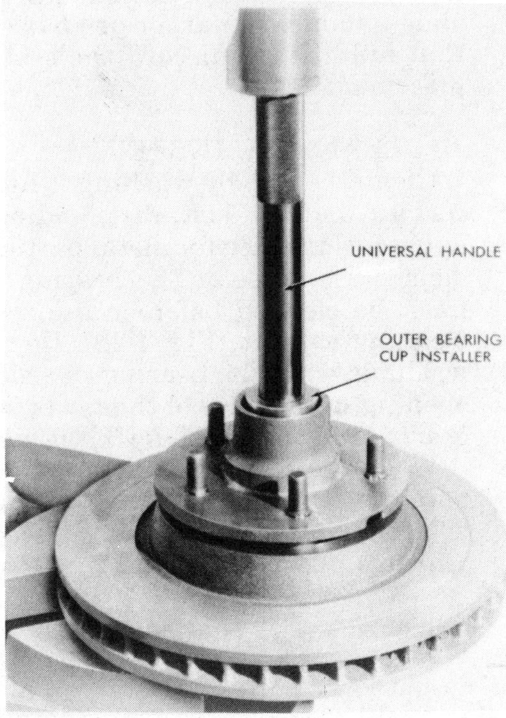

FIGURE I-31 Pressing outer bearing cup into the hub. Be certain cup is seated against lip inside hub. (Courtesy of Cadillac Motor Division)

6. REPACKING THE BEARINGS

Whenever possible use a bearing packer filled with grease that fits the manufacturer's specifications to repack the bearing. The bearing cage should be filled with grease. There should be enough grease left over to spread around the outside of the bearing. *Do not fill the hub with grease and do not mix greases.*

7. ASSEMBLING AND ADJUSTING THE WHEEL ASSEMBLY UNIT

Place inner bearing inside the hub. Use a seal driver to install the seal into the rear of the hub (Fig. I-32). Place the hub and drum assembly onto the spindle and install the outer bearing, washer, and spindle nut. Adjust the spindle nut using the auto manufacturer's recommended procedure and specifications. (See sample of specifications, page 14 or Fig. I-33.

Install a new cotter pin and cut the ends so they will not interfere with the spring static suppressors in the dust cap. Install the wheel assembly and tighten the lug nuts or bolts as specified by the manufacturer. (See sample of specifications, pages 15 and 16. Install the dust and hub caps and remove the car from jack (safety) stands. Road test the car.

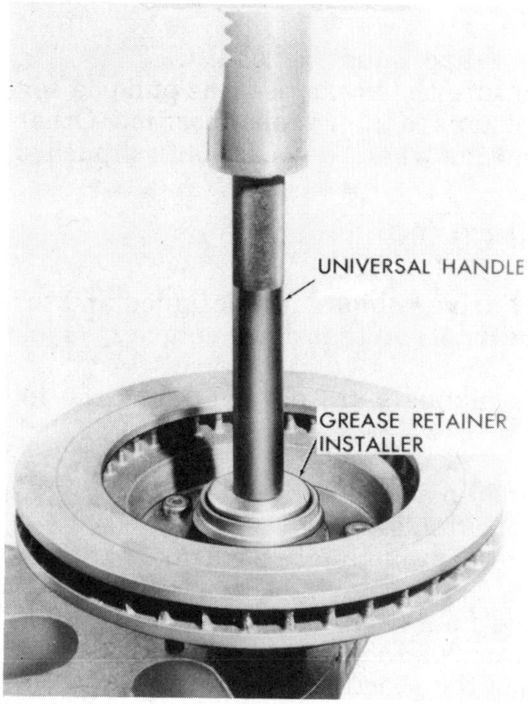

FIGURE I-32 Always use a seal driver to install grease seals. It will help protect them against damage during installation. (Courtesy of Ford Customer Service Division)

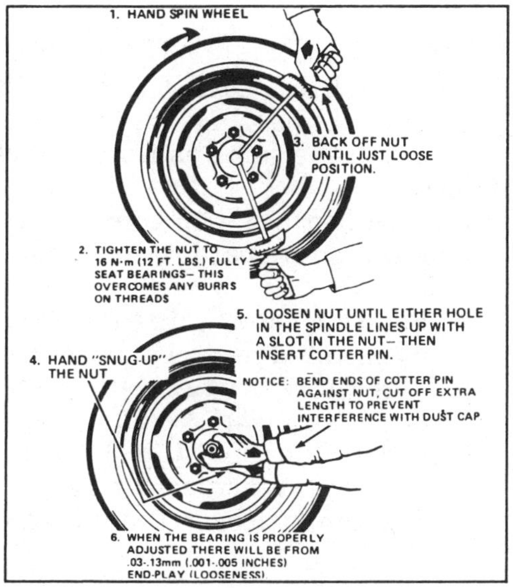

FIGURE I-33 Steps in adjusting the spindle nut. (Courtesy of Chevrolet Motor Division)

REPACKING BEARINGS ON DISC-TYPE BRAKES

The process for repacking bearings on disc-type brakes is exactly the same as on drum-type brakes, with one exception: the brake caliper must be removed before the rotor and hub assembly (Fig. I-34). This can be done by removing the caliper mounting bolts (Fig. I-35 and Fig. I-36). Each type of caliper has a different type of mount, so be sure to check the manufacturer's reference for the specific removal process recommended for each type of caliper.

In most cases, it is not necessary to remove the brake hydraulic line, but do not let the caliper hang on the rubber hose. Use a piece of wire to support the weight of the caliper.

When replacing the caliper, make certain the proper torque is applied to the caliper bolts and be sure you know which bolt you are tightening. For example: The bolt locking the caliper to the steering knuckle requires 100 to 140 ft-lb. (135–190 N-m) of torque, while the caliper-locating pins need only 25 to 35 ft-lb. (35–50 N-m). Needless to say, something would be damaged if the locating pins were torqued to 140 ft-lb. (190 N-m). All other procedures for repacking disc-type wheel bearings are the same as those used for drum brakes.

SAFETY WARNING ABOUT DISC BRAKES

1. Do not collapse the piston too much, for the brake fluid returning to the master cylinder may damage the cap seal or cause fluid leakage.
2. Collapse only one caliper at a time when packing wheel bearings; otherwise, the fluid pressure may cause the piston on the opposite side to come out of its bore.

3. If any fluid leakage is noted, the caliper must be rebuilt.
4. Before moving the vehicle, make sure the brake pedal is pumped several times to expand the caliper pistons to their proper position. Otherwise, there will be little or no braking when the pedal is first pushed.

REPACKING REAR WHEEL BEARINGS ON FRONT WHEEL DRIVE VEHICLES

Rear wheel bearings on front wheel drive vehicles are designed and serviced much the same as front wheel bearings on rear drive vehicles, as just described.

If manufacturer's or general repair manuals are not available, the following procedures are recommended:

a. Follow removal, disassembly, cleaning, packing, and assembly steps as outlined in previous bearing packing section.
b. Tighten rear wheel spindle nut to approximately 25 ft-lbs. (3.5 mkg) or 35 N-m *while rotating the wheel.*
c. Loosen the nut 1/6 to 1/4 turn and insert the cotter pin.
d. Most manufacturers specify .003 to .005 inch (.08 to .13 mm) end play. Tighten or loosen the nut to reach the specified amount.

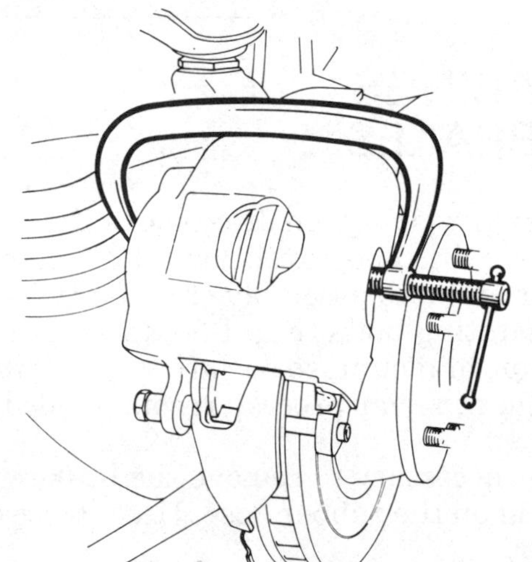

FIGURE I-35 With a single or floating caliper, the piston can be collapsed with a C-clamp. This will allow more clearance between linings to remove the caliper. Don't forget to pump the brakes before moving the vehicle. (Courtesy of Buick Motor Division)

FIGURE I-34 Typical hub and bearing assembly when used with disc rotor. (Courtesy of Chrysler Motors Corp.)

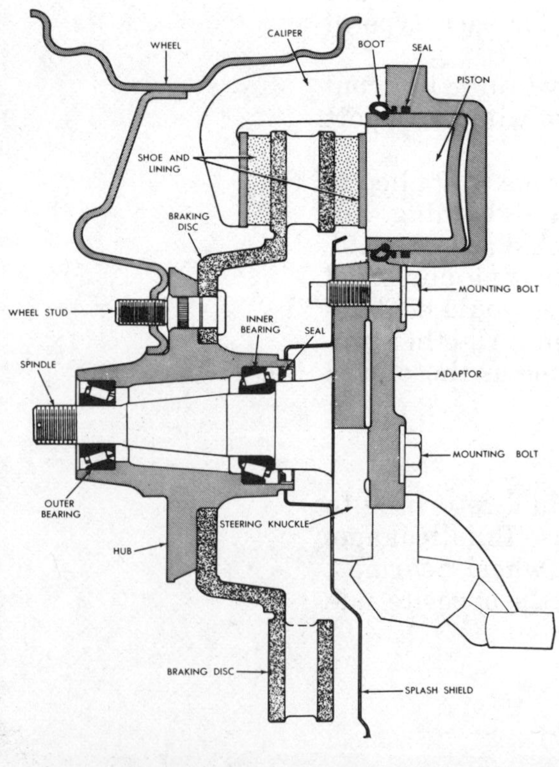

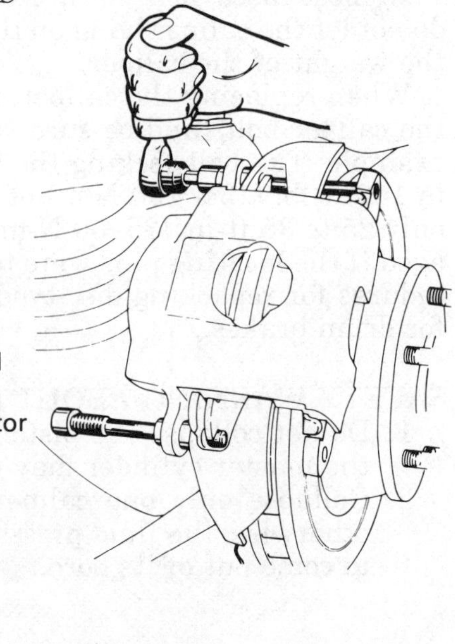

FIGURE I-36 On this Buick floating caliper, the two guide pins are removed to allow the caliper to be separated from its mounting. (Courtesy of Buick Motor Division)

REPACK FRONT WHEEL BEARINGS DRUM TYPE BRAKES

Student name________________________________Date____________________

Make of car________________Model________________Year________________

Time started______________Time finished______________Total time______________

Flat-rate time ______________

Checkup Questions:

1. The bearings were cleaned with ______________________

2. Spindle-nut torque specifications: First tightening: ______________________

   Back off: ______________________

   Second tightening: ______________________

3. Lug-nut torque specification: ______________

4. List all parts that you replaced on this job.

   a. ______________________________________

   b. ______________________________________

   c. ______________________________________

   d. ______________________________________

   e. ______________________________________

5. Did your road test indicate any problems? ______________ Yes ______________ No

If yes, what were they? ______________________________________

____

____

____

DATE COMPLETED______________________INSTRUCTOR'S SIGNATURE ______________________

REPACK FRONT WHEEL BEARINGS DISC TYPE BRAKES

Student name____________________ Date____________________

Make of car________________ Model________________ Year________________

Time started________________ Time finished________________ Total time________________

Flat-rate time ____________

Checkup Questions:

1. The bearings were cleaned with ____________________
2. Spindle nut torque specifications: First tightening: ________________

   Back off: ________________

   Second tightening: ________________
3. Caliper mounting-bolts torque: ____________________
4. Wheel-lug torque specifications are ____________
5. List all parts that you replaced on this job.
   a. ____________________
   b. ____________________
   c. ____________________
   d. ____________________
   e. ____________________
6. Did you pump the brakes before moving the vehicle? ______ Yes ______ No
7. Did your road test indicate any problems? ______ Yes ______ No

If yes, what were they? ____________________

____

____

DATE COMPLETED ____________ INSTRUCTOR'S SIGNATURE ________________

Student name____Date____________________________

Directions: Select the best answer to each question and circle the correct letter at the left.

1. Which of the following part(s) are press-fit into the hub?
   I. The cup
   II. The cone
   A. I only
   B. II only
   C. Both I and II
   D. Neither I nor II
2. The purpose of the bearing cage is to:
   I. Act as a retainer for the lubricant
   II. Separate the balls or the rollers
   A. I only
   B. II only
   C. Both I and II
   D. Neither I nor II
3. The bearing____________________must be free to creep around the spindle when in operation:
   I. Cup
   II. Cone
   A. I only
   B. II only
   C. Both I and II
   D. Neither I nor II
4. The purpose of the front-wheel grease seal is to:
   I. Seal in the grease
   II. Seal out the dirt
   A. I only
   B. II only
   C. Both I and II
   D. Neither I nor II
5. When diagnosing front-wheel bearing problems, you should:
   I. Check tires for irregular wear
   II. Drive vehicle to obtain normal operating temperature
   A. I only
   B. II only
   C. Both I and II
   D. Neither I nor II
6. When cleaning bearings, it is all right to:
   I. Wash in gasoline
   II. Spin the bearing with compressed air
   A. I only
   B. II only
   C. Both I and II
   D. Neither I nor II

7. Spindle nuts should be torqued:
   I. While rotating the wheel slowly
   II. With inch-pound torque wrench only
      A. I only
      B. II only
      C. Both I and II
      D. Neither I nor II
8. The star pattern for tightening lug nuts is used:
   I. Only when using a torque wrench
   II. Only when using an impact wrench
      A. I only
      B. II only
      C. Both I and II
      D. Neither I nor II
9. When repacking front wheel bearings:
   I. Thoroughly clean all of the old grease only when metal particles are evident
   II. Fully pack the hub with new grease
      A. I only
      B. II only
      C. Both I and II
      D. Neither I nor II
10. When repacking front wheel bearings on automobiles equipped with disc brakes:
   I. Remove the calipers
   II. Repack only the outer bearing
      A. I only
      B. II only
      C. Both I and II
      D. Neither I nor II

DATE COMPLETED ________________ INSTRUCTOR'S SIGNATURE ________________

Tires wear out before other components of an automobile; a car owner usually replaces his tires a number of times during the life of his vehicle. But to the front end technician, tires are an important part of the wheel assembly. Many front end problems are directly related to tire wear, and a careful examination of tire condition helps in diagnosing front end troubles.

OBJECTIVE I-B (written test)

After studying the preparation assignment, you will be expected to answer a minimum of seven out of ten multiple choice questions in a written test without reference materials on the following wheel assembly topics:

1. **Tire construction**
2. **Tire and wheel rim sizes**
3. **Tire load capacity ratings**

Preparation:

Study pages 27 through 53

Tires on a car provide two basic functions. First, they act as a cushion between the road and the car wheels. By reducing the force of the shocks felt from rough roads, they protect bearings and other parts of the automobile from stress, as well as providing a more comfortable ride. Second, tires are flexible and bend around obstructions so that even on a rough road a good part of the tire is making contact with the ground most of the time. The tire is the only part of the automobile to transmit engine power and brake power to the road surface where it can be useful (Fig. I-37).

Interestingly, the pneumatic (air-filled) tire was first patented in England, in 1845, by Robert William Thompson. But the following year, 1846, another Englishman named Thomas Hancock began making solid rubber tires for use on horse-drawn cabs and bicycles. The pneumatic tire sank into obscurity until the invention of what was essentially the modern automobile forced its development around the turn of the century.

FIGURE I-37 Tire and wheel rim assembly.

Today, two types of pneumatic tires are used on modern automobiles: (1) tubeless tires and (2) tires which use an inner tube. The tires are similar and function in much the same way. The tire is essentially a shell to contain the air pressure that gives the tire its tremendous strength. In the tubeless tire, the tire seals itself to the wheel rim through air pressure. The tube tire requires an inner tube, a rubber air bag, to be inserted into the cavity between the tire and wheel rim.

TIRE CONSTRUCTION

Modern tires use a combination of materials and several types of construction to contain the pressurized air (Fig. I-38). The structure (or foundation) of a tire is composed of plies (or layers) of rayon, nylon, polyester, steel, or other types of cord material within the body of the tire (see Table I-4).

These layers of cord material are impregnated with rubber, spaced properly, and formed into the number of plies necessary for the tire's designated use. The sidewall and tread material is then applied and vulcanized into place. The number of plies normally varies from two to six. Some tires have eight or more plies, depending on the intended use. The greater the load a tire is required to carry, the more plies it needs.

There are three types of tires and these three types are classified by the design of their cords or plies.

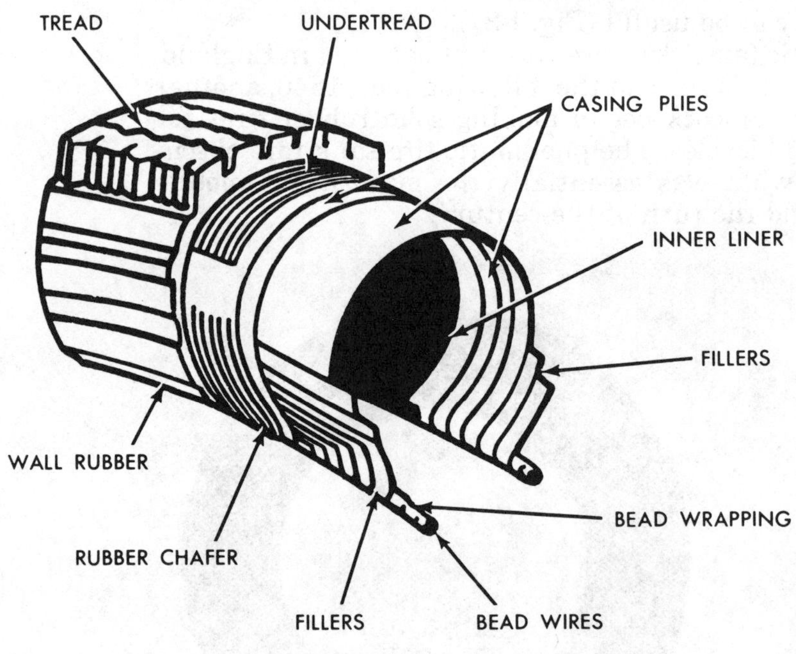

FIGURE I-38 Basic construction for the pneumatic tire. (Courtesy of Chevrolet Motor Division)

BIAS PLY: The bias-ply tire has two or four body plies that crisscross at angles between 30 and 38 degrees to the center line of the tire tread (Fig. I-39). The cord may be made of rayon, nylon, polyester, or another material. This crisscrossing of the plies adds strength to the tire sidewalls and tread. When inflated to recommended pressure, these tires give a relatively soft, comfortable ride.

BIAS-BELTED PLY: Belted tires are similar to bias-ply tires in sidewall and foundation ply construction. However, additional plies or belts of fiberglass, rayon, or fine steel cord encircle the tire under the tread. This method of construction stabilizes the tread, holds it flatter to the road, and reduces tread side motion or squirm. Belted tires give a firmer ride than the bias-ply type, but road grip is improved. The life of a belted-ply tire is usually much longer than the life of a bias-ply tire.

RADIAL BELTED-PLY: The carcass plies on radial tires are laid at right angles to the circumference of the tire. An additional two (or more) belts encircle the tire directly under the tread. The sidewalls are extremely flexible, but the tread is significantly firmer. This allows very little side motion or squirm, and results in longer tread life, improved traction, increased skid resistance, and more positive braking. Equipped with radials, the average car requires greater steering effort and has a harshness at low speeds.

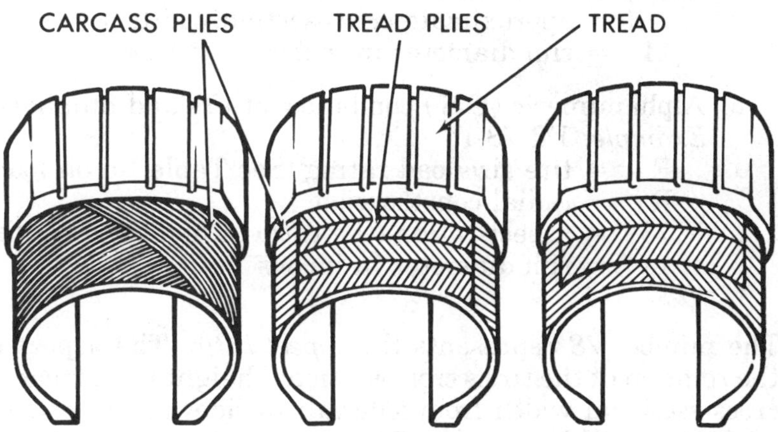

FIGURE I-39 Three types of ply construction in use today. (Courtesy of Chevrolet Motor Division)

TABLE I-4 TYPES OF TIRE CORDS

Cord Type	*Tensile Strength (psi)*	*Advantages*
Rayon	94,000	Soft ride, resilient, inexpensive
Polyester	104,000	Soft ride, more heat
Nylon	122,000	High heat resistance, excellent impact resistance, minimum flex, won't absorb water
Steel	350,000–400,000	Highest impact resistance Highest resistance to puncture
Fiberglass	407,000	Greatest strength, soft ride

PLY RATING: While there is no industry wide definition of ply rating, passenger-car tires marked "4-ply rating/2-ply" have the same load carrying capacity as 4-ply tires of the same size with the same inflation. Tires marked "8-ply rating/4-ply" likewise have the same load-carrying capacity as 8-ply rating tires, regardless of the actual number of plies. In newer tire designations, the ply rating is being replaced by a load range letter.

TIRE AND WHEEL RIM SIZES

The size of a tire is directly related to its load-carrying capacity. Since vehicles differ in size and weight, the tires for these vehicles must be selected for their appropriate load-carrying ability. A general rule is: The larger the tire, the greater the weight it can safely carry.

Currently, passenger car tires are being sold in one of four size designations: numeric, alphanumeric, metric, and P-metric.

a. Numeric is given in numbers only.
   Example: 7.75 × 14
   7.75 = approximate cross-section width
   14 = rim diameter in inches

b. Alphanumeric (U.S.) combines letters and numbers.
   Example: FR 78-14
   F = tire size/load rating (See Table I-5 on load range.)
   R = radial construction
   78 = aspect ration (the ratio of height to width)
   14 = rim diameter in inches

The number *78* represents the *aspect ratio.* The aspect ratio is defined as the relation of the tire's cross-sectional height from tread to bead to the tire's cross-sectional width from sidewall to sidewall. In other words, the height of the tire will be equal to 78 percent of the width. (See Fig. I-40.)

c. Metric (on most imported tires) is given in numbers and letters.
   Example: 195 R14
   195 = section width in millimeters
   R = radial construction
   14 = rim diameter in inches

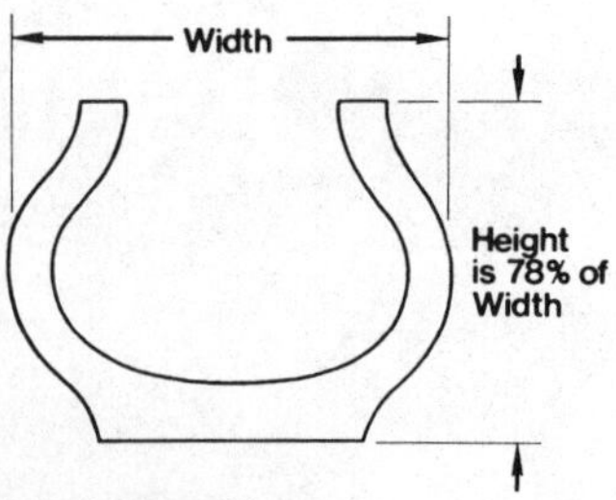

78-Series Sizes
'78' Profile Ratio

FIGURE I-40

d. P-metric, used on the newest U.S. tires, is given in numbers and letters.
   Example: P205/75 R 15 (See Fig. I-41.)
   P = passenger car tire
   205 = approximate cross-section width
   75 = height-to-width ratio
   R = radial (or B if bias-belted or D if diagonal or bias construction)
   15 = rim diameter in inches

Beside the tire size, the sidewall should tell you other important information. Molded into the sidewall rubber, at a depth of no less than 3 millimeters, will be the following information (Fig. I-42):

a. The actual number of plies in the sidewall and in the tread area and the generic name of each cord material in the plies of both sidewall and tread area. For example: TREAD 4 PLIES, 2 POLYESTER CORD, 2 STEEL CORD, SIDEWALL 2 PLIES STEEL CORD.
b. Max Load (The maximum load). This is the maximum safe-operating-load rating in both kilograms and pounds. For example: 790kg (1742 lbs). (See Table I-5)
c. Maximum permissible inflation in kilopascals and pounds per square inch. For example: 204 kPa (35 psi). (See Table I-6, Inflation Pressure Conversion Chart.)
d. Tubeless or tube-type construction.
e. Radial, if it is of radial design.
f. The code of the Department of Transportation (DOT). For example: DOT MF C5 GAN 171. *MF C5* means the tire was manufactured at the Goodyear plant in California at work station C5. *GAN* is an internal code for Goodyear and *171* means the date of production was the 17th week of 1981.
g. The country of origin is required only if the tire is imported.

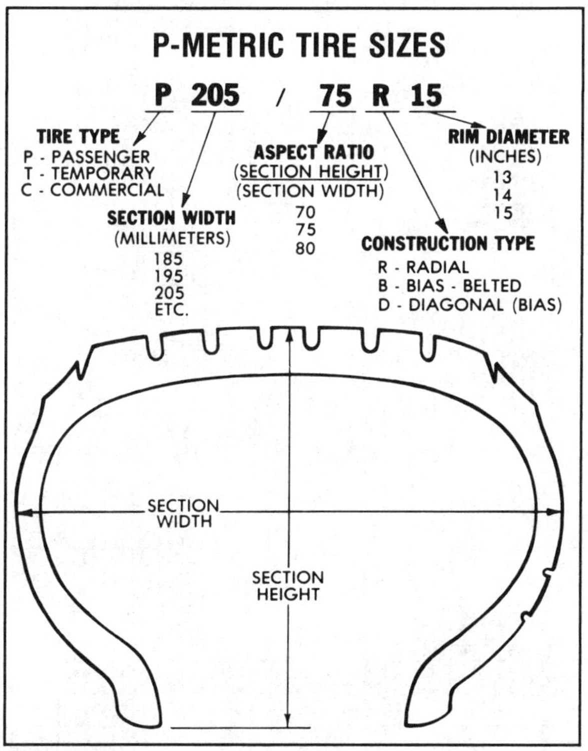

FIGURE I-41 Size designation for metric tires. (Courtesy of Chevrolet Motor Division)

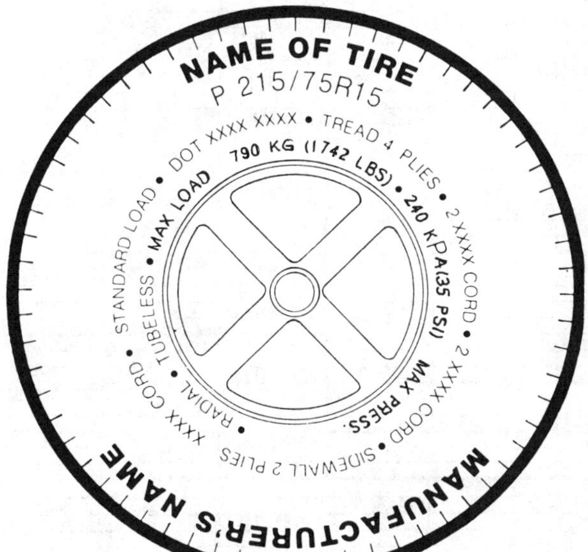

FIGURE I-42 Required sidewall information. (Courtesy Rubber Manufacturers' Association)

QUALITY GRADES

In addition to the sidewall molded markings, there is information on the tire brand label known as the DOT quality grades. These cover treadwear, traction, and temperature.

TABLE I-5 PASSENGER TIRE COMPARISON CHART & LOAD AND INFLATION TABLES

Load Limits for Tires Used On Passenger Cars

Interchangeability is not implied

Load Range B → (18–32 psi)

Standard Load → (18–35 psi)

Rim Dia	ALPHA* SIZES All Const.	EUROPEAN METRICS	P-METRIC SIZES 80 Series	75 Series	70 Series	60 Series	50 Series	Air Pressure in P.S.I./Max. Loads in Pounds 18	20	22	23	24	26	28	29	30	32	35
13 inch		6.15/155-13						575	640	700	725	750	780	820	825	850	880	
		155R13						690	730	770	790	810	845	885	900	915	950	
			P155/80R13					705	739	772	783	805	838	860	882	893	926	959
		175/70R13						820	845	865	878	890	910	935	945	955	980	
					P175/70R13			716	750	783	805	827	849	882	904	915	948	981
				P165/75R13				728	772	805	816	838	871	904	915	937	959	1003
	6.00-13	165R13						730	770	820	840	860	900	930	950	970	1010	
	A-13							770	810	860	880	900	940	980	1000	1020	1060	
			P165/80R13					783	816	860	871	893	926	959	981	992	1025	1069
					P185/70R13	P195/60R13		783	827	871	882	904	937	970	992	1003	1036	1080
		185/70R13						915	940	965	978	990	1015	1040	1053	1065	1090	
							P215/50R13	794	838	871	893	904	948	981	992	1014	1047	1091
				P175/75R13				805	849	882	904	926	959	992	1014	1025	1058	1113
	6.50-13	175R13						840	890	930	955	980	1030	1070	1090	1110	1150	
	B-13							840	890	930	955	980	1030	1070	1090	1110	1150	
						P205/60R13		849	893	937	959	981	1014	1058	1069	1091	1124	1179
			P175/80R13		P195/70R13			860	904	948	970	992	1025	1058	1080	1102	1135	1179
		195/70R13						1015	1045	1070	1085	1100	1125	1155	1168	1180	1210	
				P185/75R13				882	926	970	992	1014	1058	1091	1113	1135	1168	1224
	C-13							890	950	1000	1025	1050	1100	1140	1165	1190	1230	
							P235/50R13	926	970	1014	1036	1058	1102	1135	1157	1179	1213	1268
		185R13						930	980	1030	1055	1080	1130	1180	1205	1230	1270	
						P215/60R13		926	970	1014	1036	1058	1102	1146	1168	1179	1224	1279
					P205/70R13			937	981	1036	1058	1080	1113	1157	1179	1201	1235	1290
			P185/80R13					948	992	1036	1058	1080	1124	1168	1190	1213	1246	1301
	D-13							950	1010	1070	1095	1120	1170	1220	1245	1270	1320	
		195R13						1010	1060	1110	1140	1170	1220	1280	1300	1320	1370	
14 inch	A-14							770	810	860	880	900	940	980	1000	1020	1060	
	6.45-14	165R14						810	860	910	935	960	1000	1040	1060	1080	1120	
			P165/80R14					816	860	904	926	937	981	1014	1036	1047	1080	1124
	B-14							840	890	930	955	980	1030	1070	1090	1110	1150	
		185/70R14						960	990	1015	1030	1045	1070	1100	1115	1130	1155	
				P175/74R14				849	893	937	959	970	1014	1047	1069	1080	1124	1168
					P185/70R14			871	915	959	981	1003	1036	1080	1091	1113	1146	1201
	6.95-14							900	950	1000	1025	1050	1100	1140	1165	1190	1230	
	C-14	175R14						900	950	1000	1025	1050	1100	1140	1165	1190	1230	
		195/70R14						1055	1090	1120	1138	1155	1185	1220	1235	1250	1280	
				P185/75R14				937	981	1025	1047	1069	1113	1157	1168	1190	1235	1290
					P195/70R14			959	1003	1047	1069	1091	1135	1179	1201	1224	1257	1312
	D-14							950	1010	1070	1095	1120	1170	1220	1245	1270	1320	
						P215/60R14		981	1025	1069	1091	1124	1168	1201	1224	1246	1290	1345
					P205/70R14			992	1036	1091	1113	1135	1179	1224	1246	1257	1301	1356
		185R14						980	1040	1100	1130	1160	1210	1260	1285	1310	1360	
	E-14	205/70R14						1010	1070	1130	1160	1190	1240	1300	1325	1350	1400	
				P195/75R14				1025	1069	1124	1146	1168	1213	1257	1279	1301	1345	1400
							P245/50R14	1047	1091	1146	1168	1190	1246	1290	1312	1334	1378	1433
					P215/70R14			1069	1124	1179	1201	1224	1279	1323	1345	1367	1411	1477
						P235/60R14		1080	1135	1190	1213	1235	1290	1334	1356	1378	1422	1488
	F-14							1100	1160	1220	1250	1280	1340	1400	1425	1450	1500	
				P205/75R14				1113	1168	1224	1246	1279	1323	1378	1400	1422	1466	1532
					P225/70R14			1157	1213	1279	1301	1334	1378	1433	1455	1477	1521	1598
	G-14							1180	1250	1310	1345	1380	1440	1500	1530	1560	1620	
							P265/50R14	1190	1257	1312	1345	1367	1422	1477	1499	1521	1576	1642
				P215/75R14				1201	1268	1323	1356	1389	1433	1488	1521	1543	1587	1664
						P245/60R14		1224	1279	1345	1367	1400	1455	1510	1532	1554	1609	1675
					P235/70R14			1246	1312	1378	1400	1433	1488	1543	1565	1598	1642	1720
	H-14							1290	1360	1440	1475	1510	1580	1650	1680	1710	1770	
				P225/75R14				1301	1367	1433	1466	1499	1554	1609	1642	1664	1720	1797
					P245/70R14			1345	1411	1477	1510	1543	1598	1664	1687	1720	1775	1852
	J-14							1350	1430	1500	1540	1580	1650	1720	1755	1790	1860	
15 inch		155R15						735	780	825	845	865	905	940	960	980	1015	
			P155/80R15					772	816	860	871	893	926	959	981	992	1025	1069
				P165/75R15				805	849	893	904	926	959	1003	1014	1036	1069	1113
		165R15						820	870	910	935	960	1000	1050	1070	1090	1130	
	B-15							840	890	930	955	980	1030	1070	1090	1110	1150	
			P165/80R15					860	904	948	970	992	1025	1069	1080	1102	1135	1190
				P175/75R15				893	937	981	1003	1025	1058	1102	1124	1135	1179	1224
	C-15	175R15						890	950	1000	1025	1050	1100	1140	1165	1190	1230	
	D-15							950	1010	1070	1095	1120	1170	1220	1245	1270	1320	
		185R15						1010	1070	1130	1155	1180	1240	1290	1315	1340	1390	
	E-15	205/70R15						1010	1070	1130	1160	1190	1240	1300	1325	1350	1400	
				P195/75R15				1069	1124	1179	1201	1224	1279	1323	1345	1367	1411	1477
		195R15						1090	1150	1210	1240	1270	1330	1380	1410	1440	1490	
	F-15							1090	1160	1220	1250	1280	1340	1400	1425	1450	1500	
				P205/75R15				1168	1224	1279	1312	1334	1389	1433	1466	1488	1532	1598
		205R15						1170	1240	1300	1335	1370	1430	1490	1520	1550	1610	
	G-15							1180	1250	1310	1345	1380	1440	1500	1530	1560	1620	
					P215/70R15			1179	1246	1301	1323	1356	1411	1455	1488	1510	1554	1620
						P235/60R15		1190	1246	1312	1334	1367	1422	1466	1499	1521	1565	1642
					P225/70R15			1213	1279	1334	1367	1389	1444	1499	1521	1543	1598	1664
							P265/50R15	1246	1312	1378	1400	1433	1488	1543	1565	1598	1642	1720
				P215/75R15				1268	1323	1389	1422	1444	1510	1565	1587	1609	1664	1742
	H-15							1290	1360	1440	1475	1510	1580	1650	1680	1710	1770	
						P255/60R15		1301	1367	1433	1466	1488	1554	1609	1631	1664	1709	1786
					P235/70R15			1312	1378	1433	1466	1499	1554	1609	1642	1664	1720	1797
							P275/50R15	1334	1400	1466	1499	1521	1587	1642	1675	1698	1753	1830
		215-15							1280	1380	1430	1480	1570	1660	1710	1760	1860	
	J-15							1350	1430	1500	1540	1580	1650	1720	1755	1790	1860	
				P225/75R15				1367	1433	1499	1532	1565	1631	1687	1720	1742	1797	1874
	L-15							1430	1520	1600	1640	1680	1750	1830	1865	1900	1970	
		225-15							1370	1470	1525	1580	1670	1780	1830	1880	1980	
		230-15							1405	1550	1570	1625	1725	1825	1875	1925	2020	
				P235/75R15				1466	1543	1620	1653	1687	1753	1819	1852	1885	1940	2028
					P245/70R15			1477	1554	1620	1664	1698	1764	1830	1852	1885	1951	2028
							P295/50R15	1499	1576	1653	1687	1720	1786	1852	1885	1918	1973	2061
					P255/70R15			1499	1576	1653	1687	1731	1797	1863	1896	1918	1984	2072
						P275/60R15		1554	1642	1720	1753	1786	1863	1929	1962	1995	2050	2149

*For example, A-13 represents all aspect ratios (eg., A78-13, A70-13, AR78-13, AR70-13, etc.)

TREADWEAR

The treadwear grade is a comparative rating based on the wear rate of the tire when tested under controlled conditions on a specified government test course. For example, a tire graded 150 would wear one and a half (1-1/2) times as well on the government course as a tire graded 100. Tire performance depends upon the actual conditions of use, however, and use may depart significantly from the norm. Driving habits, service practices, and differences in road conditions and climate all affect tire performance.

TRACTION

The traction grades, from highest to lowest, are A, B, and C. These grades indicate the tire's ability to stop on wet pavement as measured under controlled conditions on specified government test surfaces of asphalt and concrete. Tires of C grade may have poor traction performance.

Note: *A tire's traction grade is based on braking (straight-line) traction tests and does not include cornering (turning) traction.*

TEMPERATURE

The temperature grades are A (the highest), B, and C (the lowest). They represent the tire's resistance to the buildup of heat. Also, they measure the tire's ability to throw off heat when tested under controlled conditions. The test is made on a laboratory test wheel. Sustained high temperature can cause the tire material to degenerate and can reduce tire life. Very high temperatures can lead to sudden tire failure. All passenger car tires must meet at least grade C performance standards as set by the federal motor vehicle safety code. Grades A and B are for the higher levels of performance and exceed the minimum performance required by law.

Note: *The temperature grade applies only to tires that are properly inflated and not overloaded. Excessive speed, under-inflation, and excessive loading, either separately or in combination, can cause heat buildup and possible tire failure.*

TABLE I-6 INFLATION PRESSURE CONVERSION CHART (KILOPASCALS TO PSI)

kPa	*psi*	*kPa*	*psi*
140	20	215	31
145	21	220	32
155	22	230	33
160	23	235	34
165	24	240	35
170	25	250	36
180	26	275	40
185	27	310	45
190	28	345	50
200	29	380	55
205	30	415	60

Conversion: 6.9 kPa = 1 psi

Note: The metric term for tire inflation pressure is the Kilopascal (kPa). Tire pressure will usually be printed in both kPa and psi. This chart converts commonly used inflation pressures from kPa to psi.

Wheel sizes are defined by three measurements: (1) rim diameter, measured at the bead surface, (2) rim width, measured to the inside edge of the flange, and (3) flange height, measured from top to bottom of the flange. The first two dimensions, rim diameter and rim width, are always in inches (Fig. I-43). The flange height is usually designated by a letter or a combination of letters. The following are samples of wheel rim designations:

J rims have 0.68-inch-high flanges
JK rims have 0.71-inch-high flanges
K rims have 0.77-inch-high flanges
L rims have 0.85-inch-high flanges

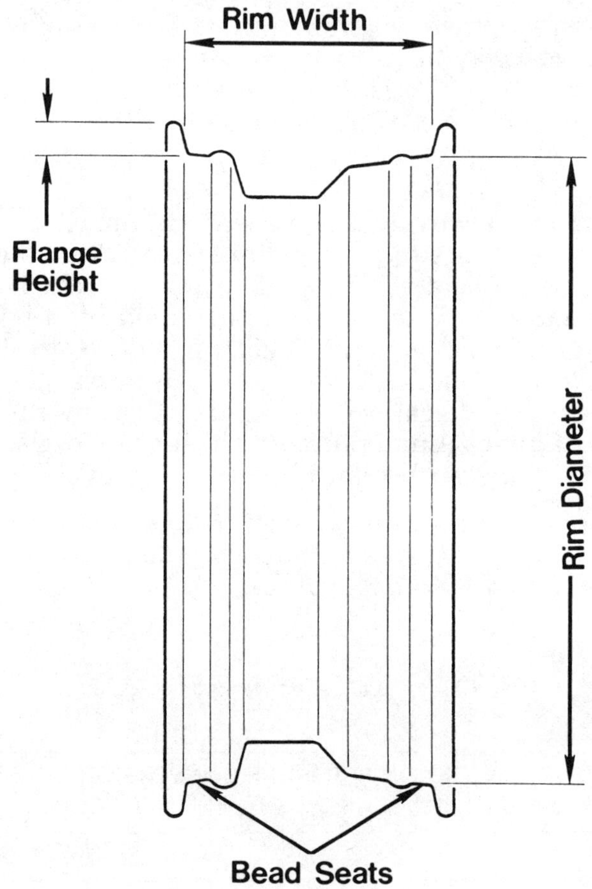

FIGURE I-43 Typical wheel rim.

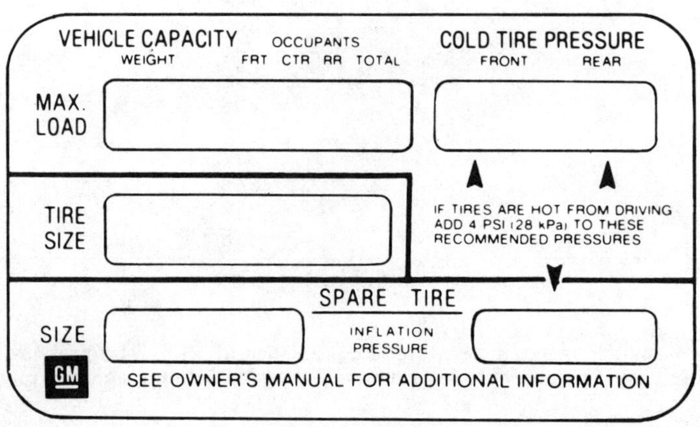

FIGURE I-44 The tire placard will be found on the doorpost of the driver's side or in the glove box. (Courtesy of Chevrolet Motor Division)

TIPS ON REPLACING TIRES

When replacing conventional-size tires such as the G 78-14 with tires of another size, be sure to check the vehicle manufacturer's recommendations. This information can be found on the tire placard located on the doorpost on the driver's side or inside the glove compartment. The placard gives the recommended original equipment tire size and inflation level (Fig. I-44).

Since the load capacity and inflation pressure are so closely related, be certain to check them when interchanging alphanumeric and P-series metric sizes. Keep in mind that there are no exact replacements as far as loads and pressures are concerned.

The tire selected should have equal or greater load-carrying capacity to the replaced tire. Do not exceed the maximum inflation as labeled on the tire.

As a general rule, do not mix different size tires on the same axle and never mix radial and nonradial tires on the same axle. Further, never install radials on the front of a vehicle with bias or bias-belted tires on the rear. Refer to the Rubber Manufacturers Association (RMA) tire application guide if tires of different size or construction are to be used on any vehicle (See Table I-7A,B).

TABLE I-7A TIRE REPLACEMENT GUIDE FOR PASSENGER CARS FROM "P" METRIC TO RADIAL PLY TIRES

IF VEHICLE TIRE PLACARD SPECIFIES A P-METRIC TIRE SIZE:
(All sizes are Load Range B or Standard Load, unless noted. Do not exceed maximum pressure marked on tire sidewall.)

P-Metric Size	*Acceptable Substitute Size*	*P-Metric Size*	*Acceptable Substitute Size*
P155/80R13	P165/75R13, P175/70R13, 175/70R13	P185/75R14	P195/70R14, P205/70R14, P215/60R14, DR78-14, DR70-14, 195/70R14
P165/80R13	P175/75R13, P185/70R13, P195/60R13, P215/50R13, AR78-13, AR70-13, 185/70R13	P195/75R14	P215/70R14, P235/60R14, P245/50R14, ER78-14, ER70-14
P175/80R13	P185/75R13, P195/70R13, P215/60R13, P235/50R13, CR78-13, CR70-13	P205/75R14	P225/70R14, P265/50R14, FR78-14, FR70-14
P185/80R13	(None)	P205/75R14 (Extra Load)	FR78-14 (Load Range D), FR70-14 (Load Range D)
P165/75R13	P165/80R13, P185/70R13, P195/60R13, P215/50R13, AR78-13, AR70-13, 175/70R13	P215/75R14	P235/70R14, P245/60R14, GR78-14, GR70-14
P175/75R13	P175/80R13, P195/70R13, P205/60R13, BR78-13, BR70-13, 185/70R13	P225/75R14	P245/70R14, HR78-14, HR70-14
P185/75R13	P185/80R13, P205/70R13, P215/60R13, P235/50R13, CR78-13, CR70-13	P185/70R14	P185/75R14, CR78-14, 185/70R14
P175/70R13	P165/80R13, P175/75R13, P195/60R13, P215/50R13, AR78-13, AR70-13	P195/70R14	P195/75R14, P215/60R14, DR78-14, DR70-14
P185/70R13	P175/80R13, P185/75R13, P205/60R13, BR78-13, BR70-13, 185/70R13	P205/70R14	P205/75R14, P235/60R14, P245/50R14, FR78-14, FR70-14
P195/70R13	P185/80R13, P215/60R13, P235/50R13, CR78-13, CR70-13	P215/70R14	P215/75R14, P265/50R14, GR78-14, GR70-14
P205/70R13	(None)	P225/70R14	P225/75R14, P245/60R14, HR78-14, HR70-14
P165/80R14	P175/75R14, BR78-14, 185/70R14	P235/70R14	JR78-14
P175/75R14	P185/70R14, BR78-14, 185/70R14	P245/70R14	(None)
		P155/80R15	P165/75R15

TABLE I-7A TIRE REPLACEMENT GUIDE FOR PASSENGER CARS FROM "P" METRIC TO RADIAL PLY TIRES

IF VEHICLE TIRE PLACARD SPECIFIES A P-METRIC TIRE SIZE:
(All sizes are Load Range B or Standard Load, unless noted. Do not exceed maximum pressure marked on tire sidewall.)

P-Metric Size	*Acceptable Substitute Size*	*P-Metric Size*	*Acceptable Substitute Size*
P165/80R15	P175/75R15, CR78-15	P235/75R15	P245/70R15, P255/70R15, P275/60R15, P295/50R15
P165/75R15	P165/80R15, BR78-15	P215/70R15	P215/75R15, P235/60R15, P265/50R15, GR78-15, GR70-15
P175/75R15	CR78-15	P225/70R15	P225/75R15, P255/60R15, P275/50R15, HR78-15, HR70-15
P195/75R15	FR78-15, FR70-15	P235/70R15	P235/75R15, LR78-15
P205/75R15	P215/70R15, P225/70R15, P235/60R15, P265/50R15, GR78-15, GR70-15	P245/70R15	P275/60R15, P295/50R15
P215/75R15	P235/70R15, P255/60R15, P275/50R15, HR78-15, HR70-15	P255/70R15	(None)
P225/75R15	JR78-15		

TABLE I-7B TIRE REPLACEMENT GUIDE FOR PASSENGER CARS FROM BIAS, BIAS BELTED AND RADIAL TO "P" SERIES TIRES

IF VEHICLE TIRE PLACARD SPECIFIES AN ALPHA-NUMERIC TIRE SIZE OR A EUROPEAN METRIC TIRE SIZE:
(All sizes are Load Range B or Standard Load, unless noted. Do not exceed maximum pressure marked on tire sidewall.)

Alpha-Numeric Size (All Series)	*Acceptable Substitute Size. Important—Add 3 PSI Above the Pressure Specified on the Vehicle Tire Placard to Assure Adequate Load Capacity.*	*European Metric Size*	*Acceptable Substitute Size. Important—Add 3 PSI Above the Pressure Specified on the Vehicle Tire Placard to Assure Adequate Load Capacity.*
AR78-13	P165/80R13, P175/75R13, P185/70R13, P195/60R13, P215/50R13, 185/70R13	155R13	P155/80R13, P165/75R13, P175/70R13, 175/70R13
BR78-13	P175/80R13, P185/75R13, P195/70R13, P205/60R13	165R13	P165/80R13, P175/75R13, P185/70R13, P195/60R13, P215/50R13, 185/70R13
CR78-13	P185/80R13, P205/70R13, P215/60R13, P235/50R13	175R13	P175/80R13, P185/75R13, P195/70R13, P205/60R13
BR78-14	P175/75R14, P185/70R14, 185/70R14	185R13	P185/80R13, P205/70R13, P215/60R13, P235/50R13
CR78-14	P185/75R14, P195/70R14, 195/70R14	175/70R13	P165/80R13, P175/75R13, P185/70R13, P195/60R13, P215/50R13
DR78-14	P195/75R14, P205/70R14, P215/60R14	185/70R13	P175/80R13, P185/75R13, P195/70R13
ER78-14	P195/75R14, P215/70R14, P235/60R14, P245/50R14	165R14	P165/80R14, P175/75R14, P185/70R14, 185/70R14
FR78-14	P205/75R14, P225/70R14	175R14	P185/75R14, P195/70R14, P215/60R14, 195/70R14
GR78-14	P215/75R14, P235/70R14, P245/60R14, P265/50R14	185R14	P195/75R14, P205/70R14, P245/50R14
HR78-14	P225/75R14, P245/70R14	185/70R14	P185/75R14, P195/70R14
BR78-15	P165/80R15, P175/75R15	195/70R14	P195/75R14, P205/70R14, P215/60R14
ER78-15	P195/75R15	155R15	P155/80R15, P165/75R15
FR78-15	P205/75R15, P215/70R15	165R15	P165/80R15, P175/75R15
GR78-15	P215/75R15, P225/70R15, P235/60R15, P265/50R15	175R15	(None)
HR78-15	P225/75R15, P235/70R15, P255/60R15, P275/50R15		
JR78-15	P225/75R15		
LR78-15	P235/75R15, P245/70R15, P255/70R15, P275/60R15, P295/50R15		

OBJECTIVE I-Bl
(written assignment)

Abnormal tire wear is an excellent indicator of front-end troubles. Working alone and with the aid of reference material, you will diagnose and report in writing to your instructor the abnormal tire wear on a minimum of four tires.

Preparation:

1. Review pages 27 through 37.
2. Study pages 37 through 43.

ABNORMAL TIRE WEAR

UNDERINFLATION

When a tire has insufficient air pressure to properly support the weight of the vehicle, the central area of the tread rises upward at the point where it contacts the road, as sidewalls flex (Fig. I-45). This concentrates tremendous stress on the outside edges of the tread, as shown in the cross-sectional view producing maximum wear on the outside tread edges and very little wear on the center portion of the tread.

If this wear is even all around the circumference of the tire, it indicates either carelessness on the part of the owner, or deliberate underinflation in an attempt to provide a softer ride.

FIGURE I-45 Under inflation. (Courtesy of Moog Automotive, Inc.)

OVERINFLATION

Excessive tire pressure does not permit the tire tread to flatten out at the point of road contact. This causes the central area of the tread to wear heavily. It also gives hard riding characteristics and reduces traction, decreasing the tire's ability to "cling" to the pavement (Fig. I-46).

When a driver keeps a vehicle's tires in an overinflated condition, it usually indicates a fear of underinflation. When occasional heavy loads are being transported, the driver sometimes inflates the tires to compensate for the added load and then forgets to reduce the air pressure to normal after the extra load has been removed.

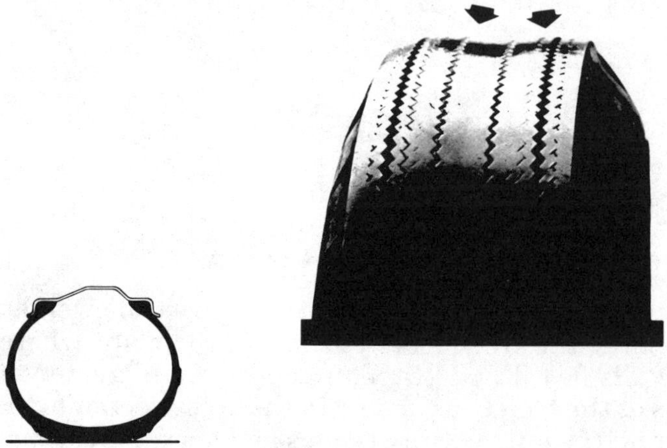

FIGURE I-46 Over inflation. (Courtesy of Moog Automotive, Inc.)

CORRECT INFLATION

Correct inflation permits the tread to contact the pavement fully with relatively even force throughout the full width of the tread. A properly inflated tire offers maximum traction and greatly extends tire life.

Notice how evenly the tread of the tire in Fig. I-47 is worn and how the original shape has been maintained at each edge of the tread.

FIGURE I-47 Correct inflation. (Courtesy of Moog Automotive, Inc.)

TOE WEAR

CONSTANT TOE WEAR (caused by incorrect toe-in or toe-out setting):

When front wheels operate with a consistent (not intermittent) toe-in or toe-out condition due to an incorrect toe setting, the tire treads will eventually appear as shown in Fig. I-48. This wear is caused by the tread continuously scuffing on the pavement. Incorrect toe-in produces excessive wear on the outside edges of tires. If the toe-in is severe, each tread rib will wear off on the outside edge and the inner edge will become sharp and ragged. Incorrect toe-out produces a similar wear pattern on the inside of the tire and on the inside of each tread rib.

The cross-sectional view shows how tread ribs are rolled over, accounting for the excessive wear on one side and the sharp edges on the opposite side of each rib.

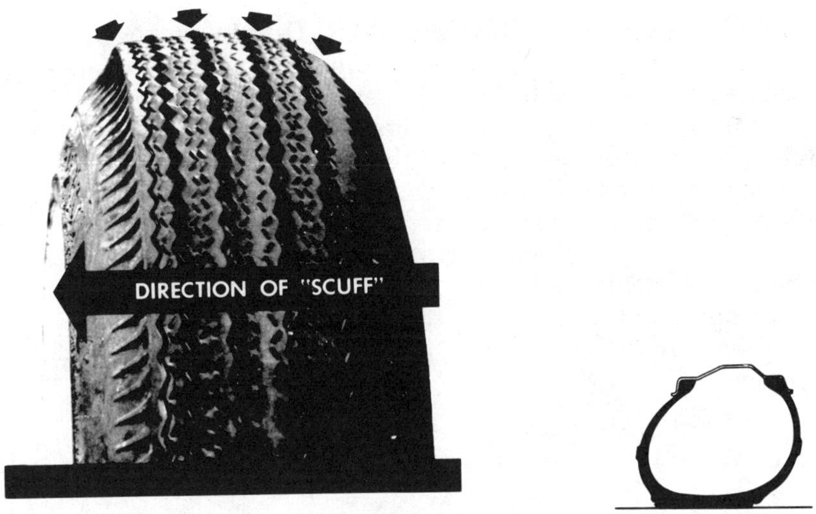

FIGURE I-48 Toe wear. (Courtesy of Moog Automotive, Inc.)

Toe-in or toe-out adjustments are usually made when the vehicle is not moving and with a fixed load height (no passengers, etc., in vehicle). Adjustments that may check out perfectly under these conditions may change intermittently when the vehicle is driven (Fig. I-49). This is due to any one or a combination of the following conditions:

1. Changes in alignment geometry
   a. Toe change resulting from load change
   b. Sagged springs—changing vehicle height
2. Changes aggravated by worn, loose, or bent parts
   a. Worn or loose ball joints
   b. Worn or loose tie-rod ends
   c. Loose idler arm bushings
   d. Loose bushings on upper and/or lower control arms
   e. Bent steering knuckles or steering arms

The illustrated portion of tire tread in Fig. I-48 shows only one of three heavily "scuffed" spots that were spaced evenly around the circumference of this particular tire. This evidence indicates that some worn part (such as a loose tie-rod end) had permitted the wheel to "swing" out of its proper toe setting, then bounce back at a frequency of three times per wheel revolution. The operator drove the car over a regular route at the same general speed most of the time (probably to and from work).

FIGURE I-49 Intermittent toe wear. (Courtesy of Moog Automotive, Inc.)

FIGURE I-50 Constant camber wear. (Courtesy of Moog Automotive, Inc.)

CAMBER WEAR

CONSTANT CAMBER WEAR (caused by incorrect camber settings):

Camber is a term used to describe the inward or outward tilt of the front wheels at the top (Fig. I-50). Outward tilt is known as *positive camber,* and inward tilt is known as *negative camber.* If too much positive camber exists, the tire will wear on the outside edge of the tread. The more positive camber there is, the greater the amount of outside edge tread wear. Conversely, if too much negative camber is present, the inside edge of the tire will show excessive wear.

The photograph in Fig. I-50 shows a typical example of camber wear. If all suspension parts are in good operating condition, a simple camber adjustment will stop the abnormal tire wear.

FIGURE I-51 Intermittent camber wear. (Courtesy of Moog Automotive, Inc.)

INTERMITTENT CAMBER WEAR (Due to Worn or Bent Parts):

Camber is the one major steering adjustment that is almost certain to be affected by worn or bent suspension parts.

Almost every steering or suspension part that is worn, loose, or bent enough to cause an undesirable camber change will also produce some degree of "intermittent toe change." If the tire "scrub" or wear is predominantly due to camber change, the tire wear will be cupped, with more pronounced irregularities toward one side of the tire (Fig. I-51). If the toe change is influenced more by a load change than by camber change, the worn spots will occur close to the center of the tread.

Major conditions which produce intermittent camber changes are as follows:

1. Worn ball joints
2. Worn bushings on upper and lower control arms
3. Bent spindle
4. Sagged springs that change vehicle height

FIGURE I-52 Incorrect turning radius. (Courtesy of Moog Automotive, Inc.)

INCORRECT TURNING RADIUS (FIG. I-53)

Incorrect turning radius will cause rapid but normal appearing tread wear that is happening too fast for the distance traveled. It is often a symptom of an incorrect turning radius angle. Wear on the tread edges may also be heavy and there may be a "rounded-off" effect across the tread. During normal driving, this condition will cause the tires to squeal on turns, even at low speeds (Fig. I-52).

Due to the angles in effect, worn tie-rod ends and loose idler-arm bushings are the chief offenders which permit "fighting" between the front wheels during turns.

Principal conditions producing incorrect turning radius are as follows:

1. Improper settings
2. Worn tie-rod ends
3. Worn idler-arm bushings
4. Worn ball joints
5. Worn control arm bushings
6. Bent steering arm

Note: *On some late-model automobiles, the angular difference between inner and outer wheel-turning radius is extremely small. This is a design feature which should not be interpreted as an error.*

The illustrations, so far, have been of tire wear conditions which were caused by one specific mechanical suspension or steering problem. It is usual, however, for more than one suspension or steering problem to cause tire wear. When this happens, tire wear patterns become complicated and difficult to analyze.

When compound wear patterns appear, the alignment technician may not be able to make an accurate diagnosis from visual observation alone. However, he can tell from unusual wear patterns that something is seriously wrong and that he should make tests to locate the worn parts and/or incorrect adjustments.

Examine carefully the following list of suspension and steering-problem combinations producing tire wear. Pay particular attention to the accompanying illustrations.

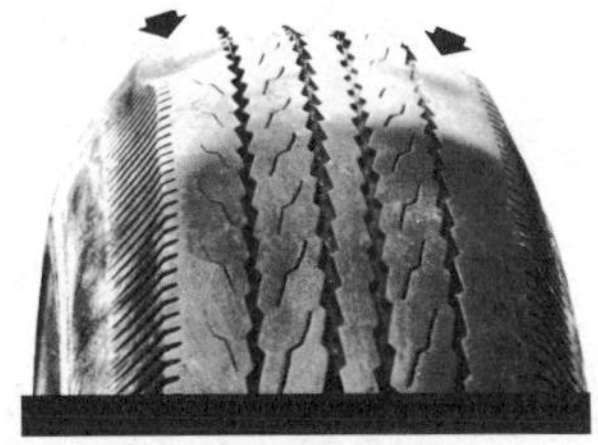

FIGURE I-53 Loose ball joint plus incorrect camber setting. (Courtesy of Moog Automotive, Inc.)

FIGURE I-54 Loose upper control arm bushings plus too much toe-out. (Courtesy of Moog Automotive, Inc.)

FIGURE I-55 Loose idler arm plus unbalance. (Courtesy of Moog Automotive, Inc.)

FIGURE I-56 Loose control arm bushings, ball joints, and idler arm. (Courtesy of Moog Automotive, Inc.)

LOOSE BALL JOINT PLUS INCORRECT CAMBER SETTING (Fig. I-53)

The tire tread has abnormal wear on the right-hand rib, which normally indicates an incorrect camber setting. In addition, numerous depressions have been worn into the tread, one of which appears in the left-hand outside rib. Such erratic wear is usually the result of some loose suspension part which allows the wheel to oscillate (wobble) in a regular pattern. A badly worn ball joint produced this particular result and was probably also responsible for the incorrect camber setting.

LOOSE UPPER CONTROL ARM BUSHINGS PLUS TOO MUCH TOE-OUT (Fig. I-54)

Heavy, erratic wear on the right-hand side plus sharp, feather edges on the left-hand tread ribs are apparent in the illustration. A toe-out condition caused the "scrubbing" action. The scrubbing put too much stress on the excessively loose control-arm bushings. The combination of excessive toe-out and loose control-arm bushings set up a vibrating action in the tire that produced the spotty wear.

LOOSE IDLER ARM PLUS UNBALANCE (Fig. I-55)

This tire was slightly unbalanced. On the road test, the tire seemed to ride normally until the automobile reached the turnpike's maximum speed limit. At high speed, the tire vibrations became apparant. The vibrations activated the loose idler arm and produced a rhythmic toe change. The result was a rather unusual tread wear pattern with maximum wear at 180-degree intervals on the tire.

LOOSE CONTROL ARM BUSHINGS, BALL JOINTS, AND IDLER ARM (Fig. I-56)

This tire tread was worn off quite rapidly (approximately three thousand miles). The vehicle in question had worn upper control arm bushings, loose upper and lower ball joints and a loose idler arm bushing. A close inspection of the tire tread wear will also reveal the effects of slight wheel unbalance, low air pressure, and the effects of incorrect camber and toe-in. **Note:** *With loose suspension parts, proper alignment settings, even if obtainable, would be impossible to retain.*

WHEEL UNBALANCE

One of the most frequent causes of tire wear is unbalanced wheels. This typically shows up as erratic or cupped tire wear as shown in Fig. I-57. This tire also shows considerable camber wear.
least one driver did. This tire also shows considerable camber wear.

Any wheel imbalance imposes severe stress on all suspension parts by forcing them into constant motion, and stress increases in proportion to increases in speed.

A very severe unbalanced condition caused this tire (Fig. I-58) to wear completely through the tread and into the breaker strip. The concentrated wear on one side also indicates an incorrect camber setting plus some degree of improper toe setting.

When vehicles are driven at sustained turnpike speeds, accurate wheel balance is of extreme importance. Two types of unbalance which should be understood are *static unbalance* and *dynamic unbalance.*

A static unbalance forces one or more portions of the tire heavily against the pavement with each revolution of the wheel. Figs. I-57 and I-58 are typical examples of static unbalance.

A dynamic unbalance (where the weight is on one side of the tire) causes the wheel to oscillate in addition to bouncing and produces a scuffing condition that soon wears the tread into a unique pattern of flat spots. The tire shown in Fig. I-59 illustrates the beginning of a typical dynamic unbalance condition. Dynamic unbalance subjects suspension and steering parts to even greater stress than static unbalance.

FIGURE I-57 Unbalance plus incorrect camber setting. (Courtesy of Moog Automotive, Inc.)

FIGURE I-59 Typical dynamic unbalance. (Courtesy of Moog Automotive, Inc.)

FIGURE I-58 Service unbalance with incorrect camber setting plus incorrect toe setting. (Courtesy of Moog Automotive, Inc.)

DIAGNOSING ABNORMAL TIRE WEAR

Student name ______________________________ Date ______________

1. Tire size: ______________ Make of car: ______________________________

   Position on car: LF______ RF______ LR______ RR______

   Tire wear pattern: ____

   ____

   Cause of abnormal wear: ____

   ____

   ____

   ____

   Shade in the area of tire wear:

FIGURE I-60

2. Tire size: ______________ Make of car: ______________________________

   Position on car: LF______ RF______ LR______ RR______

   Tire wear pattern: ____

   ____

   Cause of abnormal wear: ____

   ____

   ____

   ____

   Shade in the area of tire wear:

FIGURE I-61

3. Tire size ________________ Make of car ______________________________

Position on car: LF______ RF______ LR______ RR______

Tire wear pattern: ____

____

Cause of abnormal wear: ____

____

____

____

Shade in area of tire wear:

FIGURE I-62

4. Tire size: ________________ Make of car: ______________________________

Position on car: LF______ RF______ LR______ RR______

Tire wear pattern: ____

____

Cause of abnormal wear: ____

____

____

____

Shade in area of tire wear:

FIGURE I-63

DATE COMPLETED________________ INSTRUCTOR'S SIGNATURE ________________________

Working alone, you will check a minimum of four tires with a dial indicator for radial and lateral run out. Report in writing to your instructor the condition of each tire.

OBJECTIVE I-B2 (manipulative)

Preparation:

1. Review page 15 on wheel lug-nut torque.
2. Study pages 45 through 50.

Materials and Equipment Needed:

1. Wheel cover remover
2. Dial indicator or run-out gauge
3. Base or attaching device for dial indicator or run-out gauge
4. Wheel weight hammer
5. Jack
6. Jack (safety) stands
7. Chalk

RADIAL RUN-OUT

Radial run-out is the amount the tire or rim is out of round. Run-out can be checked by placing a dial indicator against the tread area in the center of the tire and then placing it on the inside of the wheel rim as illustrated in Figs. I-64 and I-65. Total radial run-out should not exceed 0.060 inch.

FIGURE I-65 Radial run-out check at the inside of the wheel rim. The difference between the radial run-out at the tread and this run-out is the amount the tire is out of round. (Courtesy of Cadillac Motor Car Division)

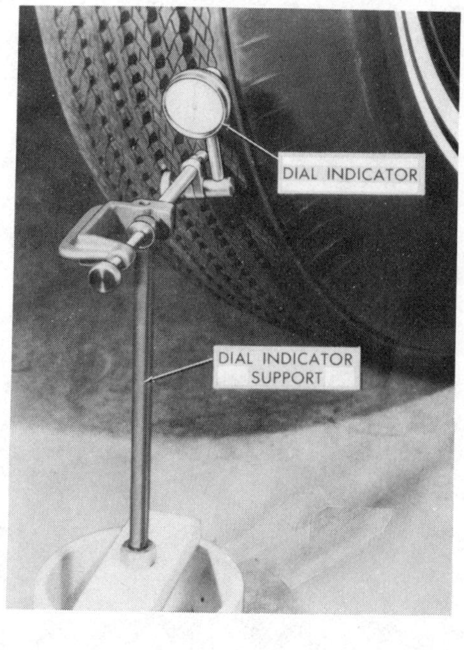

FIGURE I-64 Radial run-out check at the tread area. This is the total run-out of the tire and wheel rim assembly. (Courtesy of Cadillac Motor Car Division)

LATERAL RUN-OUT

Lateral run-out is the amount of in-and-out movement (wobble) that the tire or wheel has on one side as it rotates. The illustrations in Figs. I-66 and I-67 show the proper placement of the dial indicator to check lateral run-out. Maximum allowable lateral run-out is 0.080 inch.

When radial or lateral run-out exceeds the set limits, the position of the tire on the wheel rim should be rotated by 180 degrees and rechecked.

FIGURE I-66 Lateral run-out check at the tire sidewall. This is the total lateral run-out of the tire and wheel rim assembly. (Courtesy of Cadillac Motor Car Division)

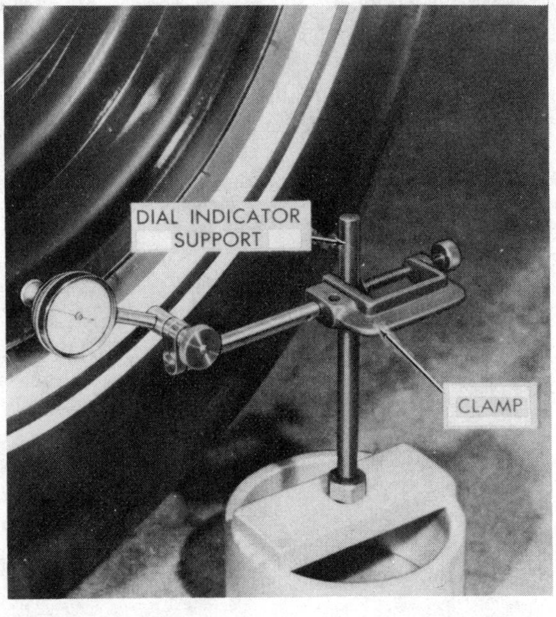

FIGURE I-67 Lateral run-out check at the side of the wheel rim. The difference between the lateral run-out at the tire sidewall and the lateral run-out at the side of the wheel rim is tire lateral run-out or amount of wobble. (Courtesy of Cadillac Motor Car Division)

Recommended Procedure:

1. Raise Automobile
   a. Lift wheels to be checked off the floor.
   b. Place jack (safety) stands under the frame of the car. (**Caution:** *Place stands so that the car will not slip or fall.)*

2. Prepare Wheels for Run-Out Check
   a. Remove wheel covers.
   b. If you are not ready to balance the wheels, mark rim with chalk for the position of each wheel weight.
   c. Remove all wheel weights, mud, and dirt from both inside and outside of wheel rim.
   d. Remove all rocks and other material from tire tread.
   e. Set dial indicator or run-out gauge as per equipment instruction.

3. Check Radial and Lateral Run-Out
   a. Using Figs. I-64 and I-65 as examples, set up equipment and check radial run-out on both the wheel rim and the tire tread for total radial run-out.
   b. Using Figs. I-66 and I-67 as examples, set up equipment and check lateral run-out on both the wheel rim and the tire sidewall for total lateral run-out.
   c. Record the findings on the run-out check report sheet.

4. Finalize Job
   a. Balance all wheels or replace the weights to their original position.
   b. Replace wheel covers.
   c. Lower car.

RUN-OUT CHECK

Student name ______________________________ Date ________________

Make of car ________________ Model ________________ Year ________________

Tire Size ____________

Run-out Check Report

Radial run-out limit specifications: ________________

Lateral run-out limit specifications: ________________

Source or reference for specifications: ______________

Left Front Wheel

Radial run-out

At wheel rim: ________________

At tire tread: ________________

Difference between readings: ____________

Lateral run-out

At wheel rim: ____________

At tire sidewall: __________

Difference between readings: ____________

Describe tire and wheel condition: ____

____

____

Right Front Wheel

Radial run-out

At wheel rim: ________________

At tire tread: ________________

Difference between readings: ____________

Lateral run-out

At wheel rim: ____________

At tire sidewall: __________

Difference between readings: ____________

Describe tire and wheel condition: ____

____

____

Left Rear Wheel

Radial run-out

At wheel rim: ______________________

At tire tread: ______________________

Difference between readings: ______________________

Lateral run-out

At wheel rim: ______________________

At tire sidewall: ______________________

Difference between readings: ______________________

Describe tire and wheel condition: ______________________

____

____

Right Rear Wheel

Radial run-out:

At wheel rim: ______________________

At tire tread: ______________________

Difference between readings: ______________________

Lateral run-out

At wheel rim: ______________________

At tire sidewall: ______________________

Difference between readings: ______________________

Describe tire and wheel condition: ______________________

____

____

DATE COMPLETED: ______________________ INSTRUCTOR'S SIGNATURE ______________________

Working alone and without reference material, you will draw a diagram illustrating the correct procedure of tire rotation for each of the following.

1. Bias and bias-belted tires using only four tires.
2. Bias and bias-belted tires using five tires.
3. Radial tires using only four tires.
4. Radial tires using five tires.

OBJECTIVE I-B3 (written assignment)

Preparation:

1. Review material for abnormal tire wear, pages 37 through 44.
2. Study pages 51 through 53.

TIRE ROTATION

Tires should be rotated and run at different wheel positions in accordance with car manufacturer's recommendations or every 5,000 miles. Rotation equalizes wear, thereby increasing mileage, handling, and riding comfort. Air pressure should be checked after each rotation to insure the proper front and rear tire pressure relationship.

Both traditional bias and bias belted tires can use the original cross-width rotation with a slightly different pattern. The tire diagram at the left of Figure I-68 will illustrate the two different ways to rotate tires with bias and bias belted type cords. In the past, the general practice was to move front tires straight back to the rear while cross-switching the rear tire to the front. As shown in the illustration, the recommended procedure has changed. The two front tires are now cross-switched to the rear, while the rear tires are moved straight forward.

Cross-switching is not recommended for radial ply tires. In four tire rotation of radial tires, the front tires are moved straight back and the rear tires are moved straight forward. See Figure I-68 for the proper procedure for rotating tires using both the four and five tire method.

Cross-switching is not recommended on radial tires because after being mounted on one side of the car and being driven for a period of time, the tire develops a radial set in the cords. When the tire is switched to the other side of the car the radial set must be reversed by driving. This will cause a temporary out-of-round condition, that can be very uncomfortable to the driver and dangerous to the life of the tire.

Caution: *Do not cross switch radial tires.*

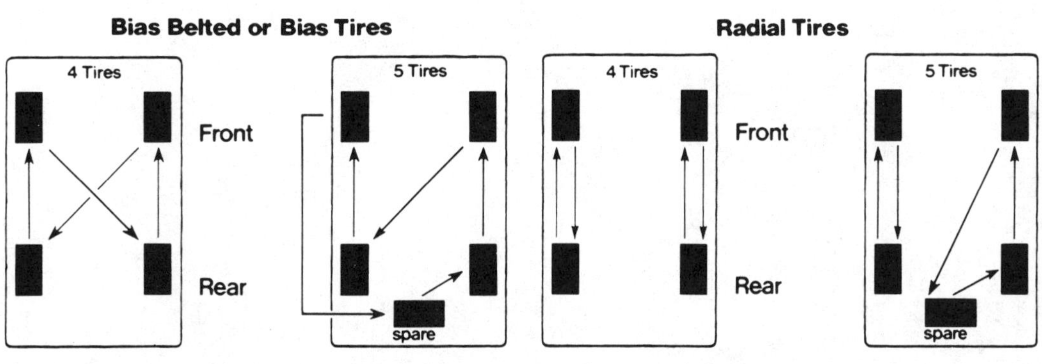

FIGURE I-68 Recommended tire rotation methods for bias, bias-belted, and radial tires

TIRE ROTATION

Student name______________________________Date____________________

Draw lines with arrows between the tires to indicate rotation pattern in each illustration below:

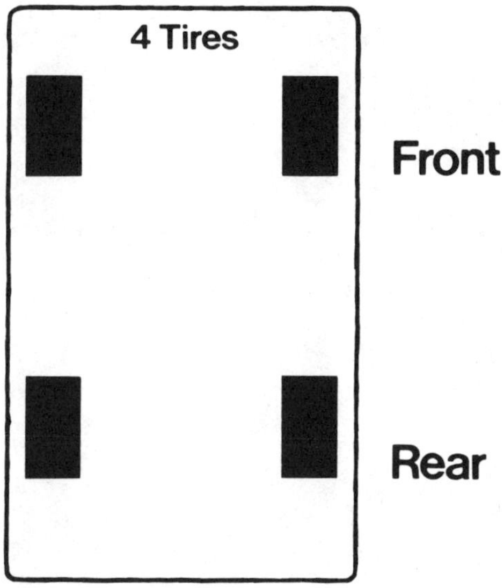

a. Four tire bias and bias-belted rotation.

FIGURE I-69

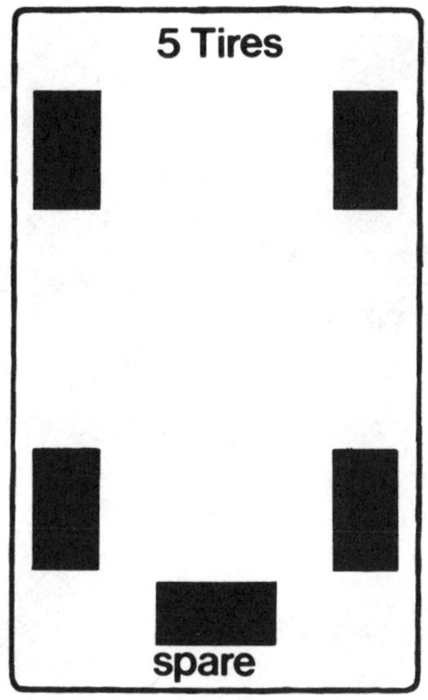

b. Five tire bias and bias-belted rotation.

FIGURE I-70

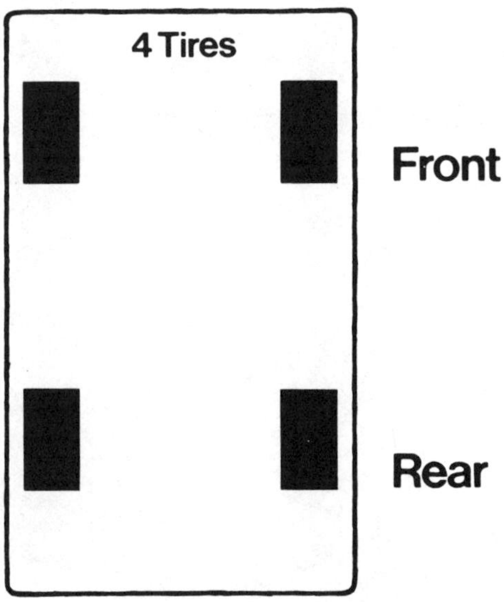

c. Four tire radial rotation.

FIGURE I-71

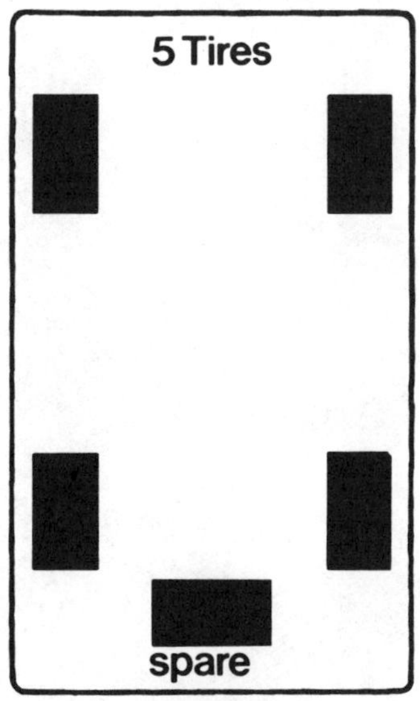

d. Five tire radial rotation.

FIGURE I-72

DATE COMPLETED ____________________INSTRUCTOR'S SIGNATURE ____________________

Student name____________________Date________________

Directions: Select the best answer to each question and circle the correct letter at the left.

1. A tire with the plies crisscrossing from sidewall to sidewall is a:
   A. Bias ply
   B. Bias belted
   C. Bias radial
   D. Radial
2. A tire with plies that crisscross from side to side and encircle the tire under the tread is a:
   A. Bias ply
   B. Bias belted
   C. Bias radial
   D. Radial belted
3. A tire with plies laid at right angles to the circumference of the tire is a:
   A. Bias ply
   B. Bias belted
   C. Bias radial
   D. Radial
4. A bias-ply tire has the cords criss-crossed from side to side at a:
   I. 50-degree angle
   II. 30-degree angle
      A. I only
      B. II only
      C. Both I and II
      D. Neither I nor II
5. A tire that is a 78 series has the:
   I. Height 78 percent of the width
   II. Width 78 percent of the height
      A. I only
      B. II only
      C. Both I and II
      D. Neither I nor II
6. The label P 205/75 R 15 on a tire sidewall means it is a
   A. Bias-belted tire
   B. Metric tire
   C. Tire used on 75 mm wheel rims
   D. 25 inch tire
7. The label FR 78-14 is a (an):
   I. Alphanumeric tire
   II. Tire with a 14 percent aspect ratio
      A. I only
      B. II only
      C. Both I and II
      D. Neither I nor II

8. Wheel rim diameter is measured to the:
   I. Edge of the flange
   II. Bead surface of the rim
      A. I only
      B. II only
      C. Both I and II
      D. Neither I nor II
9. Radial tires may be used on the car in:
   I. Sets of four only
   II. Pairs only
      A. I only
      B. II only
      C. Both I and II
      D. Neither I nor II
10. The highest air pressure recommended for a tire with a load range of B is:
   A. 28 psi
   B. 32 psi
   C. 36 psi
   D. 38 psi

DATE COMPLETED______________________INSTRUCTOR'S SIGNATURE ______________________

OBJECTIVE I-C (written test)

After studying the preparation assignment you will be expected to answer a minimum of seven out of ten multiple choice questions in a written test, without reference materials, on the following topics:

1. Wheel balance
2. Radial and lateral run-out
3. Static wheel balance
4. Dynamic wheel balance

Preparation:

1. Review pages 45 through 50.
2. Study pages 57 through 90.

WHEEL BALANCE

An out-of-balance wheel assembly is one of the major factors in vibration and instability of the front end at medium to high speeds (40 to 60 mph or 64–97 km/h). To eliminate vibration and instability, the wheel must be balanced to a point where the weight distribution around the circumference of the wheel and between the inside and outside halves of the wheel assembly is equal.

The wheel assembly that we are concerned with is the part that rotates on the spindle.The tire, wheel, rotor or drum,hub, and at least part of the front wheel bearings make up the rotating wheel assembly.

A P205/75R14 tire traveling at 60 mph (97 km/h) makes approximately 800 revolutions per minute. At these speeds an out-of-balance tire disturbs the centrifugal force created by the spinning wheel assembly and may cause a bounce or wobble. The driver can lose control of the automobile because of this bounce or wobble effect and even if he doesn't, damage to suspension and steering parts of the automobile may occur.

Two types of out-of-balance conditions cause trouble. A *static out-of-balance* wheel has a heavy spot somewhere around the circumference of the wheel assembly. The result can be a wheel assembly that hops up and down as the wheel rotates at certain speeds (Fig. I-73). This same condition occurs at all speeds when the wheel assembly has too much radial run-out. A *dynamic out-of-balance* wheel has a heavy spot that is not centered in the tread area but is probably on the outer edge of the wheel assembly. The result can be a shimmy or wobble. This same shimmy or wobble can occur because of excessive lateral run-out. It is possible that a wheel assembly may be in perfect static balance but still be out of dynamic balance.

STATIC WHEEL BALANCE

The term "static," as used in connection with wheel balance, means "having no motion" or "at rest." A tire that is in perfect static balance will stand motionless or at rest on its own axis. The wheel assembly illustrated here is mounted on an off the car dynamic wheel balancer (Fig. I-74). The wheel will turn on the machine spindle as a result of the smallest out-of-balance condition. If a wheel assembly is in static balance, it will remain in any position on the spindle without rotating.

The wheel will hold in that position if it is in static balance. When the wheel assembly is out of balance, the tire will rock back and forth until the heavy spot comes to rest at the bottom of the wheel.

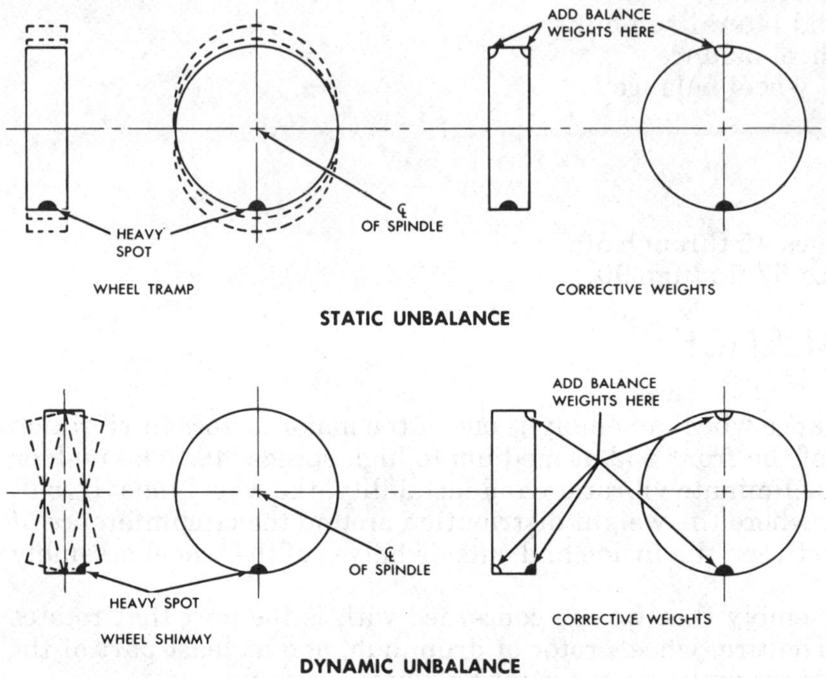

FIGURE I-73 Dynamic and static out-of-balance conditions. (Courtesy of Chevrolet Motor Division)

FIGURE I-74 If this wheel will hold in any position in which it is set without rotating on its own axis, it can be said that the wheel is in static balance.

The most common static wheel balancer in use is usually referred to as a bubble balancer because it uses a level bubble to indicate the light side of a wheel (Fig. I-75).

Complete instructions on the use of static (bubble) balancer can be found on pages 60 through 65. Follow the instructions carefully or you may create a dynamic wheel balance problem simply by placing weights improperly on the wheel rim.

FIGURE I-75 Typical static (bubble) wheel balancer (Courtesy of FMC Corp., Automotive Service Equipment Division)

DYNAMIC WHEEL BALANCE

A wheel assembly may be in perfect static balance, but at the same time it may be out of dynamic balance. The meaning of the word "dynamic" is "of or pertaining to energy, force, or *motion* in relation to force." Stated in more simple terms, it means balanced while in motion to correct side to side motion of the wheel.

A wheel that is out of dynamic balance will shimmy or wobble when in motion.

When a tire is out of balance, the overweight condition may be evenly distributed across the tread area, as illustrated in Fig. I-76, or it may be off to one side. Both of these conditions cause a static out-of-balance condition. Static balance is achieved simply by placing wheel weights 180 degrees around the wheel from the overweight spot. These wheel weights must be equal to the amount of the overweight spot, then split, with half of the weight on the inside and half on the outside of the wheel rim. This will place the wheel in static balance but not necessarily in dynamic balance. Where the overweight spot is off to one side of the tread, the spot will try to reach the true plane of rotation (the center of the tread) on every half revolution, thus causing the wheel to shimmy or wobble while in motion (Fig. I-77).

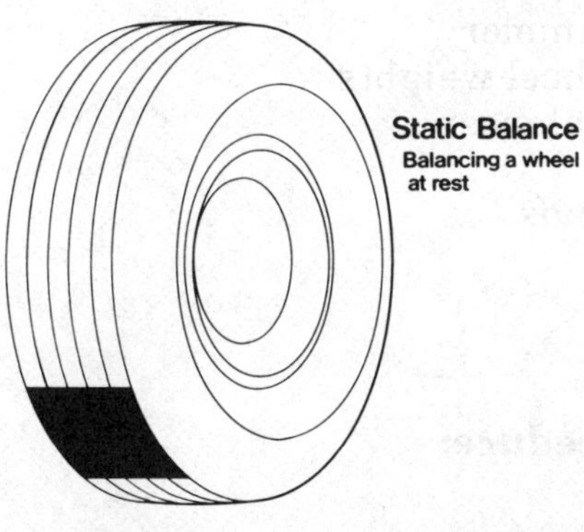

FIGURE I-76 The wheel is out of static balance when the weight is not evenly distributed around the circumference of the wheel assembly.

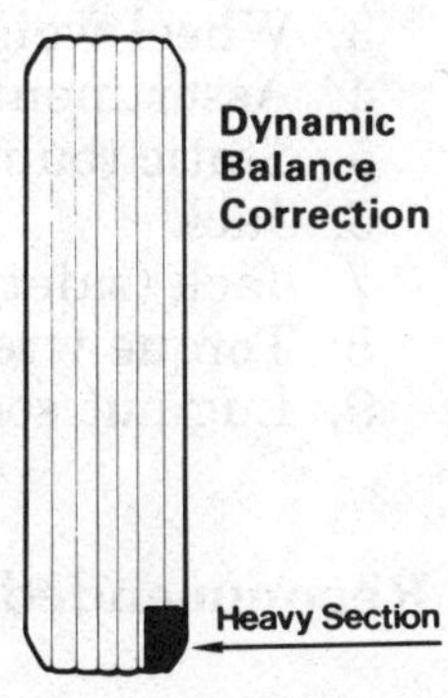

FIGURE I-77 A wheel out of balance dynamically will have the heavy spot off to one side.

The type of balancing weight used on each wheel will vary, depending on the type of wheel and the type of wheel cover used. Three basic designs of wheel weights are shown in Fig. I-78 with an explanation of where each is used.

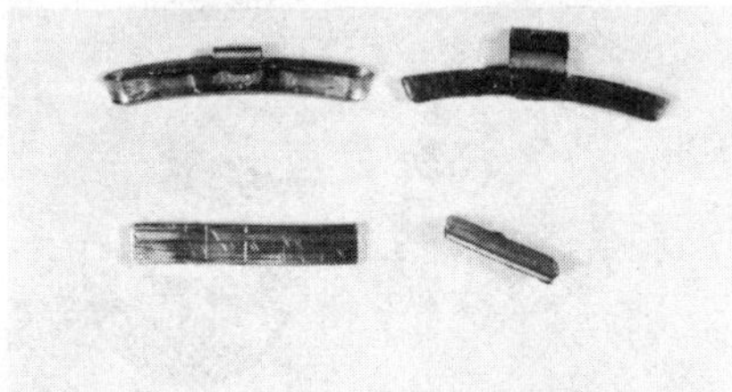

FIGURE I-78 Typical wheel weights. Top left—regular. Top right—full hub cap type. Two lower weights—stick on weights for mag wheels.

OBJECTIVE I-C1 (manipulative)

Working alone with the aid of the material in this text, you will correctly static balance a minimum of four wheels using a static type, off-the-car (bubble) wheel balancer. Maximum time limit: 20 minutes per wheel.

Preparation:

1. Review pages 57 through 60.
2. Study pages 60 through 65.
3. Carefully observe instructor's demonstration on static wheel balance procedures.

The method and illustrations used for this instruction are those for the Acra-speed Static Balancer manufactured by the FMC Corporation. However, the basic principles explained here apply to most static balancers in commercial use.

Materials and Tools Needed:

1. Hubcap remover
2. Lug wrench
3. Wheel weight hammer
4. Assortment of wheel weights
5. Static (bubble) balancer
6. Jack
7. Jack (safety) stands
8. Torque wrench
9. Lug nut socket

Recommended Procedure:

1. *Raise automobile*
   a. Lift wheels to be balanced off the floor.

b. Place jack (safety) stands under frame of the car. (**Caution:** *Place stands so that the car will not slip or fall.)*

2. *Remove wheels and prepare for balancing*
   a. Remove the wheel cover.
   b. Remove lug nuts. (Place inside the wheel cover for safekeeping.)
   c. Remove wheel. (Identify wheel with chalk as to position on car (e.g., RF for "right front".)
   d. Remove rocks and other material from tire tread.
   e. Remove all weights from wheel rim.
   f. Clean any mud or dirt from wheel rim.
   g. Level the balancer by adjusting one or all of the three adjustable legs while the handle is in the "off" position (Fig. I-79).

Note: *Since your position to the balancer is important for its most efficient operation, always stand at the front or handle side of the machine. This places the handle and wheel weights within easy reach at all times.*

Once the wheel is off the car, proceed with the following steps:

3. *Balance wheels*
   a. With the handle in the "off" position, place the wheel on the balancer, with the outside of the wheel facing up and the valve stem in line with the black arrow on the level face (Fig. I-80). Be certain the wheel is centered on the leveling flange.
   b. Move the handle to the "on" position. Steady the wheel by lightly placing your hands on the tread of the tire. Remove your hands and allow the wheel to come to rest. When the bubble stands still, it will

FIGURE I-79 Adjustment for leveling static wheel balancer.

FIGURE I-80 Place wheel with valve stem in line with arrow on balancer head.

have moved to the light side of the wheel. Rotate the wheel until the bubble is directly in front of you. This will place the heavy side of the wheel away from you and directly in line with the bubble (Fig. I-81).

c. Holding the wheel stationary with one hand, move the handle to the "off" position and then back to the "on" position. This recenters the balancing head and allows it to float freely on its pivot point (Fig. I-82). It is important to recenter the balancing head each time the wheel is bumped or turned during the balancing process.

d. With the handle in the "on" position and the wheel assembly (bubble) at rest, you are ready to select the amount of weight needed to balance the wheel. Use the three illustrations in Figs. I-83 through I-85 to help select the proper size weight.

In Figure I-83 the bubble is close to the cross hairs and will possibly need four ¼- or ½-ounce (10 g or 20 g) weights. Figure I-84 has the bubble just touching the inside ring. This wheel may need four ¾- or 1-ounce (20 g or 30 g) weights. The bubble in Figure I-85 is tight against the outside of the level and might take four 2-ounce or larger weights. *(Notice that you always use four weights of equal size on the wheel.)*

e. With the handle in the "on" position, place one weight of each pair on the rim flange. These weights should be evenly spaced from the center line of the bubble (Fig. I-86). Place the other weight of each pair directly behind the first weight. Move the pairs of weights together or apart around the rim, maintaining equal distance from the center of the original light point.

If in moving your weights for proper balance you have to go beyond a 90-degree angle, you have selected weights that are too heavy.

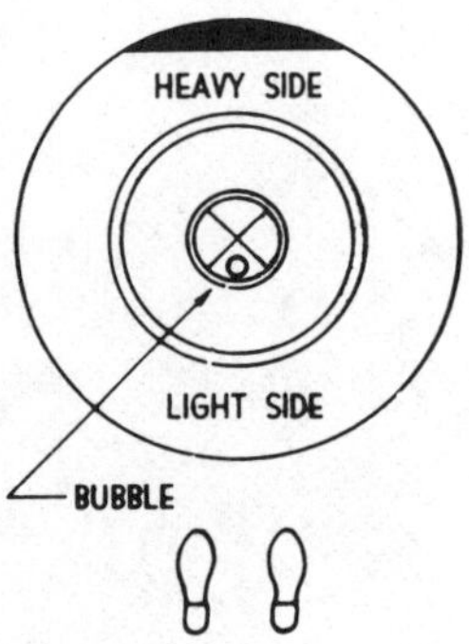

FIGURE I-81 Turn the wheel assembly so that the light spot of the wheel is on the operator's side of the machine. (Courtesy of FMC Corporation, Automotive Service Equipment Division)

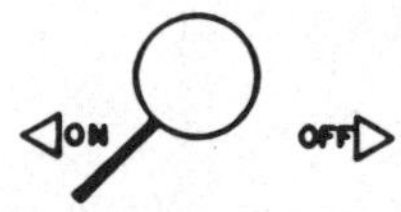

FIGURE I-82 Turn handle to the "off" position, then to the "on" position to recenter balance after turning the wheel. (Courtesy of FMC Corporation, Automotive Service Equipment Division)

FIGURE I-83 With the bubble close to the cross hairs, the wheel will require 2 pair of ¼- or ½-ounce weights. (Courtesy of FMC Corporation, Automotive Service Equipment Division)

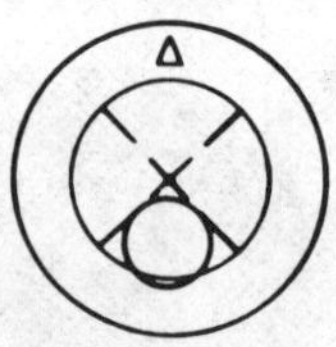

FIGURE I-84 With the bubble in this position, it will require two pair of ¾- or 1 ounce weights. (Courtesy of FMC Corporation, Automotive Service Equipment Division)

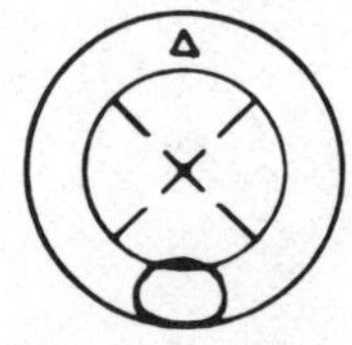

FIGURE I-85 With the bubble touching the outside ring, it will require two pair of 2-ounce or larger weights. (Courtesy of FMC Corporation, Automotive Service Equipment Division)

Exchange the four weights for four equal but lighter weights and try again. If the weights are closer than 6 inches (154 mm) from center to center, the four weights are not heavy enough to balance the tire and should be exchanged for four equal but heavier weights.

f. With the handle moved to the "off" position, draw a chalk line on the tread that is in line with the center of each pair of weights (Fig. I-87). To accurately locate these lines, sight through the center of the weights to the center of the balancer.

g. With the handle still in the "off" position, remove both pairs of weights from the wheel and lift the wheel from the balancer. Attach two of the weights to the inside or underside of the rim. Each weight should be attached directly in line with one of the chalk lines on the tread of the tire (Fig. I-88).

h. Replace the wheel on the balancer with the outside facing up. Then place one of each of the remaining weights on the rim in line with the chalk marks (Fig. I-89).

i. Move the handle to the "on" position and check for proper balance. If necessary, move the two weights closer together or further apart until the bubble centers on the cross hairs (Fig. I-90). Move the handle to the "off" position and attach the two remaining weights to the outside

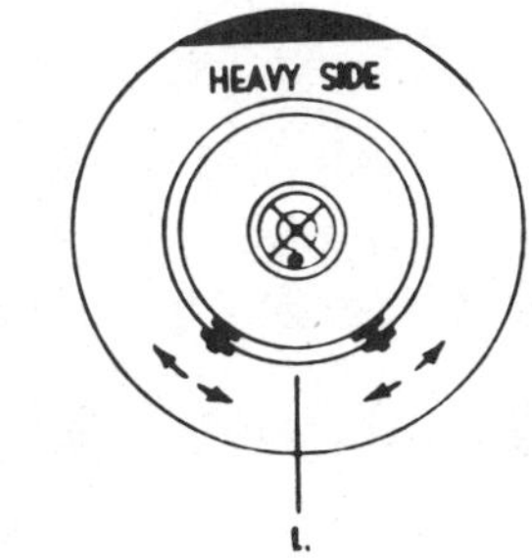

FIGURE I-86 Position pairs of weights on rim and tire. Move the pairs either together or apart around the rim, maintaining equal distance from the light point. (Courtesy of FMC Corporation, Automotive Service Equipment Division)

4. *Replace wheels*
   a. Tighten lug nuts to proper torque using a torque wrench.
   b. Replace wheel covers.
   c. Lower vehicle.

5. *Road test*

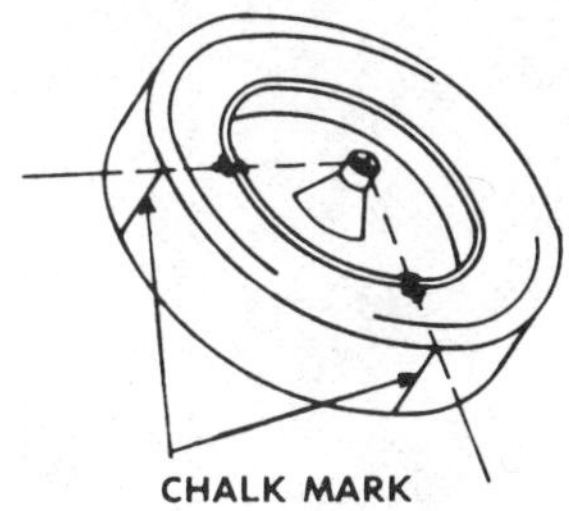

FIGURE I-87 Draw a vertical line with chalk on the tire tread to show position of weights on the backside of the rim. (Courtesy of FMC Corporation, Automotive Service Equipment Division)

FIGURE I-88 Secure weights on back side of the wheel rim, in line with the chalk marks.

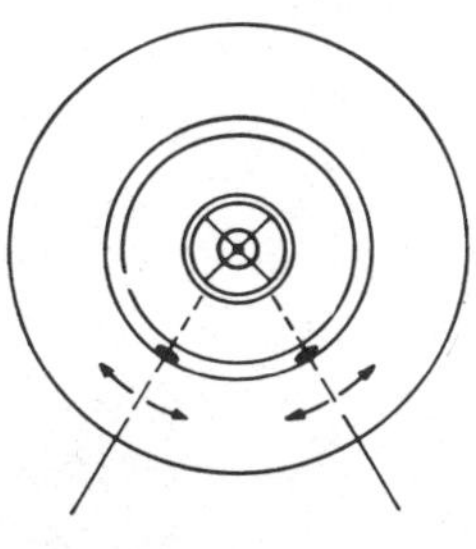

FIGURE I-89 Place two remaining weights on the wheel rim and move slightly together or apart to center bubble. (Courtesy of FMC Corporation, Automotive Service Equipment Division)

FIGURE I-90 With the handle in the "off" position, attach the two outside weights.

STATIC (BUBBLE) BALANCE

Student name ______________________ Date ____________

Make of car ______________ Model ______________ Year ________

Time started ______________ Time finished ______________ Total time

Flat-rate time ______________________________

Checkup Questions

1. How many weights did you use on each wheel? ____________
2. Were the weights for each wheel all the same size? ____________ Yes No
3. Record the size of weights used on each wheel.

   a. Left front: ____________ b. Right front: ____________

   c. Left rear: ____________ d. Right rear: ____________
4. Did your road test indicate any problems? ____________ Yes No

   If so, what were they? ______________________________

DATE COMPLETED ____________ INSTRUCTOR'S SIGNATURE ____________

OBJECTIVE I-C2 (manipulative)

Working alone with the aid of the material in this text, you will static and dynamic balance a minimum of four wheels on an instructor assigned automobile. You will use an on the car, spin-type balancer. Time recommended: 20 minutes per wheel.

Preparation:

1. Review pages 57 through 60.
2. Study pages 67 through 77.

The procedure used here is for an electronic strobe light type of on-the-car spin balance. **Note:** *If the balancing equipment in your shop is not the same type used in this text, your instructor will provide you with instructions for that particular balancer. Place that information in the space provided on pages* 75 *through* 77 .

Materials and Tools Needed:

1. Wheel cover remover
2. Wheel-weight hammer
3. Assortment of wheel weights
4. Wheel spinner (Fig. I-91).
5. Balancing head and attachment or strobe light
6. Jack and jack (safety) stands
7. Chalk or crayon

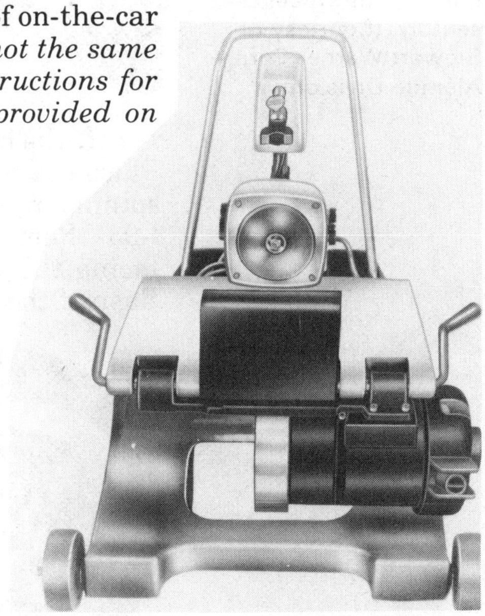

FIGURE I-91 Electronic on the car balancer capable of balancing a wheel with both static and dynamic problems. (Courtesy of Stewart Warner Corp., Alemite Division)

PREPARATION FOR BALANCING

Place a jack under the center of the front cross member and raise both front wheels about 2 inches (50 mm). Position jack (safety) stands under the frame so that if the jack slips, the tire will not touch the ground. Remove all wheel weights and dirt from both inside and outside of the wheel. Check for brake drag by rotating wheel. On drum brakes, it is possible to reduce drag by tapping on the backing plate to center the brake shoes in the drum. If this does not work, use a brake spoon to back off the brake shoes. When the car is equipped with disc brakes, spread calipers by using the method recommended by the manufacturer. Wheel bearings must be checked, and replaced or adjusted as necessary. Stones or other objects that could fly off during spinning should be removed from the tire tread.

RUN-OUT CHECK

Check for both radial and lateral run-out using the procedures outlined earlier in this unit. Make corrections on any excessive run-out before continuing. These corrections may be made by rotating the tire 180 degrees on the wheel rim and rechecking for run-out. If rotating the tire does not correct the problem, replacement of the tire or wheel rim will be necessary.

POSITIONING EQUIPMENT

The electronic pickup unit is placed flat on the ground with the pickup magnet extended to a flat surface on the lower control arm (Fig. I-92). Tighten the thumb screw to secure the magnet in place.

FIGURE I-92 Correct position of electronic pickup for up-and-down movement of the wheel assembly. (Courtesy of Stewart Warner Corp., Alemite Division)

Lift the back end of the spinner by the handle and roll it towards the tire. Place the brake plate squarely on the tire tread and lower the handle. The spinner motor should then be 2 inches (50 mm) from the tire tread. (Fig. I-93). Remove the strobe light from its holder and position it on the outside, facing the wheel. Tap the wheel with your hand. The strobe light should flash if the pickup is positioned properly.

FIGURE I-93 Spinner should be one or two inches from the tire tread. (Courtesy of Stewart Warner Corp., Alemite Division)

REFERENCE MARK

Apply a reference mark anyplace on the outside of the tire (Fig. I-94). Yellow tire-marking crayon works very well.

CHECKING FOR BALANCE

Depress the spinner control handle using just enough pressure to spin the wheel. Point the strobe light at the wheel and watch the meter (Fig. I-95). If the wheel is out of balance, the needle will rise as the wheel picks up speed.

When the needle reaches its peak rise on the scale, it will start to drop. At this point release the spinner control and continue to watch the meter on the strobe light.

As the wheel speed decreases, the needle will again rise. When the needle starts to drop off this time, look at the tire and note the position of the reference mark. Raise the brake control handle and apply just enough pressure to bring the wheel to a smooth stop. **Caution:** *Don't jam the brake pad against the tires.*

FIGURE I-94 Place a wide chalk mark on the tire. (Courtesy of Stewart Warner Corp., Alemite Division)

FIGURE I-95 Spin the wheel and watch for the highest meter reading. (Courtesy of Stewart Warner Corp., Alemite Division)

POSITIONING WEIGHT

A simple way to remember the position of the reference mark is to compare its position to the hour hand on a clock. For example, if the reference mark appeared at the top, it would be twelve o'clock (Fig. I-96). At the bottom it would be six o'clock.

Turn the wheel and place the reference mark in the same position it was in when the needle on the strobe light meter started to drop for the second time.

The heavy spot of the wheel assembly is now on the bottom. With the wheel held in this position, place two equal weights on the rim both at exactly twelve o'clock (Fig. I-97). One must be placed on the inside of the rim and the other on the outside of the rim. The two weights must equal the total shown on the meter.

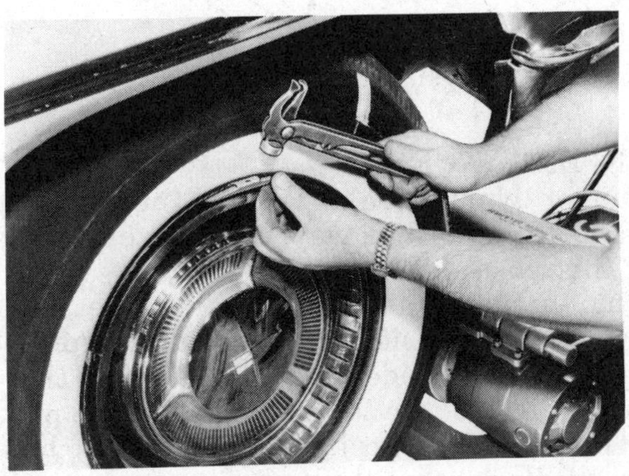

FIGURE I-97 Rotate the chalk mark to the position where the meter reads the highest, and place the proper size weight at twelve o'clock. (Courtesy of Stewart Warner Corp., Alemite Division)

FIGURE I-96 Visualize the hands of a clock on the wheel. (Courtesy of Stewart Warner Corp., Alemite Division)

When the needle on the strobe light rises above the center of the scale, use two 1½ ounce (40 g) weights. The first line on the strobe light meter represents ½-ounce (10 g); the distances between the other lines represent 1 ounce (30 g) each. Do not use more than 3 ounces (80 g) of weight on the initial balance attempt.

RECHECKING FOR BALANCE

Recheck the balance of the wheel by spinning the wheel again while pointing the strobe light at the tire. Watch the meter. If the needle remains in the green section of the meter, the wheel assembly is in both static and dynamic balance.

However, if the needle moves out of the green section, it indicates there is still a slight unbalance. Note the position of the reference mark on the tire when the needle starts to drop after it reaches its highest point with the wheel coasting.

Stop the wheel and place the reference mark at the same position noted. Check the position of the weight and refer to Fig. I-98. Reposition or change the amount of weight as indicated on the manufacturer's chart. Repeat this operation until the needle remains in the green section on the strobe light meter.

Note: *If after several attempts the needle does not stay in the green section, proceed to dynamic-balance.*

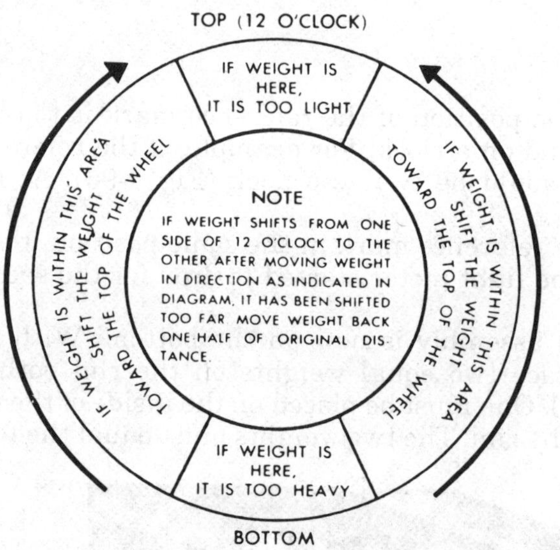

FIGURE I-98 Instructions for repositioning of weights after rechecking for up-and-down vibrations. (Courtesy of Stewart Warner Corp., Alemite Division)

DYNAMIC BALANCE

To set up for dynamic balance, turn the wheel outward at the front about halfway and leave it that way. (It may be necessary to lock the steering wheel in this position.) Place the pickup so that the pickup magnet will contact the backing plate at the front edge as close to the center line as possible (Fig. I-99). Position the spinner as previously outlined. With the strobe light on the outside but pointed at the tire, spin the wheel and let it coast slowly. Tighten the thumb screw on the pickup arm. This will set the pickup in place. If the wheel is out of dynamic balance, the needle will rise. When the needle reaches its peak rise on the scale and as it begins to drop off, release the spinner and allow the wheel to coast.

FIGURE I-99 Electronic pickup position for checking wobble or side-to-side movement of the wheel assembly. (Courtesy of Stewart Warner Corp., Alemite Division)

As the wheel speed decreases, the needle will rise again. This time, when the needle begins to drop off, note the position of the reference mark on the tire.

Bring the wheel to a smooth stop with the brake on the spinner, and position the reference mark as noted while the wheel was coasting. Loosen the thumb screw on the pickup arm before attaching the dynamic weights.

POSITIONING DYNAMIC WHEEL WEIGHTS

Two weights are used to correct the dynamic unbalance condition. Each weight is equal to the amount indicated on the strobe light meter (Fig. I-100). For example, if the meter showed 1 ounce (30 g), two 1-ounce (30 g) weights would be used.

Place the wheel with the reference mark in the same position that it was in when the meter reached its highest point while coasting. Place one weight on the inside of the rim at the front of the wheel, in line with the pickup magnet, and the second weight on the outside of the rim at the rear of the wheel, 180 degrees away from the first weight.

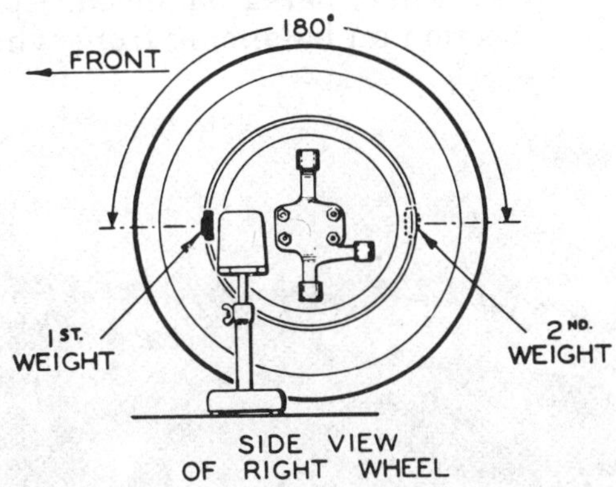

FIGURE I-100 Placing weights for correction of dynamic out-of-balance condition. (Courtesy of Stewart Warner Corp., Alemite Division)

RECHECKING FOR DYNAMIC BALANCE

Spin the wheel again to check the balance. (Remember to tighten the thumb screw on the pickup while the wheel is coasting.)

With the wheel coasting, point the strobe light at the wheel and watch the meter. If the needle stays in the green section, the wheel is in dynamic balance.

If the needle moves out of the green section, note the position of the reference mark at the highest meter reading. Stop the wheel and position the reference mark.

Check the location of the weights and refer to the charts in Figs. I-101 and I-102. Reposition or change the amount of weight as indicated on the chart.

Repeat this operation until the meter needle will stay in the green section while the wheel is coasting.

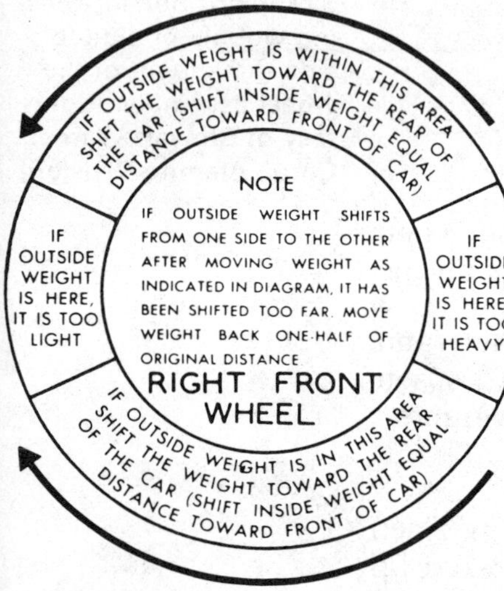

FIGURE I-101 Chart for repositioning dynamic weights on right front wheel. (Courtesy of Stewart Warner Corp., Alemite Division)

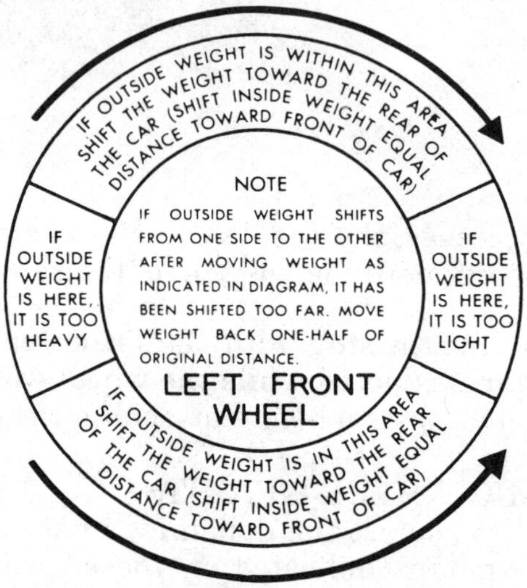

FIGURE I-102 Chart for repositioning dynamic weights on left front wheel. (Courtesy of Stewart Warner Corp., Alemite Division)

BALANCING REAR WHEELS

(See special instruction for limited slip differential.)

1. Remove all weights.
2. Block the front wheels.
3. Lift only one rear wheel by placing a jack under the frame rail and raising the wheel approximately 2 inches off the floor.
4. Position the pickup magnet under the rear axle as close as possible to the wheel being balanced. Apply a reference mark as outlined in the section on balancing front wheels (Fig. I-103).

FIGURE I-103 To balance rear wheels, lift one wheel off the ground and place electronic pickup on rear axle as close to the backing plate as possible. *(Note: When rear axle assembly is a limited-slip differential, see Fig. I-104.)*

5. Have driver start the engine, shift into high gear or drive position, and accelerate slowly. Find the speed at which the meter reading on the strobe light is the highest. *Do not exceed 40 mph on the speedometer, since the wheel is running at twice the speedometer reading.*
6. Apply weights as in front-wheel balancing.
7. Recheck your work. *(There is no way to balance the rear wheels for dynamic balance.)*

SPECIAL INSTRUCTIONS FOR LIMITED-SLIP DIFFERENTIALS

Limited-slip differentials can be identified in one of several ways:

A. A metal identification tag attached to the filler plug.
B. A color-coded filler plug or painted area around the plug (usually red).
C. Code letter(s) stamped into the carrier, cover plate, or axle tube or code letters in the vehicle identification number (VIN).
D. With the engine *off* and the transmission in PARK on an automatic transmission or reverse on a manual transmission, jack both rear wheels off the ground. Now attempt to turn one wheel while watching the other. If the wheel cannot be turned, the differential is the limited-slip type. With a common differential (nonlimited slip), the wheel on the opposite side will turn when the wheel nearest you is turned.

Warning: *Do not block either of the rear wheels at any time when balancing rear wheels of cars with limited-slip differentials.*

1. Place a jack under center of differential (Fig. I-104). Raise rear of car and take off the opposite wheel from the one to be balanced. Be sure to replace all wheel lugs or nuts securely.
2. Follow the normal procedure for balancing of rear wheels.
3. After balancing the first wheel, replace the opposite wheel and balance it. It is not necessary to remove the balanced wheel.
4. When completed readjust brakes where necessary.
5. Road test.

FIGURE I-104 For a limited slip differential, lift both wheels and remove one wheel while you balance the wheel on the other side.

JOB REPORT SHEET I-C2

ON-THE-CAR SPIN BALANCING

Student name______________________________Date______________________

Make of car______________Model______________Year______________________

Time started______________Time finished______________Total time______________

Flat-rate time ______________

The following space is provided for recording instructions, supplied by your instructor, that differ from those given in the text. These instructions are applicable only in cases where the types of equipment used in your shop differ from those listed in this text. However, it should be remembered that most general instructions will apply.

1. *Positioning Equipment*

____

____

____

____

____

____

2. *Balance Front Wheels*
   a. Preparation for balancing:____

   ____

   ____

   ____

   b. Static balance (or balance for up-and-down movement): ______________________

   ____

   ____

   ____

c. Dynamic balance (or balance for side movement). **Note:** *When necessary and when equipment will permit:* ______

______

______

______

______

______

d. Final check for accuracy: ______

______

______

3. *Balance Rear Wheels*
   a. Preparation for balance: ______

   ______

   ______

   ______

   b. Use engine to drive rear wheels. This may be accomplished with remote speed control if available with your equipment. When not available, have assistant control speed from inside the automobile. (**Caution:** *Do not exceed speedometer readings of 40 mph.*)
   c. Static balance (or balance for up-and-down movement): ______

   ______

   ______

   ______

   ______

   ______

   d. Final check for accuracy: ______

   ______

   ______

4. *Lower Automobile*
   a. Readjust brakes where necessary.
   b. Replace hub caps.
5. *Road Test*

Checkup Questions:

1. Describe the number and amount of each weight used on each wheel.

a. Left Front

b. Right Front

c. Left Rear

d. Right Rear

2. Were weights placed on both the inside and the outside of the rim?

_______ Yes _______ No

3. Did your road test indicate any problems? _______ Yes _______ No

If so, what were they? _______________

DATE COMPLETED_______ INSTRUCTOR'S SIGNATURE:_______

Working alone with the aid of the material in this text you will static and dynamic balance a minimum of four wheels on an instructor assigned automobile using an off-the-car dynamic balancer. Time recommended: Maximum time limit of 20 minutes per wheel.

OBJECTIVE I-C3 (manipulative)

Preparation:

1. Review pages 57 through 59.
2. Study pages 79 through 88.

The procedure used is for the Sun 1730 or Nortron 7300 off-the-car balancer.
Note: *If the balancing equipment in your shop is not the same as the* Sun *or* Nortron *balancing equipment, your instructor will provide you with instructions for the particular balancing equipment used. However, it should be remembered that most general instructions will apply.*

Materials and Tools Needed:

1. Wheel cover remover
2. Lug wrench
3. Socket and ratchet for spindle nut
4. Wheel-weight hammer
5. Torque wrench
6. Assortment of wheel weights
7. Off-the-car dynamic wheel balancer
8. Jack
9. Jack (safety) stands

Recommended Procedure:

1. *RAISE AUTOMOBILE*
   a. Lift off the floor those wheels that are to be balanced.
   b. Place jack (safety) stands under frame of car. **Caution:** *Place stands so that car will not slip or fall.*

2. *REMOVE WHEEL AND PREPARE FOR BALANCING*
   a. Remove the wheel cover.
   b. Remove the lug nuts.
   Note: *Certain makes of automobiles have left-hand threaded studs and lug nuts (Fig. I-105) on the left side.*
   c. Inflate the tire to the proper air pressure.
   d. Remove any matter, such as stones or pebbles, caught in the tire thread.
   e. Remove all the old weights, including the stick-on type for mag wheels.
   f. Place the power switch in the OFF position before plugging the power cord of the balancer into the proper AC outlet (Fig. I-106).
   g. Turn the power switch to ON (Fig. I-107).

FIGURE I-105 Left-hand threaded studs and lug nuts

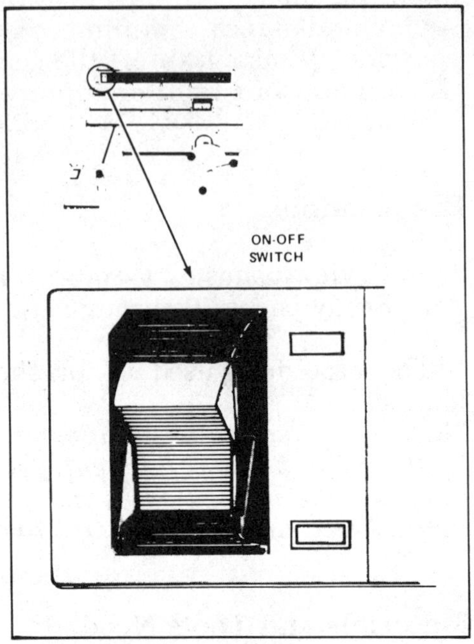

FIGURE I-107 On-off Switch (Courtesy of Sun Electric Corp.)

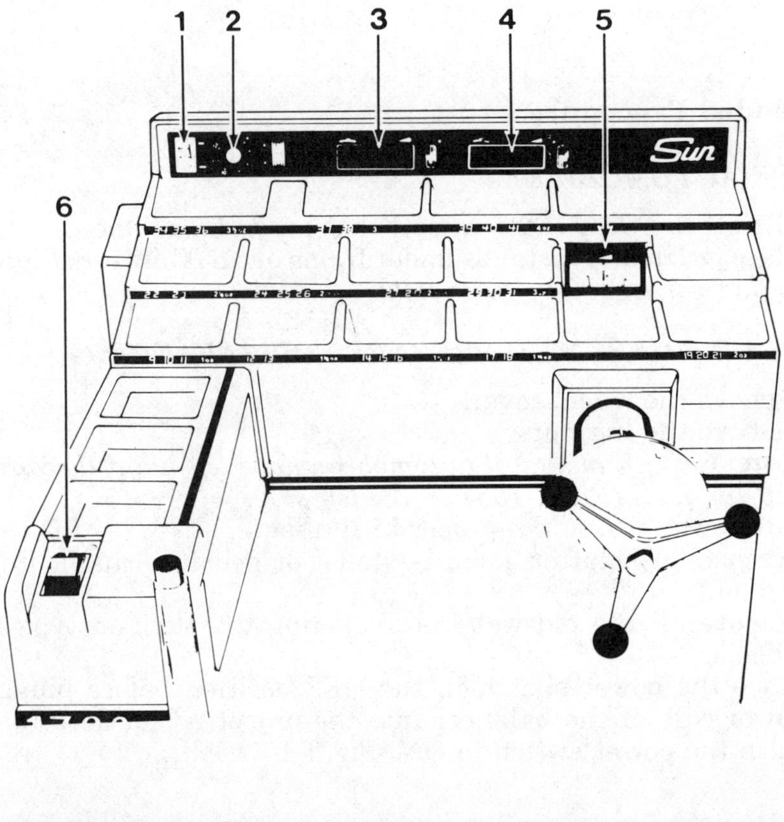

FIGURE I-106 Parts of a Sun 1730 or Nortron 7300 balancer.
1. On-off switch
2. Rim diameter control
3. Outer weight and position control
4. Inner weight and position control
5. Position window
6. Start button
(Courtesy of Sun Electric Corp.)

3. *MOUNTING THE WHEEL*

a. Place the cone-set mounting spring on the balancer shaft with the large end against the faceplate and small end toward the threads (Fig. I-108).
b. Select the cone that best fits the center of the wheel to be balanced. For specially designed wheels on which adapters will not fit, refer to the mounting chart (Fig. I-109).
c. Put the wheel on the cone and hold it in position to prevent it from falling onto the shaft and causing nicks or damaged threads.
d. Place the drum hub-nut assembly on the shaft and thread it tightly against the wheel.
Note: *Be certain the hub nut engages the threads at least four full turns. If necessary, the hub nut may be reversed. Also, use the rubber bumper only when you want to protect custom wheel finishes. When the rubber bumper is used, an extra effort is needed to tighten the hub nut adequately.*
For extended center-style wheels, use the hub nut without the retaining drum.

FIGURE I-108 Standard method of mounting wheels (Courtesy of Sun Electric Corp.)

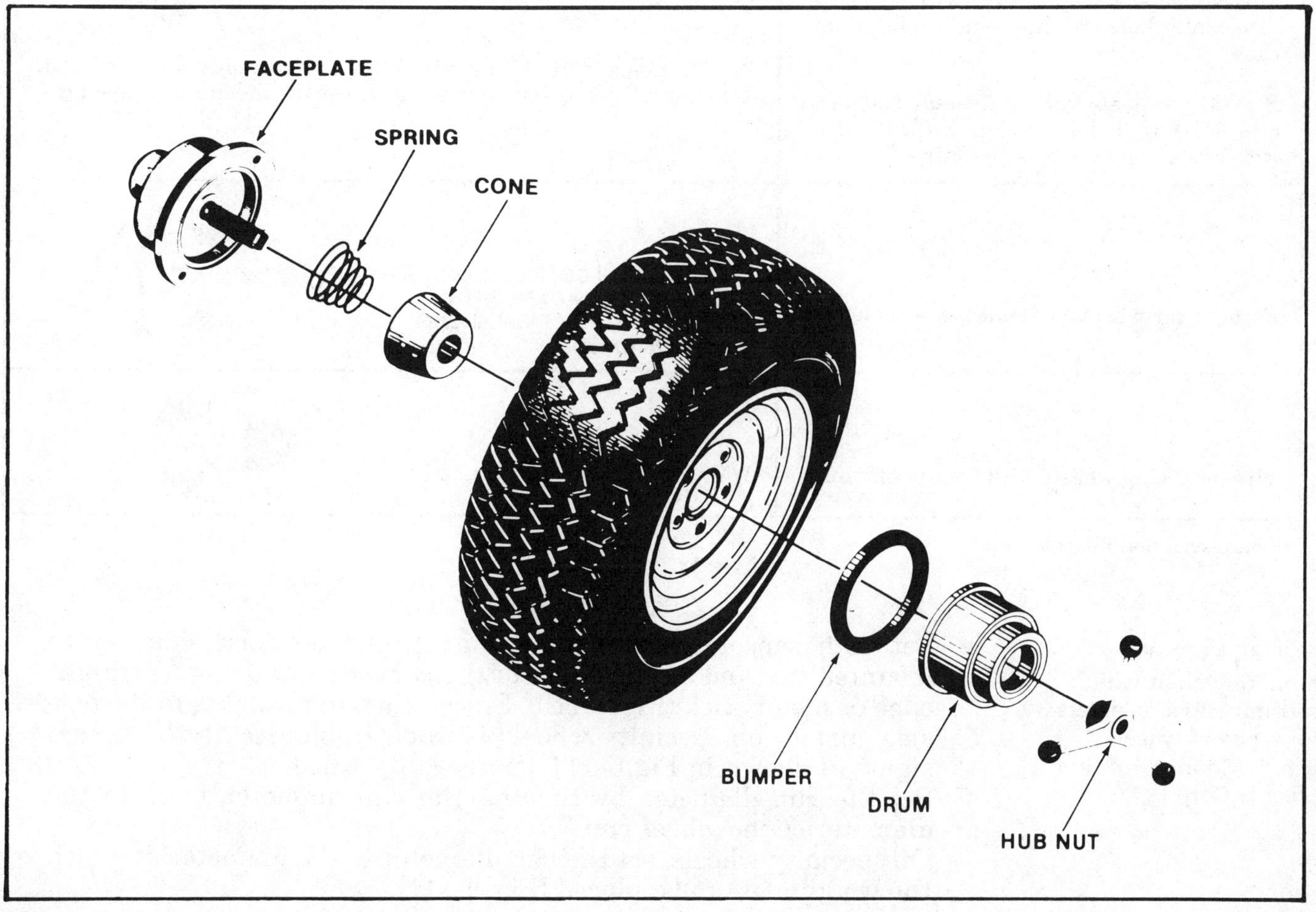

MOUNTING CHART

Large and small reversible cones and hub nut give the quickest mounting and usually provide more accurate results. Older wheels may require the adapters, but for the highest efficiency, if the wheel looks good, use the cones. Wheels with obviously inaccurate center holes require the adapters.

WHEEL TYPE	MOUNTING
1966 and later vehicle wheels 1965 and older vehicle wheels with worn lug holes Mag styled, specialty wheels with accurate center hole Specialty wheels with universal bolt circles, Uni-lug, K-lug, or slots and with accurate center hole	
1966 or later vehicle wheels with worn, damaged, or inaccurate center hole. Wheels with bolt circle centering. Mag styled specialty wheels without accurate center hole or center hole too large for cone Specialty wheels with universal bolt circles, Uni-lug, K-lug, or slots without accurate center hole or hole too large for cone	UNIVERSAL OR COMBINATION LUG BOLT CIRCLE ADAPTER 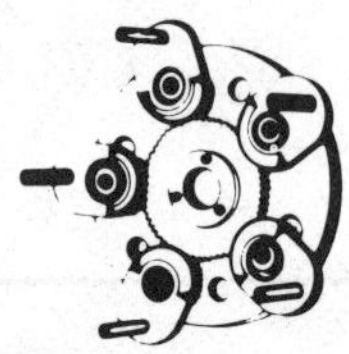 NOTE: Use adapters from wheel kit, European lug nuts, or a combination of fittings to ensure wheel runs true before balancing
Foreign car wheels without center holes	UNIVERSAL OR COMBINATION LUG BOLT CIRCLE ADAPTER WITH THREADED STUD OF BALANCER SHAFT REMOVED 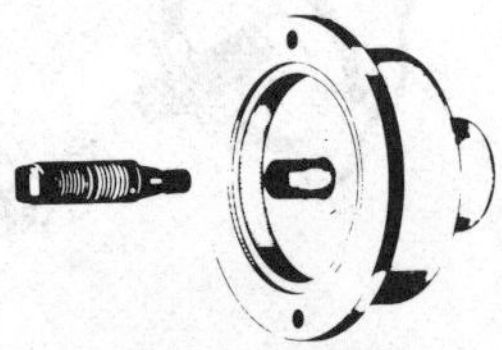
Wire or racing wheels with "knock-off" hub	WIRE WHEEL ADAPTER (OPTIONAL) 

Courtesy of Sun Electric Corp.

FIGURE I-109 Wheel-mounting chart with recommended adapters for types of wheels listed. (Courtesy of Sun Electric Corp.)

e. Set both sensor sights by making contact on each side of the mounted tire and outside edge of the sensor sight using a straight edge or meter stick (Fig. I-110). Since attaching weights to the outside surface on specialty wheels is undesirable, locate the sensor sights as shown in Fig. I-111 for specialty wheels.

f. Set the rim diameter by turning the rim diameter knob to the diameter of the wheel rim.
On specialty wheels, set the rim diameter to the diameter at which the weights are to be placed (Fig. I-111).

4. *BALANCING OPERATIONS*

a. *By hand,* spin the wheel clockwise (the normal direction of machine rotation) and check the lateral and radial run-out by eye. *Lateral* run-out is the side-to-side or wobble motion of the wheel and tire. *Radial* run-out is the up-and-down motion of the tire and wheel. If either type of run-out is more than ±1/16 inch (1.6 mm), check the mounting setup of the balancer, then check for bent wheels and/or defective tires. If the excessive run-out cannot be corrected, do not balance the wheel. Notify your instructor or supervisor about the problem.

FIGURE I-110 Use line of sight or a meter stick to align sight sensors in locating positions for weights. (Courtesy of Sun Electric Corp.)

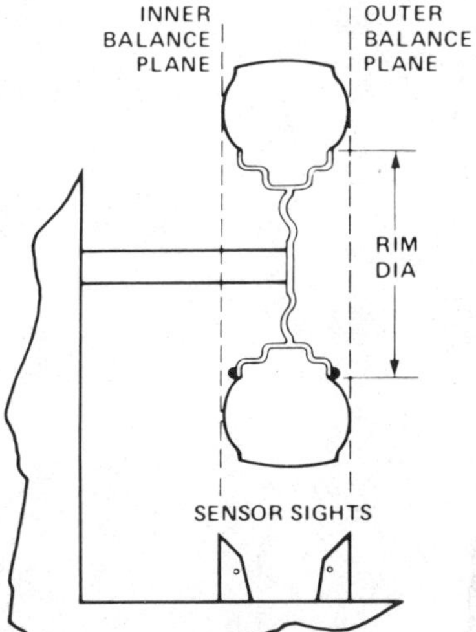

FIGURE I-111A Typical steel wheel sensor and rim diameter positions.

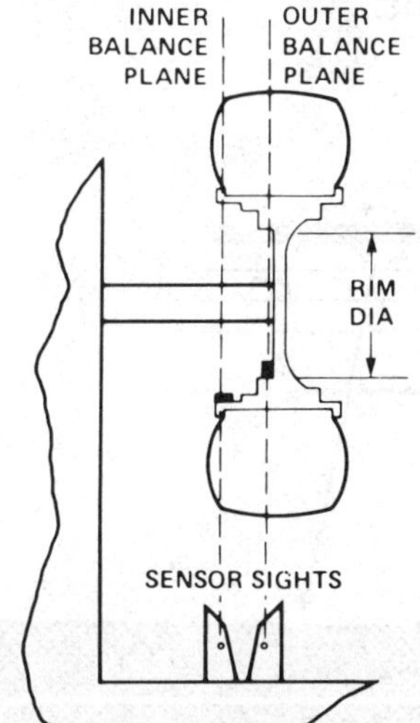

FIGURE I-111B For specialty wheels, align sensors and rim diameter as shown.

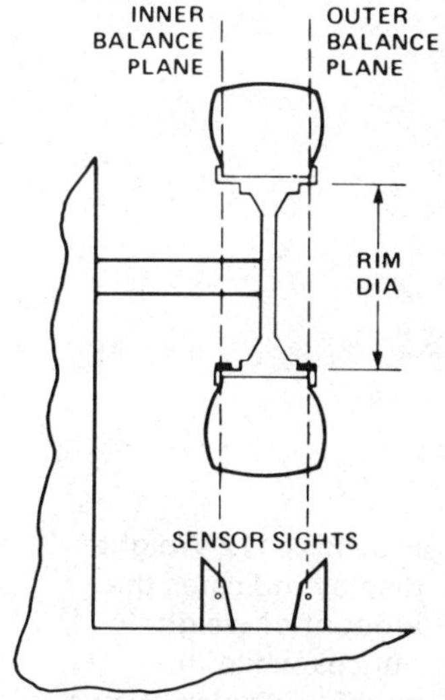

FIGURE I-111C Follow procedures for mag wheels when attaching weights. (Courtesy of Sun Electric Corp.)

b. While the wheel and tire are still spinning, lower the safety hood and push the START button. The safety hood must be in the full down position before the balancer will spin the tire.
c. Check the weight and position readings on the display panel at the end of the spin cycle. Lift the safety hood and rotate the wheel until the position number called for by the outer position display (see Fig. I-112) appears in the outer readout window.
d. Attach the weight indicated by the outer weight display on the outer edge of the rim at top-dead center. Be sure the weight is attached so it aligns exactly with the position number in the window.
e. Repeat the last two steps to get the inner weight and position.
Note: *The more accurate you are in selecting and placing the weight in the position recommended, the more often you will get the correct balance in one spin.*
f. After both outer and inner weights are attached and the safety hood is closed, push the START button and the balancer will complete the second spin cycle.
g. Both position and weight displays should read zero. If small amounts of weight (less than 1 ounce or 30 grams) should be added, do so. If excessive weight is shown on the displays, remove all weights and begin again with the first spin.
h. Mag-type wheels that require so-called stick-on weights are balanced the same way as original equipment wheels except for the positioning of the sight sensors and the weight attachment. After the first spin of the wheel, select the appropriate amount of regular clip-on weight. Using masking or duct tape, place the weights in the recommended positions and secure them with tape. BE SURE THE TAPE WILL HOLD THE WEIGHT WHILE IT SPINS. When the wheel is balanced (both displays read zero), remove the clip-on weights and tape and apply the exact same amount of stick-on weight.

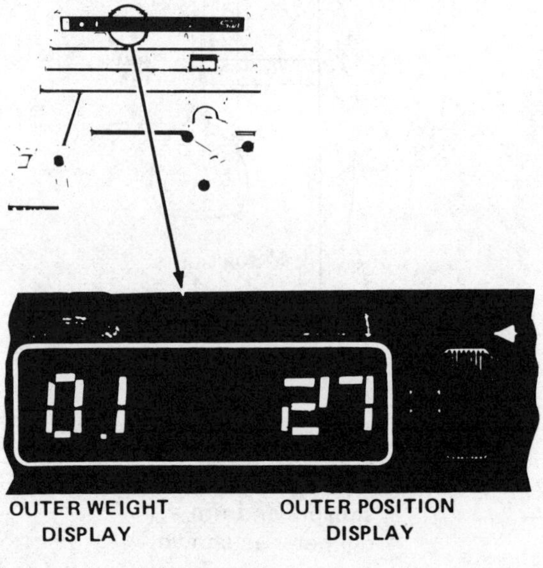

FIGURE I-112 Weight display indicates the amount of weight in ounces while the position display shows one of fifty positions on the wheel where the weight can be attached. (Courtesy of Sun Electric Corp.)

MOUNTING WHEELS WITH COMBINATION ADAPTER

If it is difficult to mount the wheel with the cone adapters, the combination adapter can be used (Fig. I-113). Lug hole patterns of 3, 4, 5, 6 or 8 holes for 10- to 17-inch diameter wheels can be balanced with this adapter. Six and 8 lug hole wheels would use 3 and 4 lug swivels respectively.

Set up the combination adapter as follows:

a. Install a lug swivel in the common hole (marked $\begin{smallmatrix}3\\4\\5\end{smallmatrix}$ in Fig. I-114).
b. Line up the index mark on the lug swivel to match the dot above the numbers 3, 4, and 5.
c. Insert the Allen head swivel bolt from behind the adapter plate and push it through. *Do not tighten it.*
   Note: *This first lug will be used with all lug bolt circle patterns.*
d. Install additional lug swivels to match the number of lug bolt holes to be used. These are: 3 for 3 and 6 lug holes, 4 for 4 and 8 lug holes, 5 for 5 lug holes.
   Note: *Be sure that index marks on all installed swivels line up with the appropriate "3", "4", or "5" mark on the center gear of the adapter.*

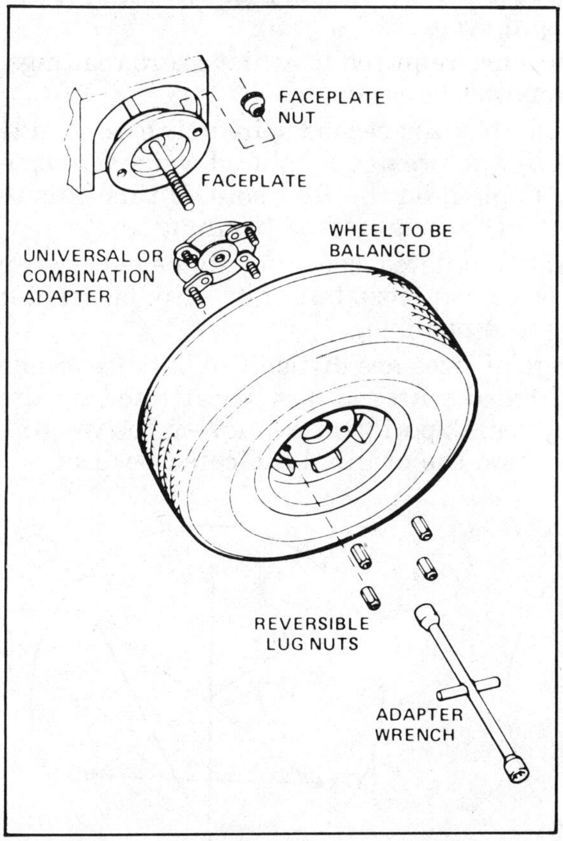

FIGURE I-113 Use the lug wrench with the machine to tighten the nuts. First tighten lug nuts, then tighten lug swivel bolts with an allen wrench. (Courtesy of Sun Electric Corp.)

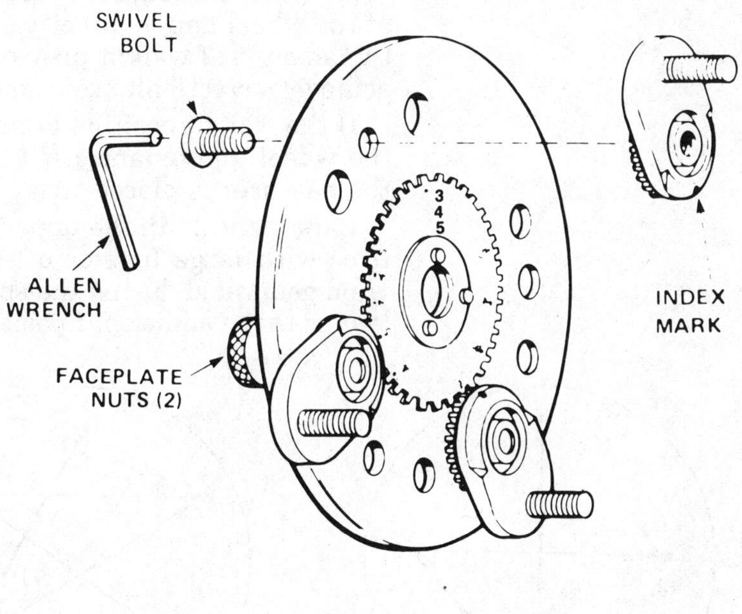

FIGURE I-114 Universal or combination adapter. Be certain the index marks of all lug swivels link up to the number of lug holes to be used. (Courtesy of Sun Electric Corp.)

e. Install the combination adapter on the wheel to be balanced. Hand tighten the lug nut supplied with the adapter, then tighten the lug nuts with the adapter wrench supplied with the machine. Use a criss-cross or star pattern when tightening (Fig. I-115).

f. Lug nuts must be centered and threaded on the studs at least four full turns. Use the lug wrench on the left front of the balancer to tighten the lug nuts. *Do not use other lug wrenches or impact tools to tighten lug nuts.*

g. Tighten the Allen head swivel bolts with an Allen wrench while holding the lug swivels.

h. Install the wheel and adapter on the balancer shaft. Line up the two studs with their holes in the faceplate. Install and firmly hand tighten the two faceplate nuts.

i. The combination adapter should remain on the balancer faceplate for additional wheels with the same bolt hole circle.

HELPFUL HINTS:

A perfectly balanced wheel with excessive run-out can still cause vehicle vibration.

Fine-tune balancing can be achieved by holding down the START button continuously during the spin cycle.

If multiple spins (more than two) are required to achieve zero readings, weight amount and placement may not be accurate.

If weight placement is indicated after the second spin on the same side of the wheel and is directly opposite the previously placed weight, reduce the amount of weight previously applied on the first spin. If this fails to achieve correct balance, remove all the weights and begin again.

If the weight position or amount fluctuates, check for slippage between the wheel and balancer shaft. Look for a loose hub nut. Also, be certain the balancer is placed firmly on its support legs.

Large wheels in the upper range of sizes are difficult to balance. Some tires with large lugs or off-road tread patterns may be affected by the wind generated during the spin cycles. Experiment by moving the weights two or three additional positions from the original indicated position.

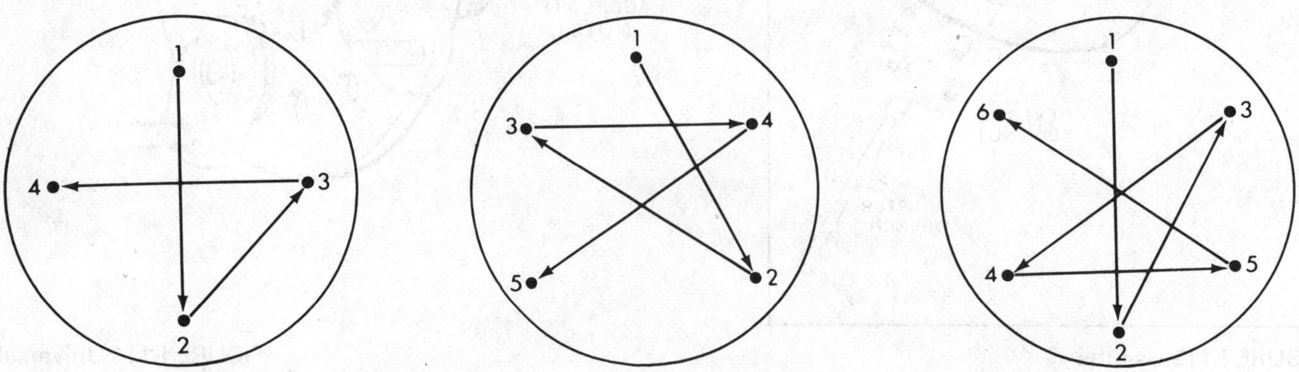

FIGURE I-115
Recommended lug nut tightening sequence to minimize wheel run-out.

OFF-THE-CAR DYNAMIC WHEEL BALANCING

Student name__________ Date__________

Make of car__________ Model__________ Year__________

Time started__________ Time finished__________ Total time__________

Flat-rate time __________

The following space is provided for recording instructions not included in the text.

The procedure for setup and balancing will vary, depending on equipment available in your shop.

Balance Wheels

1. Mounting of wheel to balance (front wheels): __________

(rear wheels): __________

2. Procedure for static balance or combined static-dynamic balance: __________

3. Procedure for dynamic balance: __________

4. Final check for accuracy: __________

Checkup Questions

1. What is the maximum acceptable wheel and tire run-out? ____________________

____

____

2. Explain the placement of the dynamic weights on each wheel.

   a. Left front outside: ____________________

      inside: ____________________

   b. Right front outside: ____________________

      inside: ____________________

   c. Left rear outside: ____________________

      inside: ____________________

   d. Right rear outside: ____________________

      inside: ____________________

3. Did your road test indicate any problems? __________ Yes __________ No

If so, what were they? ____________________

____

____

____

DATE COMPLETED____________ INSTRUCTOR'S SIGNATURE: ____________

TEST I-C

Student name________________________________ Date____________________

Directions: Select the best answer to each question and circle the correct letter at the left.

1. A wheel that is out of round has:
   I. Radial run-out
   II. Lateral run-out
      A. I only
      B. II only
      C. Both I and II
      D. Neither I nor II
2. Total run-out is the total of:
   I. Radial and lateral run-out
   II. Wheel-rim and tire run-out
      A. I only
      B. II only
      C. Both I and II
      D. Neither I nor II
3. When the weight of the wheel assembly is equally distributed around the circumference of the tire, it will be in:
   I. Dynamic balance
   II. Static balance
      A. I only
      B. II only
      C. Both I and II
      D. Neither I nor II
4. A wheel assembly that bounces at any speed:
   I. Is out of static balance
   II. Has too much radial run-out
      A. I only
      B. II only
      C. Both I and II
      D. Neither I nor II
5. A wheel that bounces at 40 to 60 mph only:
   I. Is out of static balance
   II. Has too much radial run-out
      A. I only
      B. II only
      C. Both I and II
      D. Neither I nor II
6. A wheel that wobbles at all speeds:
   I. Is out of dynamic balance
   II. Has too much lateral run-out
      A. I only
      B. II only
      C. Both I and II
      D. Neither I nor II

7. When the overweight spot on a tire is evenly spread across the tire tread, the wheel is:
   I. Out of dynamic balance
   II. Out of static balance
      A. I only
      B. II only
      C. Both I and II
      D. Neither I nor II
8. When the overweight spot is only on one side of the tire tread, the wheel may be:
   I. Out of dynamic balance
   II. Out of static balance
      A. I only
      B. II only
      C. Both I and II
      D. Neither I nor II
9. When balancing a wheel assembly with equipment that is capable of static and dynamic balance:
   I. Check for excessive wheel and/or tire run-out before spinning
   II. Remove all old weights
      A. I only
      B. II only
      C. Both I and II
      D. Neither I nor II
10. It is possible for the wheel assembly to be in:
   I. Static balance but not dynamic balance
   II. Static balance but continue to bounce because of excessive radial run-out
      A. I only
      B. II only
      C. Both I and II
      D. Neither I nor II

DATE COMPLETED____________________ INSTRUCTOR'S SIGNATURE ____________________

unit II

Suspension Systems

This unit explains the parts of a suspension system. These parts are the front suspension, the rear suspension, and the shock absorber. Included in the subjects covered are: independent and I-beam front suspension, ball joint construction and inspection procedures, springs, and control arm assemblies. All service procedures discussed are currently used by the automotive repair trade.

Notice: *Objectives II-A through II-E2 are part of the overall Objective II-F, which involves rebuilding the entire front suspension of an automobile. It is suggested that the student use Job Report Sheet II-F, page 189, to record progress, whether or not the required objectives are completed on one or more vehicles. The individual Job Report sheets should be turned in as each job is completed.*

Your instructor will assign the number of objectives required for the course in which you are enrolled.

FRONT SUSPENSION

The front suspension systems of the earliest automobiles were copied or modified from horse-drawn wagons. A typical early system used a rigid straight axle as the main support. Even today many trucks and light utility vehicles still use the beam or rigid-axle suspension because of its good load-carrying ability.

The problem with the rigid axle was that it was not comfortable to ride on. Aware of this fact, auto makers looked for a suspension system that would improve the ride. The result was the independent front suspension (Fig. II-1). Three versions of the independent design are in use today. They are: (1) the long and short control arm type, with a spring or torsion bar acting on the lower arm, (2) the long and short control arm type, with the spring or torsion bar acting on the upper arm, and (3) the MacPherson strut type.

OBJECTIVE II-A (written test)

After studying the preparation assignment you will be expected to correctly answer a minimum of seven out of ten multiple-choice questions in a written test without reference materials on ball joint construction and inspection procedures.

Preparation:

Study pp. 91 through 109.

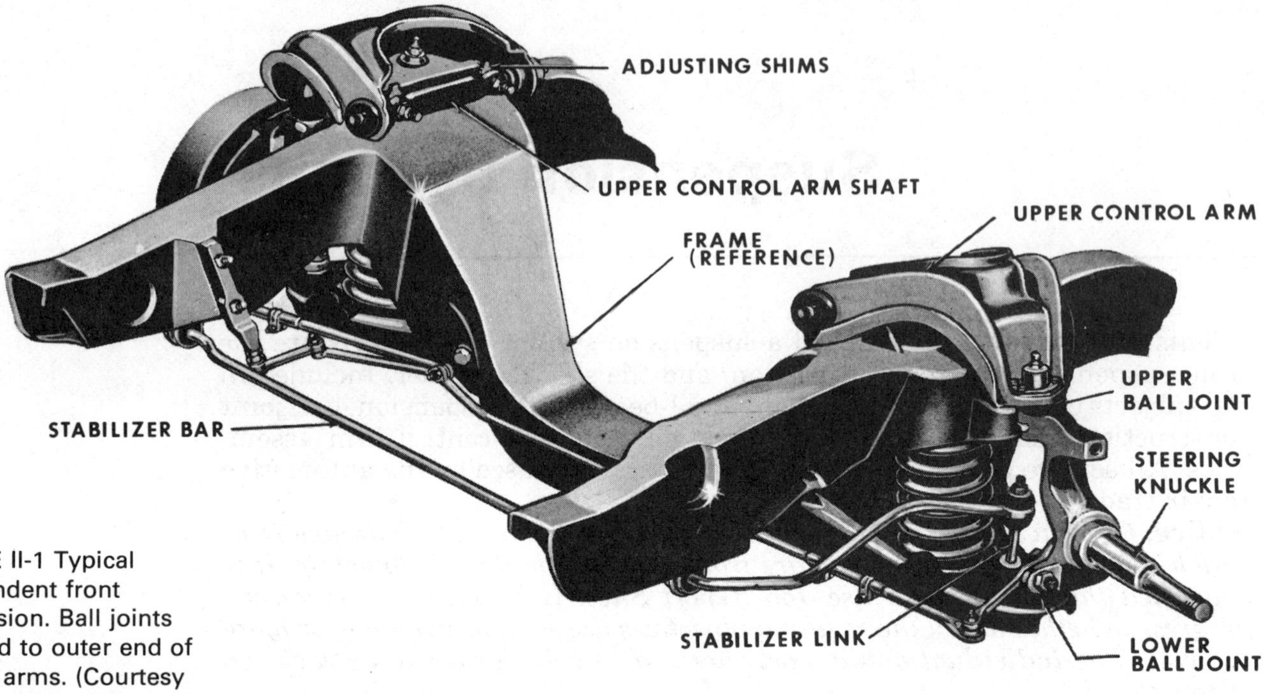

FIGURE II-1 Typical independent front suspension. Ball joints attached to outer end of control arms. (Courtesy of Buick Motor Division)

BALL JOINTS

The independent or ball joint suspension consists of an upper and a lower ball joint (**Note:** *The MacPherson strut suspension has only a lower ball joint)*. The ball joints are mounted between the upper and lower control arms and the steering knuckle. They permit both the up-and-down movement that an independent front suspension should have and the turning motion required of the front-wheel assembly. (See Fig. II-1.)

In effect, the ball joints and the control arms form the outer hinges of the front suspension system.

Ball joints are identified in two ways: (1) by their internal design which is either compression-loaded or tension-loaded and (2) placement in a suspension system which is either load-carrying type or follower type, which is a nonload carrier.

A compression-loaded ball joint takes its name from the fact that the joint is being compressed from two directions. The vehicle weight is pushing down on one end of the joint while the counterforce of the road is pushing up from the other direction. This design is shown in Figs. II-2, II-4, and II-5.

In the tension-loaded joint design, the weight of the vehicle is carried by the lower control arm and ball joint. This is countered by the force of the road pushing up on the upper control arm. This type is shown in Fig. II-3.

NOTE: *A ball joint will have a compression-loaded or a tension-loaded design. In addition it also will be a load-carrying or follower joint.*

The control arm that carries the weight of the vehicle is the load-carrying arm. The joint to which it is attached is the load-carrying joint. The supportive arm and joint are called the follower or nonload carrying arm. See and compare Figs. II-2, II-3, II-4, and II-5.

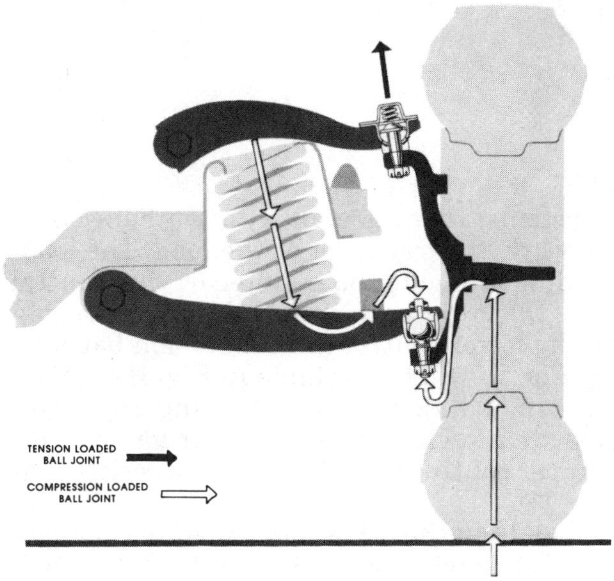

FIGURE II-2 Compression loaded lower ball joint. Weight of vehicle is pushing downward on lower control arm and ball joint. The road is pushing upward on the wheel steering knuckle and ball joint. (Courtesy of Moog Automotive, Inc.)

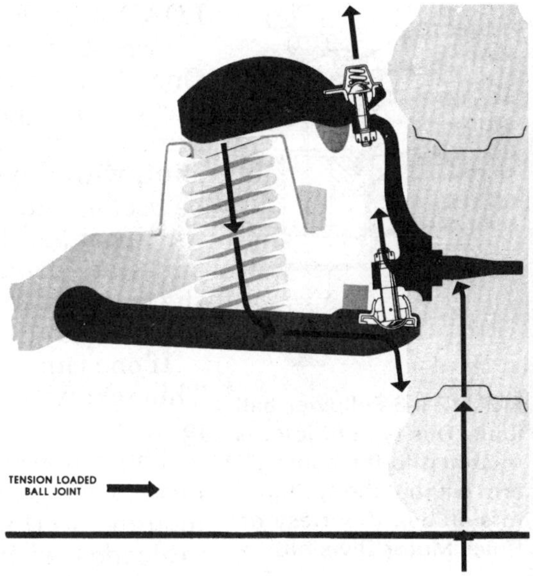

FIGURE II-3 Tension-loaded lower ball joint. Weight of vehicle is pushing downward on lower control arm and ball joint. (Courtesy of Moog Automotive, Inc.)

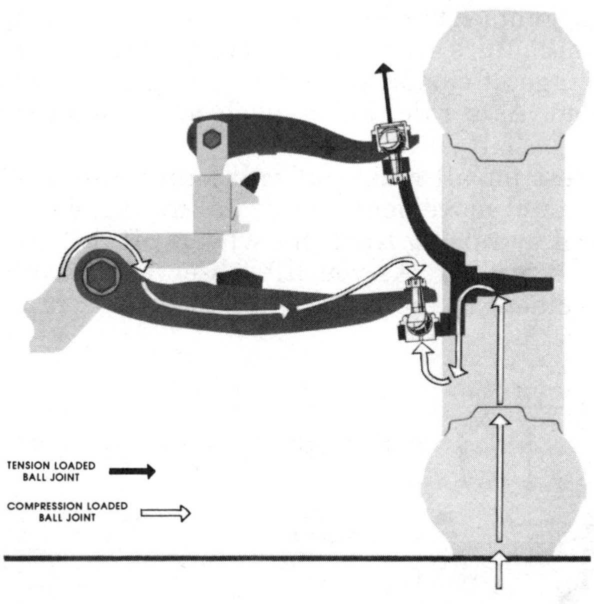

FIGURE II-4 Load-carrying lower ball joint on a torsion-bar suspension. This example is typical of Chrysler Corporation cars. (Courtesy of Moog Automotive, Inc.)

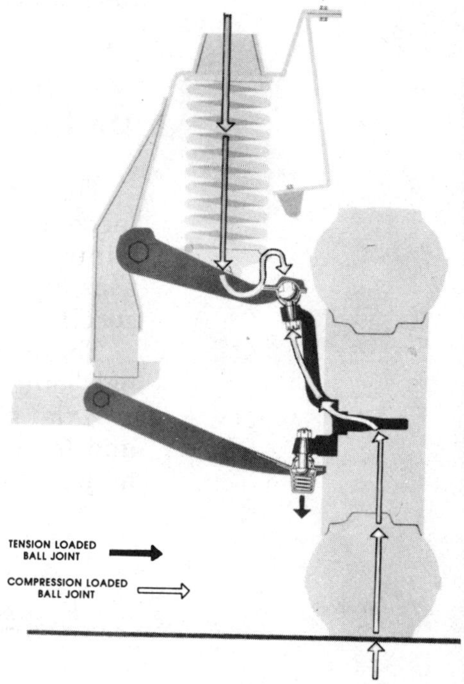

FIGURE II-5 Load-carrying ball joint on upper control arm. This example is typical of many intermediate and compact models of Ford, American Motors, and General Motors. (Courtesy of Moog Automotive, Inc.)

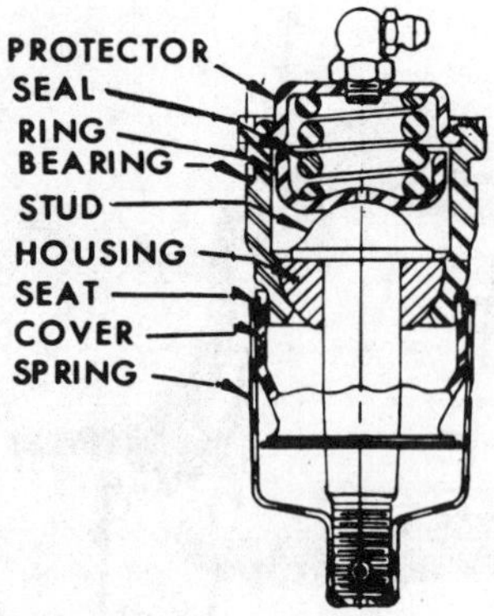

FIGURE II-6 Follower ball joint. This type of joint is mounted to the control arm without the spring or torsion bar. (Courtesy of Buick Motor Division)

LOAD AND FOLLOWER JOINTS

Except for MacPherson strut systems, there are always two ball joints. One of the two will do most of the work or carry most of the load. The load-carrying ball joint is always the joint mounted to the spring-carrying control arm. The example in Fig. II-3 shows the lower ball joint as the load-carrying joint, with the spring mounted to the lower control arm. A torsion bar may be used in place of the coil spring (Fig. II-4). The example in Fig. II-5 shows the upper ball joint as the load-carrying ball joint. Load-carrying upper ball joints are usually found on compact cars with unibody construction, with the spring mounted between the upper control arm and a spring pocket that is built into the auto body.

If one joint carries most of the load, then the other joint must be a helper. This text will use the name *follower joint* when referring to the helper ball joint.

The follower joint is always mounted to the control arm that does not have a coil spring or a torsion bar mounted to it. In other words, it is mounted to the unloaded arm. The ball stud of the follower joint is usually preloaded with a steel spring so that the joint forms a snug connection between the steering knuckles and the control arm (Fig. II-6). The preload on the ball joint gives a turning resistance that contributes to the handling and steering stability of the automobile.

BALL JOINT INSPECTION

The right time to replace ball joints has been a subject argued among mechanics for some time. Many mechanics are cautious about changing ball joints. Others have taken advantage of customers by replacing ball joints that still had a lot of life in them. As a rule, follow each manufacturer's specifications on the replacement of ball joints.

There is no one or universal test for all makes of ball joints. However, ball joints should be checked for axial movement and radial movement.

Axial movement can be checked by raising the front wheels off the floor and lifting the tire upward with a pry bar. As you lift, observe or have a helper look for up-and-down movement in the joint. See Fig. II-7.

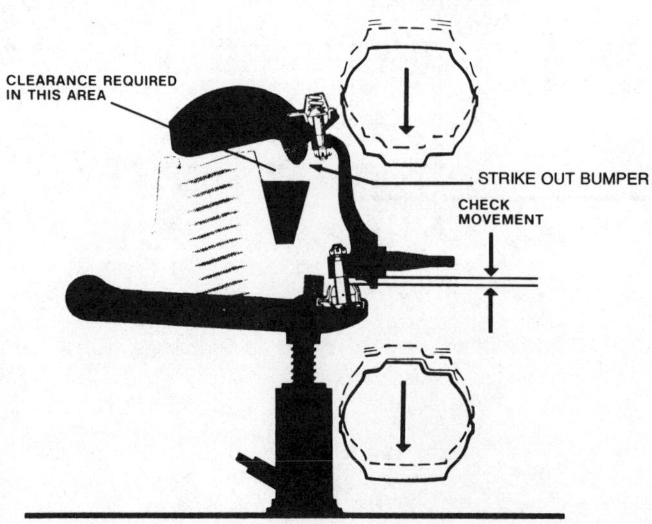

FIGURE II-7 Axial movement is the amount of vertical movement indicating ball joint wear. (Courtesy of Moog Automotive, Inc.)

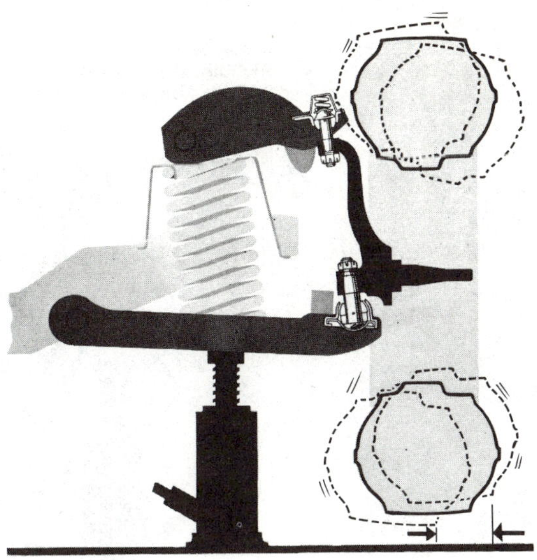

FIGURE II-8 Radial movement is tire-sidewall movement indicating ball joint wear. *Note: Loose wheel bearings can cause the same symptoms.* (Courtesy of Moog Automotive, Inc.)

Radial movement can be checked by raising the front wheels off the floor and turning them to the straight ahead position. With one hand at the top of the tire and the other at the bottom of the tire, alternately pull with one hand while pushing with the other. Radial movement will be any in-and-out movement in the joint (Fig. II-8). Loose wheel bearings will show movement in the wheel assembly, so make certain that you are checking ball joint movement and not wheel assembly movement. If the spindle nut is loose, the brake drum or disc will move but the backing plate or splash shield will not.

BALL JOINT WEAR INSPECTION

(**Note:** *The Motor Vehicle Manufacturer's Association recommends the following procedure for general use. Follow this method if manufacturer's procedure and specifications are not available for the vehicle being checked.)*

Except for ball joints that have wear indicators, the load-carrying ball joint (the one mounted to the spring or torsion-bar carrying control arm) must be unloaded during the test. This is done by jacking the vehicle as shown in Fig. II-9 or II-10. The difference in jack placement is determined by spring position. Torsion-bar suspension would be jacked the same as shown in Fig. II-9. MacPherson strut suspension would be jacked as shown in Fig. II-10.

OBJECTIVE II-A1 (manipulative)

Working with one other student as a team and with the aid of Job Report Sheet II-A1, you will check for wear in the ball joints on a minimum of two instructor assigned automobiles. Manufacturer's recommended procedure and specifications must be followed. Time recommended: Maximum of 20 minutes per automobile.

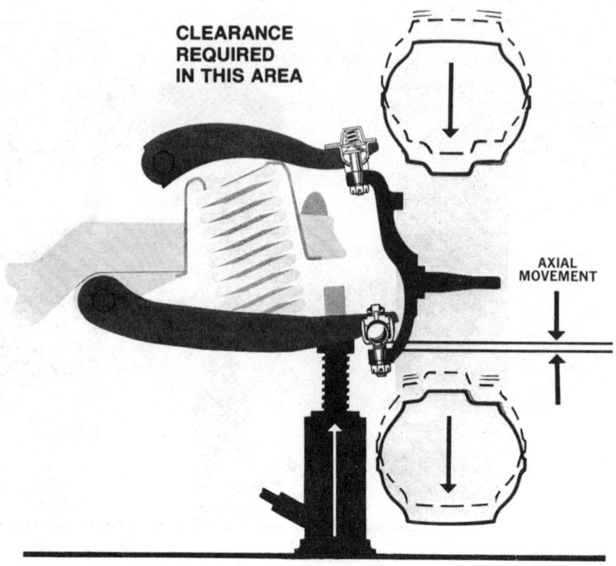

FIGURE II-9 Illustration shows proper jack position for checking lower ball joint axial movement on a coil-spring suspension system with spring between the control arms (Type 1).

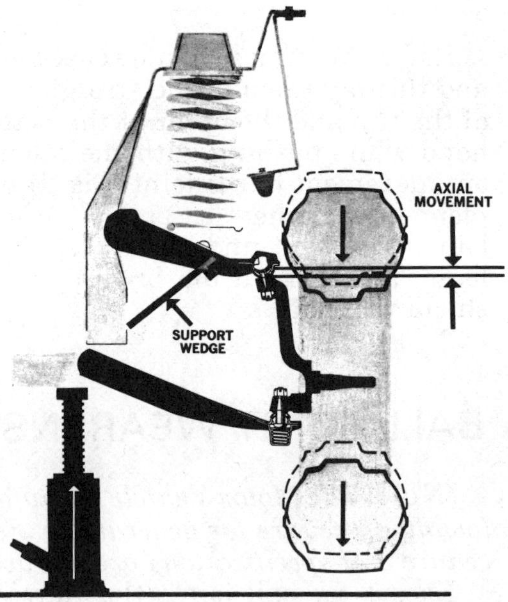

FIGURE II-10 Proper jacking position for ball joint check on coil-spring suspension system with spring above the upper control arm (Type 2).

NOTE: Wedge support must be installed before jacking.

Preparation:

1. Review pages 91 through 101
2. Read manufacturer's or general repair manual for correct specifications and procedures.
3. Carefully observe instructor's demonstration on ball joint inspection.

Materials and Tools Needed:

Since inspection procedures vary considerably, read the manufacturer's or general repair manual before assembling tools and materials.

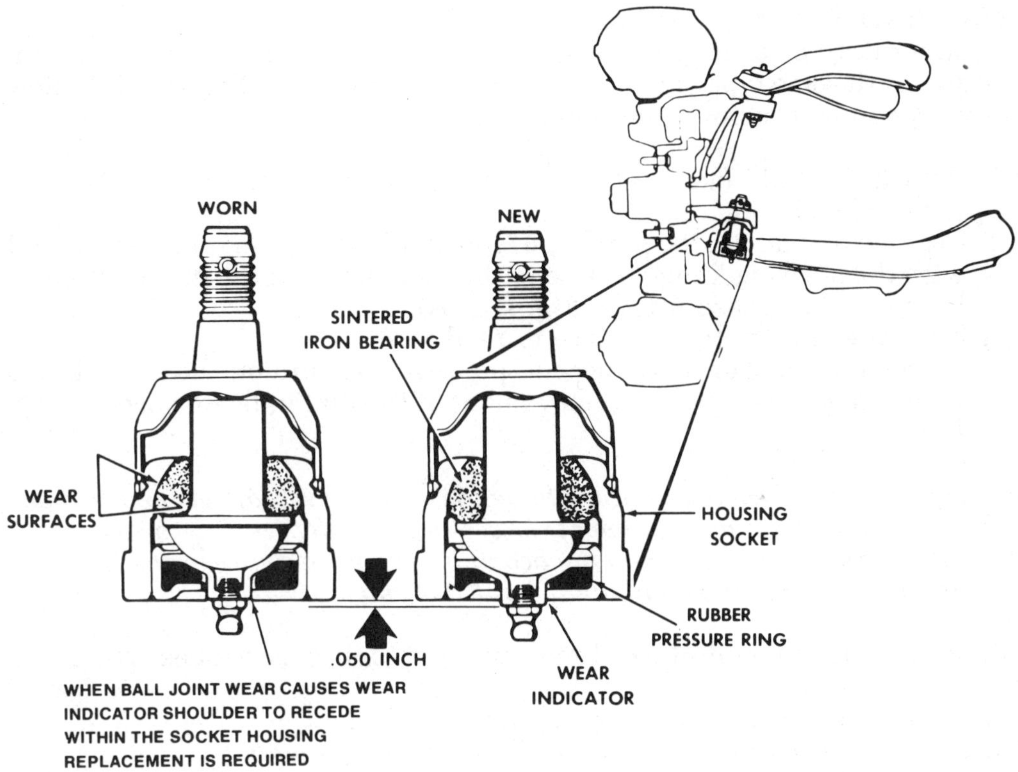

FIGURE II-11 Indications of ball joint wear. Example on left shows a worn joint. Joint at the right is new. (Courtesy of Chevrolet Motor Division)

TYPE 1 VEHICLES

The load-carrying ball joint for the Type 1 suspension may be one of two kinds: (1) joints with wear indicators, and (2) joints without wear indicators.

A ball joint with a wear indicator is shown in Fig. II-11. It can be recognized by the small, round shoulder, or boss, holding the grease fittings on the opposite side of the stud and nut. This round boss is the wear indicator. It projects .050 to .070 in. (1.27 to 1.78 mm) beyond the surface of the cover when the joint is new or unworn.

In checking for wear, the Type 1 wear-indicator joint (the lower joint) must remain loaded. That is, the wheels must remain on the ground with the suspension at curb height so that all wear surfaces inside the joint will remain in contact with each other.

Ball joints of Type 1 design have been standard on many General Motors' vehicles since 1973 and on the Ford Zephyr and Fairmont models with the modified MacPherson strut suspension system since the 1978 models.

Lower load-carrying ball joints without wear indicators must be unloaded to check for looseness from wear. This is done by raising the vehicle and placing jack (safety) stands under the lower control arms as close to the wheel as possible. When the lower joint is properly unloaded, there should be some clearance at the strike-out bumper.

Note: *Always look for clearance at the strike-out bumper when checking lower load-carrying ball joints since this is the only way to ensure that all spring or torsion bar support has been relieved (unloaded). If there is no clearance, move the jack (safety) stand closer to the wheel or raise only one side of the vehicle at a time.*

AXIAL AND RADIAL MOVEMENT

To accurately measure the amount of axial and radial movement, a dial indicator with some type of holding device is required (see Fig. II-12). Follow these steps in measuring axial movement.

1. Attach the dial indicator to the control arm that connects with the ball joint to be checked.
2. Position and adjust the plunger of the dial indicator so it makes contact with the steering knuckle close to the stud nut or to the ball joint housing near the grease fitting (Fig. II-13A).
3. Turn the ring on the dial face to read zero.
4. Using a pry bar or large piece of pipe, pry upward on the tire while a helper reads the amount of movement of the dial indicator needle (Fig. II-14).

Note: *If the wheel bearing appears to be loose enough to affect the accuracy of the ball joint wear measurement, have someone depress the brake pedal inside the vehicle. This will stop the bearing movement between the hub and steering knuckle and allow only ball joint movement, if any.*

5. Record the amount of axial movement on Job Report Sheet II-A1.

Follow this procedure in measuring radial movement.
(**Note:** *Radial measurement is not required for all joints.)*

1. Attach the dial indicator to the control arm that connects with the ball joint to be checked.
2. Position and adjust the plunger of the indicator against the edge of the wheel rim nearest the joint to be checked (Fig. II-13B).
3. Turn the ring on the dial face to read zero.
4. Alternately push and pull the top and the bottom of the tire-wheel assembly (see Fig. II-8). Note the amount of looseness on the indicator dial.
5. Record the amount as radial movement on Job Report II-A1.

TYPE 2 VEHICLES

In Type 2 vehicles, the coil spring or torsion bar acts directly on the upper control arm. The upper ball joint in Type 2 vehicles is the load carrier and must be checked for wear and looseness. To unload the upper joint, place a support wedge between the frame and upper control arm *before* the vehicle is raised. To raise the vehicle, jack it from the frame or crossmember. Lower the vehicle so its frame or crossmember is resting on jack (safety) stands. Follow this procedure in measuring axial movement:

1. Install wedge supports (shown in Fig. II-10), raise the vehicle, and place it on jack (safety) stands.
2. Check to see there is clearance between the upper control arm and the strike-out bumper.
3. Attach the dial indicator to the control arm that is being checked.
4. Position and adjust the plunger of the dial indicator so it makes contact with the steering knuckle close to the stud nut or to the ball joint housing near the grease fitting (see Fig. II-13A).
5. Turn the ring on the dial indicator face to read zero.

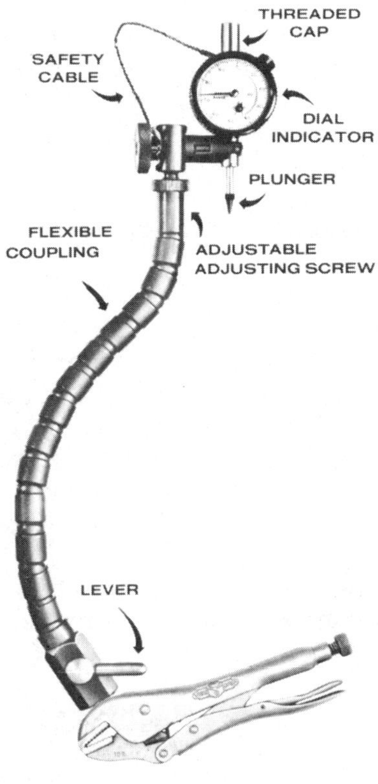

FIGURE II-12 This dial indicator is a precision instrument and should be handled carefully to prevent damage. The mounting procedure for the checking tool may vary depending on the style of ball joint used on the vehicle. Manufacturer's tolerances may be axial (vertical), radial (horizontal) or both. (Courtesy of Moog Automotive, Inc.)

FIGURE II-13A For a radial check, attach dial indicator to the control arm of the ball joint being checked. Position and adjust plunger of dial indicator against edge of wheel rim nearest to ball joint being checked. Slip dial ring to zero marking. Move the wheel in and out and note the amount of ball joint radial looseness registered on the dial. (Courtesy of Moog Automotive, Inc.)

FIGURE II-13B For an axial check, first position the dial indicator on the control arm, then clean off the flat on the spindle next to the ball joint stud nut. Position the dial indicator on the spindle flat and depress the plunger approximately .250 of an inch. Turn lever to tighten indicator in place. Use a pry bar between the floor and tire and record the reading. (Courtesy of Moog Automotive, Inc.)

FIGURE II-14 If the ball joint looseness reading on the indicator exceeds specifications, the ball joint should be replaced. (Courtesy of Moog Automotive, Inc.)

6. Using a pry bar or a large piece of pipe, pry upward on the tire while a helper reads the amount of movement of the dial indicator needle.
7. Record this figure as the axial movement or looseness on Type 2 suspensions.

For radial measurement, follow the procedures outlined for radial measurement for Type 1 ball joints (Fig. II-13B).

MACPHERSON STRUT

MacPherson strut systems have only one typical ball joint. It is attached between the bottom of the strut (steering knuckle) and the lower control arm, which is sometimes called a track arm (Fig. II-15).

Since the main weight of the vehicle is placed on the spring and strut and passes through the wheel, the load is carried through to the top of the strut assembly. This makes the lower ball joint a follower and it should be checked as such.

An exception to the lower ball joints on MacPherson struts acting as followers is the modified MacPherson strut design used on some Ford models. Ford has a modified strut without a surrounding spring. Therefore, it does not carry the load but merely takes the place of the upper control arm. The Ford design has a coil spring mounted separately and a lower ball joint that carries the load. At the present time, all Ford-made modified struts have wear-indicator joints installed at the factory. They should be checked according to procedures outlined on page 98.

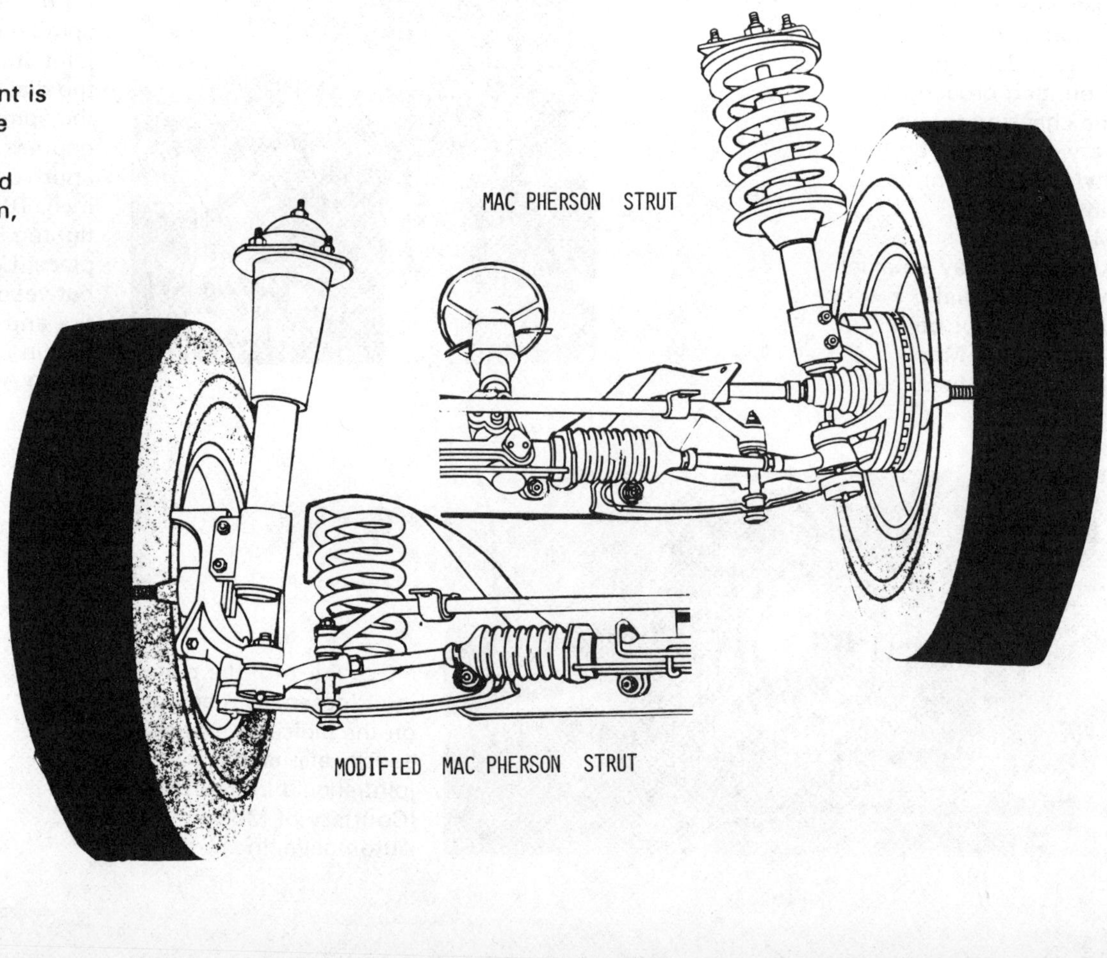

FIGURE II-15
MacPherson strut systems' one ball joint is attached between the bottom of the strut (steering knuckle) and the lower control arm, which is sometimes called a track arm.

JOB REPORT SHEET II-A1

BALL JOINT INSPECTION

ASSIGNMENT 1

Student name__________Date__________

Make of car__________Model__________Year__________

Time started__________Time finished__________Total time__________

Flat rate time __________

Specifications:

1. Maximum allowable axial movement: __________.

   Actual axial movement: __________.

2. Maximum allowable radial movement:__________.

   Actual radial movement: __________.

Checkup Questions:

1. Is the lower ball joint ☐ compression ☐ tension loaded?
2. Is the upper ball joint ☐ compression ☐ tension loaded?
3. Which ball joint is the load-carying__________, follower__________?
4. How are ball joints attached to control arms? ☐ Pressed ☐ Riveted ☐ Threaded
5. Any evidence of collision damage?__________If yes, explain: __________

   ____
6. Other suspension defects noticed: __________

   ____
7. Reference materials used: __________

   ____

DATE COMPLETED__________INSTRUCTOR'S SIGNATURE __________

JOB REPORT SHEET II-A1

BALL JOINT INSPECTION

ASSIGNMENT 2

Student name____________________Date____________________________________

Make of car____________________Model____________________Year____________________

Time started____________________Time finished____________________Total time________________

Flat rate time ____

Specifications:

1. Maximum allowable axial movement: ________________________________.

   Actual axial movement: ____.

2. Maximum allowable radial movement:________________________________.

   Actual radial movement: _______________________________________.

Checkup Questions:

	compression	tension	
1. Is the lower ball joint	☐	☐	loaded?
2. Is the upper ball joint	☐	☐	loaded?

3. Which ball joint is the load-carying____________________, follower____________________?

	Pressed	Riveted	Threaded
4. How are ball joints attached to control arms?	☐	☐	☐

5. Any evidence of collision damage?____________________If yes, explain: ____________________

   ____

6. Other suspension defects noticed: ____________________________________

   ____

7. Reference materials used: ____

   ____

DATE COMPLETED____________________INSTRUCTOR'S SIGNATURE ____________________________

BALL JOINT SERVICE

Before replacing a ball joint, the technician should secure the manufacturer's service manual whenever possible. The manufacturer's manual offers the most specific directions for replacement of ball joints. When the manual is not available, upper and lower ball-joint service can still be accomplished with the directions given here.

Be certain to review the following list of safety procedures before beginning actual work.

Safety Warning

- Wear suitable eye protection at all times.
- Use adequate lighting.
- Following the ball joint installation, all vehicles require a front end alignment. Refer to Unit IV for alignment instructions.

OBJECTIVE II-A2 (manipulative)

Working with one other student as a team, you will remove and replace at least two ball joints on an instructor assigned automobile. Use the Job Report Sheet II-A2 and manufacturer's recommended specifications as a guide. Time recommended: Twice the flat-rate time allowed in Motors or Chilton Flat Rate manuals. This objective may be completed in conjunction with Objective II-F, a complete front end rebuild.

Preparation:

1. Review pages 91 through 101.
2. Study pages 103 through 109.
3. Study manufacturer's or general repair manuals for automobile to be used.
4. Carefully observe instructor's demonstration on ball joint removal and replacement.
5. Review steps for completion of this job with your instructor prior to beginning work.

Materials and Tools Needed:

1. Service (floor) jack and jack (safety) stands
2. Lug or impact wrench and socket
3. Socket set with handles
4. Assorted box/open-end combination wrenches
5. Diagonal pliers
6. Spreader tool (taper breaker) or two large hammers
7. Pneumatic chisel, or press tool, or special ball joint socket
8. Torque wrench
9. Cotter pins
10. Chassis grease
11. Hand towels or wiping rags
12. Manufacturer's or general repair manual

As explained earlier, the load-carrying joint is the one most often in need of replacement. However, always check both joints for looseness. It is common to replace ball joints without removing the control arms to which they are attached. If, after thorough inspection, other defects are discovered requiring the removal of the control arm, see the control arm section starting on page 123. With the control arm removed, continue with ball joint replacement instructions from step 7.

Procedure:

1. Raise the vehicle on a hoist or place it on jack (safety) stands so the weight of the car will keep the lower control arm in position when the ball joint is disconnected.
2. Mark the lug stud nearest the valve stem and then remove the tire and wheel assembly (Fig. II-14).
3. Remove the cotter pin from the ball stud.
4. Loosen the ball stud nut approximately one turn (Fig. II-16).

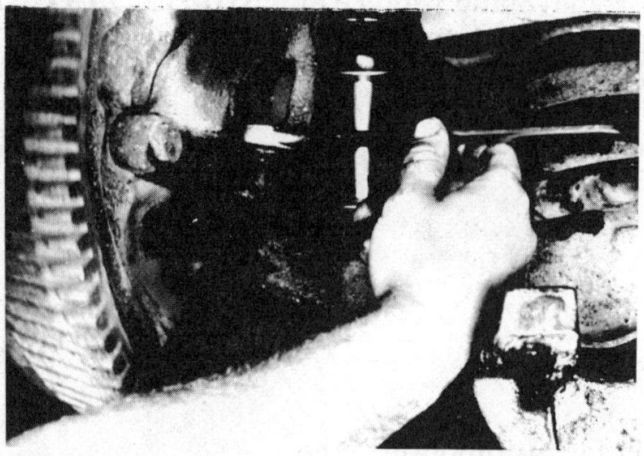

FIGURE II-16 Loosening a ball stud nut with an open-end wrench. (Courtesy of Moog Automotive, Inc.)

FIGURE II-17 Taper breaking tools are used to separate tapered ball studs from the steering knuckle. (Courtesy of Moog Automotive, Inc.)

5. With a taper breaking tool (spreader), break the ball stud loose from the steering knuckle (Fig. II-17). If the ball stud resists, place the spreader in position and apply tension. Then strike the taper area of the steering knuckle with a hammer to loosen the ball stud.
6. To remove a ball stud without the taper breaker, strike the taper area heavily while backing the opposite side of the knuckle taper with another hammer (Fig. II-18).
7. Removal of rivet-type ball joint:
   a. If the ball joint is held in position with rivets, grind off the rivet heads with a grinder, or chisel off with a sharp bladed pneumatic tool (Fig. II-19).
   b. Punch out rivet stems and remove joint.
8. Removal of press-type ball joint:
   a. To remove a press-type ball joint on the car, a press tool such as the Moog T-468 is required (Fig. II-20). Remove the dust boot and place the tool in position. If a press tool is not available, the control arm must be removed. To do this, see the section on control arms, page 123.
   b. Press out the old joint by operating the hydraulic jack. If an unusual amount of force seems to be needed, check the position of the adapters against the joint for proper alignment. If the adapters are located properly and the joint still resists, remove the control arm and use an arbor press or hydraulic press to remove the joint.

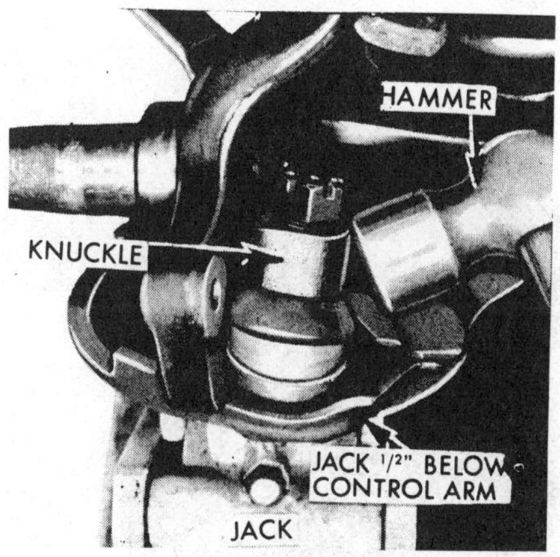

FIGURE II-18 Illustration shows proper procedure for striking the taper area of the steering knuckle to separate ball stud and knuckle. (Courtesy of Buick Motor Division)

FIGURE II-19 A pneumatic chisel is being used to remove rivet heads of original equipment ball joints. (Courtesy of Moog Automotive, Inc.)

FIGURE II-20 Portable ball joint press, used to remove and replace press type ball joints. (Courtesy of Moog Automotive, Inc.)

9. Removal of threaded-type ball joint:
   a. Clean the surface area near the grease fitting or plug.
   b. Secure the proper socket to fit the specific size of joint (Fig. II-21). With a long breaker bar attached, turn the joint counterclockwise to loosen and remove it.

FIGURE II-21 This is a special ball joint socket used to remove and replace threaded-type ball joints. (Courtesy of Moog Automotive, Inc.)

FIGURE II-22 When inspecting tapered hole of steering knuckle, check for distortion or cracks. Clean hole before reassembly.

10. No matter the type of joint, thoroughly clean and inspect the taper opening in the steering knuckle and ball joint opening in the control arm (Fig. II-22). Cracks, bends, or other irregular surfaces are reasons to replace these parts (Fig. II-23).
11. Secure a replacement joint and compare the size and the fit of both the new and old units.
12. Replacement of rivet type:
    a. In place of rivets, bolts are used to install the new part. Tighten the nuts to the recommended torque (Fig. II-24).
    b. Insert the ball stud into the taper hole of the knuckle and tighten the nut to the recommended torque (Fig. II-25).
    c. Insert a new cotter pin. When the pin hole does not line up, *further* tighten the nut to the next hole. Do not loosen the nut to insert a new pin.
13. Replacement of press-type ball joint:
    a. With press tool and correct adapters, install the joint to its seated position (Fig. II-26).

FIGURE II-24 Installing high strength bolts in place of original rivets. Torque to recommended specifications. (Courtesy of Moog Automotive, Inc.)

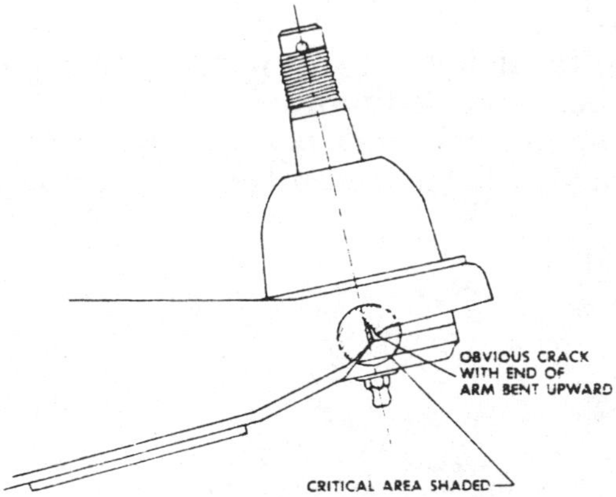

FIGURE II-23 Cracks, bends, or other irregular surfaces in ball joint area are reasons to replace parts.

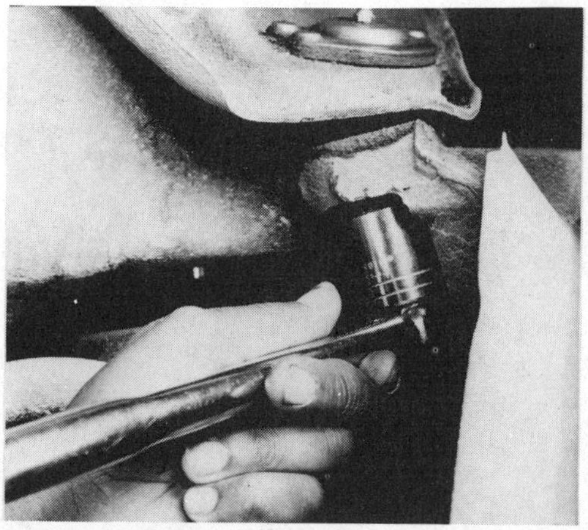

FIGURE II-25 With a torque wrench and socket, tighten ball stud nut to specifications.

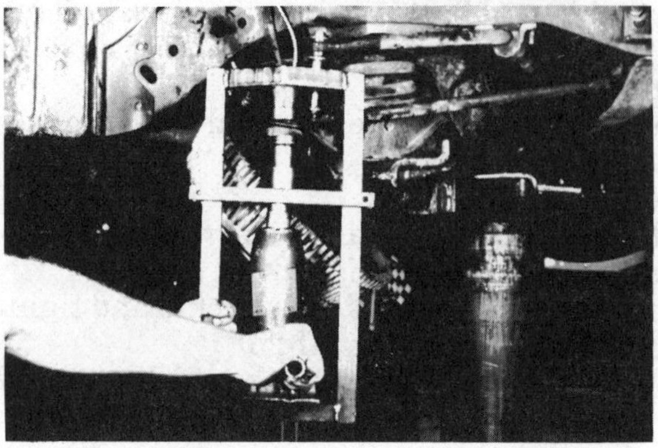

FIGURE II-26 Use a ball joint press tool to install ball joints. Maximum press capacity is 5 tons. (Courtesy of Moog Automotive, Inc.)

b. Some replacement joints for press ball joints are threaded and require the use of a socket and handles for installation. This type of joint will cut its own threads as it moves to the seated position.
c. Insert the ball stud into the taper hole of the knuckle and tighten the nut to the recommended torque. Make sure both surfaces are clean and free of irregularities.
d. Insert a new cotter pin. When pin hole does not line up, *further* tighten nut to the next slot. Never loosen the nut to insert the cotter pin.

14. Replacement of threaded-type ball joints:

Note: *Replacement ball joints for GM cars originally equipped with threaded joints have been superseded by press-fit units. Follow the installation instructions with the new joints (Fig. II-27).*

a. With the threads in the control arm clean, tighten the joint until it is properly seated (Fig. II-21).
b. Insert the ball stud into the taper hole of the steering knuckle and tighten the nut to the recommended torque.
c. Insert a new cotter pin. If the pin hole does not line up, *further* tighten the nut to the next slot. Never loosen the nut to insert the cotter pin.

15. Perform alignment procedures.
16. Road test.

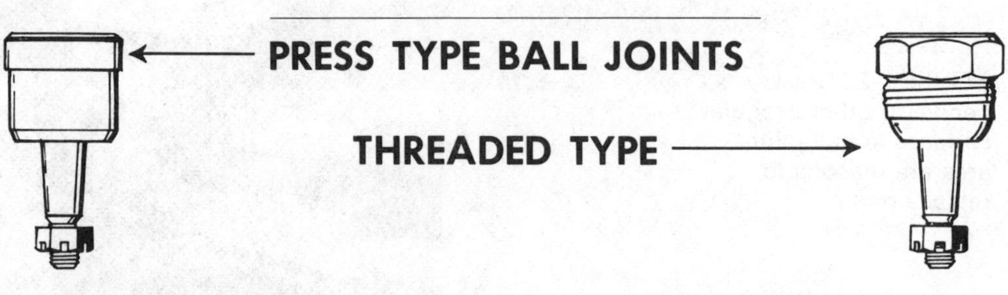

FIGURE II-27 OEM replacement ball joints for all GM cars originally equipped with threaded joints have been superseded by press fit units. Most after market manufacturers have also followed this policy.

JOB REPORT SHEET II-A2

BALL JOINT REPLACEMENT

Student name________________________Date________________

Make of car____________Model____________Year________________

Time started____________Time finished____________Total time____________

Flat-rate time ________________________________

Specifications:

1. Torque required for ball-stud nut: __________ft-lb.

2. Torque required for replacement bolts: _______ft-lb.

Notes:

- Wear suitable eye protection at all times.
- Use adequate lighting.
- Align all vehicles following ball joint installation. Refer to Unit IV.

Checkup Questions:

1. How are ball joints attached to control arms?

2. Were all safety precautions observed during completion of this job? _______ Yes _______________ If not, why?

____

3. Other suspension defects noticed: ________________________________

____

4. Reference materials used to complete this job: ________________________

____

____

DATE COMPLETED________________INSTRUCTOR'S SIGNATURE ________________

Student name________________________________ Date____________________

Directions: Select the best answer to each question and circle the correct letter at the left.

1. The most common type of ball joint wear is:
   I. Taper wear
   II. Radial
      A. I only
      B. II only
      C. Both I and II.
      D. Neither I nor II
2. The load carrying ball joint is attached to the:
   I. Spring carrying control arm
   II. Follower joint
      A. I only
      B. II only
      C. Both I and II.
      D. Neither I nor II
3. Compression-loaded ball joints:
   I. May have wear indicators
   II. Are found on most MacPherson strut systems
      A. I only
      B. II only
      C. Both I and II.
      D. Neither I nor II
4. When checking ball joints, a loose spindle nut will show the same wear as:
   I. Axial movement
   II. Radial movement
      A. I only
      B. II only
      C. Both I and II.
      D. Neither I nor II
5. The follower ball joint is mounted to the:
   I. Steering knuckle
   II. Unloaded control arm
      A. I only
      B. II only
      C. Both I and II.
      D. Neither.I nor II
6. The ball joint most often needing replacement is the:
   I. Follower
   II. Load carrying
      A. I only
      B. II only
      C. Both I and II.
      D. Neither I nor II

7. Without the use of a spreader tool, the ball joint taper can be broken with a:
   I. Hammer
   II. Press tool
      A. I only
      B. II only
      C. Both I and II.
      D. Neither I nor II
8. When tightening ball stud nuts, what procedures should be used?
   I. Tighten with a torque wrench
   II. Further tighten until cotter pin can be inserted
      A. I only
      B. II only
      C. Both I and II.
      D. Neither I nor II
9. Replacement ball joints for threaded control arm may:
   I. Require heat to press in new joint
   II. Require special strength bolts
      A. I only
      B. II only
      C. Both I and II.
      D. Neither I nor II
10. Vehicles with torsion-bar front suspension have:
   I. No ball joints
   II. Only one ball joint
      A. I only
      B. II only
      C. Both I and II.
      D. Neither I nor II

DATE COMPLETED____________________ INSTRUCTOR'S SIGNATURE ____________________

OBJECTIVE II-B (written test)

After studying the preparation assignment you will be expected to correctly answer a minimum of seven out of ten multiple-choice questions in a written test without reference materials on kingpin suspension systems.

Preparation:

Study pages 113 through 121.

KINGPINS

Prior to 1950, the kingpin suspension was the most common type of suspension used in passenger cars. In this type of suspension, the kingpin holds the spindle to the steering knuckle, permitting the wheel assembly to turn from side to side. A kingpin suspension for a passenger car is shown in Fig. II-28. Upper and lower outer control arm bushings are used with kingpins to allow the up-and-down movement needed by the suspension.

In the early 1950s, the kingpin design was replaced with ball joint construction for cars with independent front suspension (Fig. II-29). However, kingpins are still used with a solid or straight axle on light and heavy trucks.

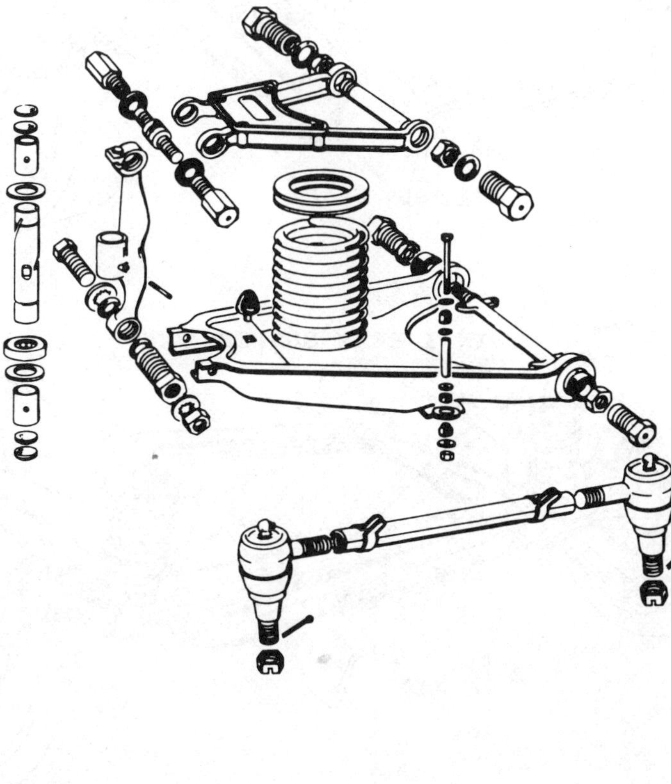

FIGURE II-28 Early kingpin suspension. Kingpin, bushings, and seals shown on the left. (Courtesy of Moog Automotive, Inc.)

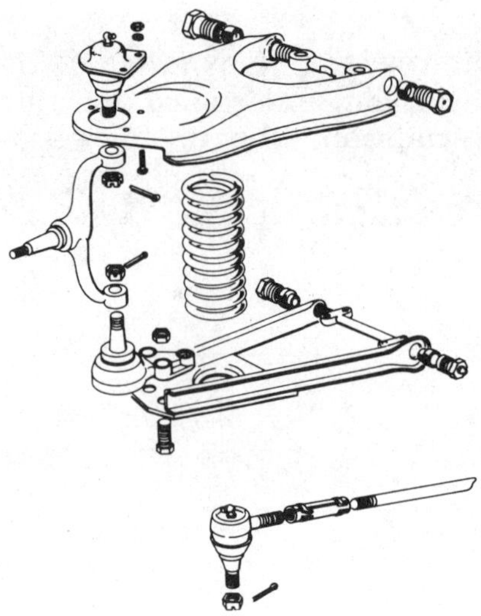

FIGURE II-29 Typical independent front suspension that uses control arms, ball joints, and steering knuckle spindle combination. (Courtesy of Moog Automotive, Inc.)

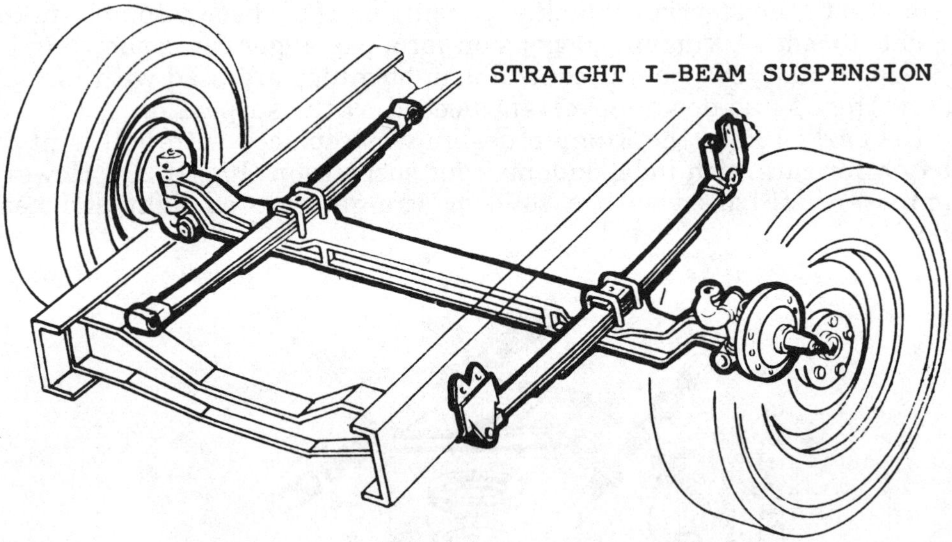

FIGURE II-30A Straight I-beam suspension

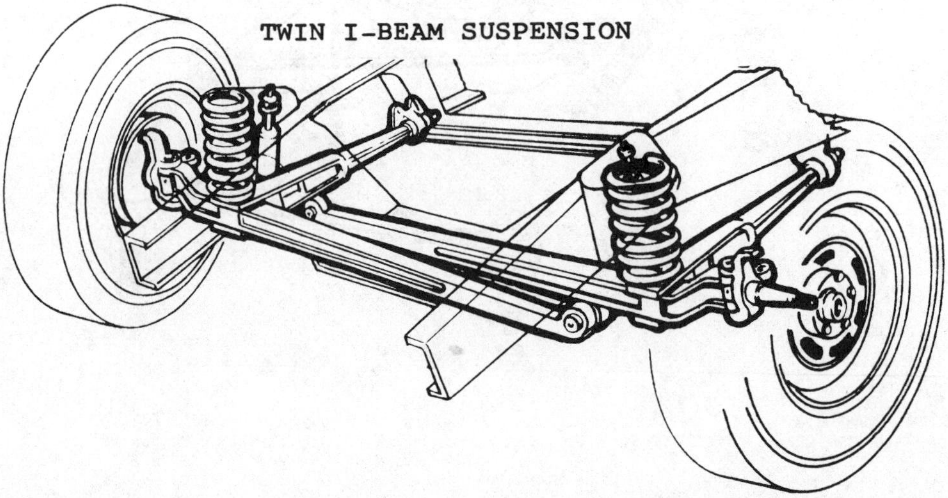

FIGURE II-30B Twin I-beam suspension

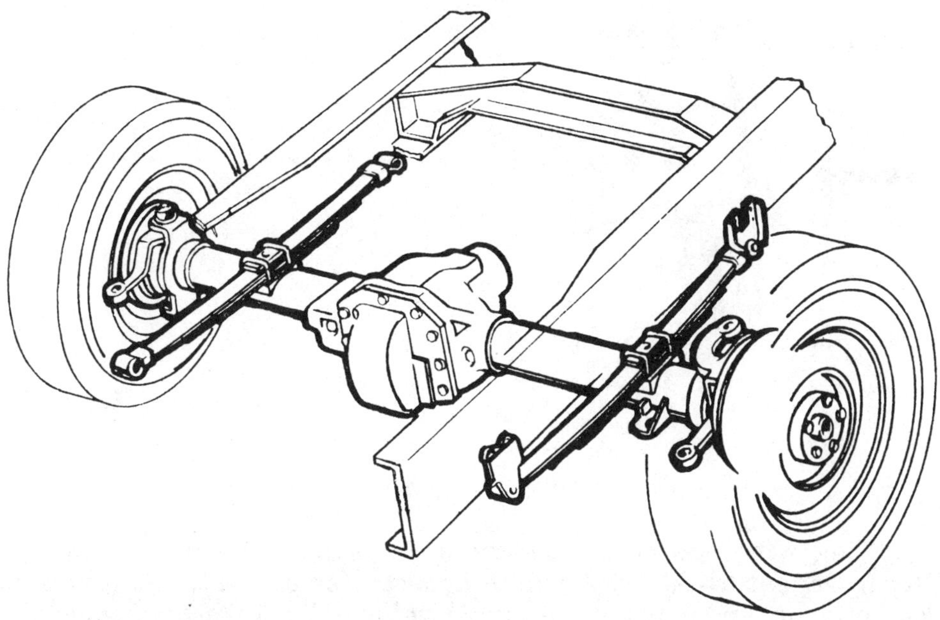

FIGURE II-30C Four-wheel-drive suspension

Fig. II-30 shows three variants of the solid-beam axle using kingpins. The straight I-beam axle (Fig. II-30A) has a solid axle and parallel leaf springs to support the vehicle weight. The springs are mounted to the frame by shackle bushings. Kingpins are used for pivoting the wheels and are held in place by a locking bolt or draw key (Fig. II-31).

Twin I-beam vehicles (Fig. II-30B) have the coil spring mounted between the frame and the I-beam. This design gives an independent front suspension using no upper control arms. One end of each I-beam is attached to the frame with a rubber bushing and the other end is connected to the spindle. The spindle is held in place on the I-beam by a kingpin. Delrin (nylon) or bronze bushings are used in the upper and lower ends of the spindle.

Four-wheel-drive suspension systems (Fig. II-30C) have either leaf or coil springs on a rigid driving axle up front. Both ball joints and knuckle pins or kingpins are used to provide pivot to the front wheels.

KINGPIN WEAR CHECK

Kingpins can be checked for wear by raising the axle off the ground. Then place one hand on the top of the tire and the other on the bottom and alternately move the wheel in and out at the top and bottom. The degree of kingpin wear is indicated by the amount of movement in the wheel and brake assembly. Before starting repairs, make certain that the looseness is in the kingpin and not just an improperly torqued wheel spindle nut. Ordinarily, repairs should be made when the movement at the brake backing plate exceeds 1/32 inch (.79 mm). If the wheel and drum move but the backing plate does not, then torque the wheel spindle nut and recheck for movement in the backing plate.

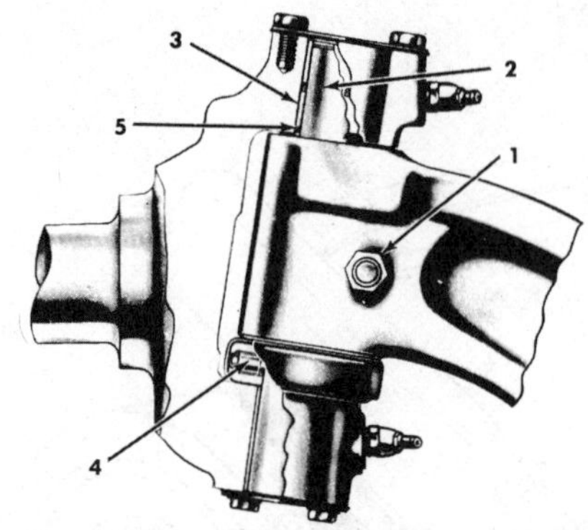

FIGURE II-31 Detailed view of a kingpin front-suspension assembly. (Courtesy of Chevrolet Truck Division)

OBJECTIVE II-B1 (manipulative)

Working with one other student as a team and with the aid of Job Report Sheet II-B1, you will diagnose, remove, fit, and replace the kingpins on an instructor assigned vehicle. Time recommended: Twice the flat-rate time allowed in Motors or Chilton Flat Rate manual. This objective may be completed as part of Objective II-F—a front end rebuild.

Preparation:

1. Review pages 113 through 116.
2. Study pages 116 through 119.
3. Study manufacturer's or general repair manual for automobile to be used.

Materials and Tools Needed:

1. Secure individual tools and materials as described in instruction sheet enclosed with the new parts or found in a repair manual.

Since kingpin service is not a large part of the work performed by a front end specialist, the following procedures will serve only as a general outline. Refer to directions supplied with the new parts and manufacturer's service manuals.

KINGPIN SERVICE (SINGLE I-BEAM OR TWIN I-BEAM)

1. Jack up the vehicle and place jack (safety) stands under the frame.
2. Remove the wheel, tire, hub, and drum assembly.
3. Remove the backing plate bolts and lift the brake and plate assembly away from the axle. Use wire to suspend the brake backing plate assembly away from the axle. This precaution protects the brake and backing plate assembly and the brake hose.

4. Remove the lock pin or bolt holding the kingpin.
5. Remove the upper and lower seals on the spindle just above and below the kingpin. This may be a small metal expansion plug.
6. Drive the kingpin down through the spindle or the support. Keep any shims found on the top of the beam.
7. Drive or press the old bushings out of the spindle or the support and inspect the kingpin opening for out of round or cracks (Fig. II-32).
8. Install the new bushings. Press them in place. Do not hammer them. Be sure that the bushing grease holes are lined up with those in the steering knuckles.
9. Ream or hone the bushings to fit the new kingpin (Fig. II-33).
10. Install the spindle on the beam. The shims go above and the bearing goes below (cup side down) the beam before placing the kingpin in position (Fig. II-31).
11. Attempt to fit an .005″ (0.127 mm) feeler gauge between the shims and the spindle. If the gauge can be inserted, add shims until it will not fit.
12. After tightening the bolt to correct torque, install new seals above and below the kingpin.
13. Lubricate the kingpin, and check to see if the chassis grease flows through the bearing.
14. Reinstall the brake backing plates, hubs, drums, and wheel assemblies. Be certain wheel bearings are repacked and assembled as outlined in Objective I-A1.
15. Have your instructor check your work before installing the drum or rotor.
16. Remove jack (safety) stands and align front suspension.
17. Road test.

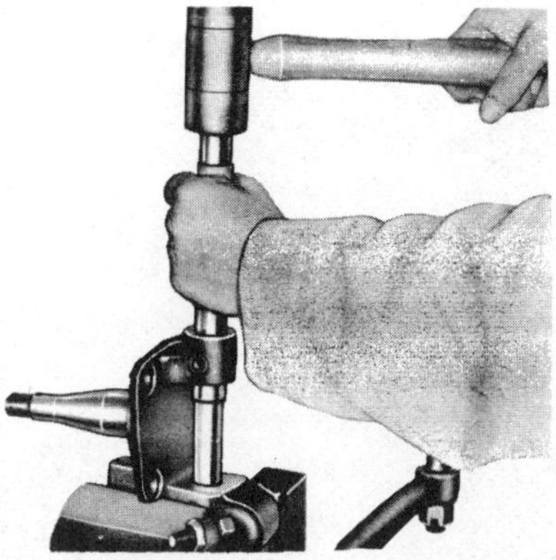

FIGURE II-32 Driving out bronze bushings from spindle. (Courtesy of Ford Customer Service Division)

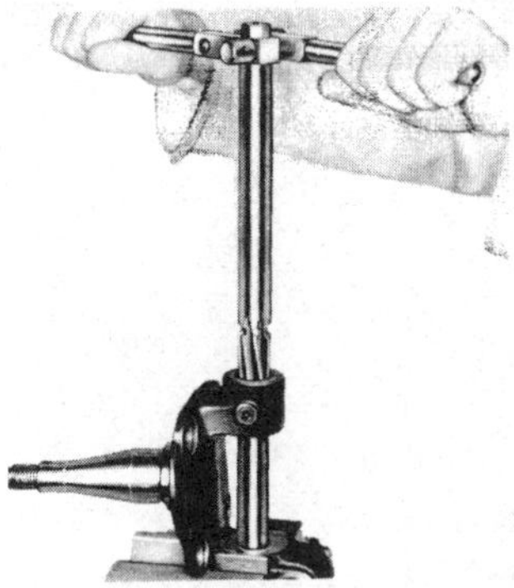

FIGURE II-33 Use a reamer to finish new bushings for proper fit with new kingpin. (Courtesy of Ford Customer Service Division)

KINGPIN REPLACEMENT

Student name____________________Date____________

Make of car____________Model____________Year____________

Time started____________Time finished____________Total time____________

Flat-rate time ____________

Specifications:

Follow specifications and procedures as outlined in instruction sheet accompanying new parts or repair manuals.

Check-up Questions:

Kingpin size? ____________________ inch (mm).

What additional suspension defects were noted? ____________________

____

Was vehicle front end aligned?____________

Was vehicle road tested?____________

DATE COMPLETED____________INSTRUCTOR'S SIGNATURE ____________

TEST II-B

Student Name ______________________ Date ______________

Directions: Select the best answer to each question and circle the letter at the left.

1. In early passenger cars, the kingpin was used to hold the:
   I. Spindle to its bearings
   II. Axle to its spring
   A. I only
   B. II only
   C. Both I and II.
   D. Neither I nor II
2. When checking the wear of kingpins, the following procedure(s) is acceptable:
   I. Jack vehicle, move wheel alternately top to bottom, watching movement of backing plate
   II. With vehicle on ground, turn steering side to side. Measure amount of movement at backing plate
   A. I only
   B. II only
   C. Both I and II.
   D. Neither I nor II
3. When checking for kingpin wear, if the drum moves but the backing plate does not, you should:
   I. Tighten spindle nut and recheck
   II. Lubricate wheel bearing and axle, then recheck
   A. I only
   B. II only
   C. Both I and II.
   D. Neither I nor II
4. Kingpin repairs should be made when the movement at the backing plate exceeds:
   I. 1/32 inch (.79 mm)
   II. Movement of spindle
   A. I only
   B. II only
   C. Both I and II.
   D. Neither I nor II
5. The kingpin is held solidly to the axle by:
   I. A lock bolt
   II. A pin
   A. I only
   B. II only
   C. Both I and II.
   D. Neither I nor II
6. To check for kingpin wear, the axle must be:
   I. Disassembled
   II. Cleaned
   A. I only
   B. II only
   C. Both I and II.
   D. Neither I nor II

7. When replacing new kingpin bushings:
   I. Align grease holes of bushings with those of the knuckle
   II. Fit pin to recommended clearance
      A. I only
      B. II only
      C. Both I and II.
      D. Neither I nor II
8. The preferred method(s) of bushing installation is:
   I. Pressing
   II. Shrinking
      A. I only
      B. II only
      C. Both I and II.
      D. Neither I nor II
9. When the backing plate is removed it should be tied up with wire or rope to:
   I. Keep it from interfering with further work
   II. Prevent brake-hose damage
      A. I only
      B. II only
      C. Both I and II.
      D. Neither I nor II
10. Among the final required operations in kingpin service is:
   I. Alignment of front suspension
   II. Torquing spindle nut
      A. I only
      B. II only
      C. Both I and II.
      D. Neither I nor II

DATE COMPLETED____________________INSTRUCTOR'S SIGNATURE ____________________

After studying the preparation assignment you will be expected to correctly answer a minimum of seven out of ten multiple-choice questions in a written test without reference materials on upper and lower control arms, and upper and lower control arm bushings.

OBJECTIVE II-C (written test)

Preparation:

1. Study pages 123 through 135.
2. Review manufacturer's service manual or general repair manual if necessary.

CONTROL ARMS

Automotive designers, in testing various types of "soft ride" suspensions, developed the independent front suspension. The most common design consists of two control arms, usually termed *upper* (short) and *lower* (long) (Fig. II-34). (Together, they are sometimes called LASA, for "long arm, short arm.")

The shorter suspension control arm is mounted on the top of the frame and the longer or lower arm is mounted to the front cross member of the frame. The arc created by the movement of the upper (short) arm is smaller than the arc created by the movement of the lower (long) control arm. When there is up-and-down movement of the car, the upper arm moves through its arc much faster than the lower arm. As a result, the further up the wheel assembly moves in relation to the body of the car, the further the wheel is pulled inward at the top. Check the upper suspension movement and see how it affects the position of the tire (Fig. II-35).

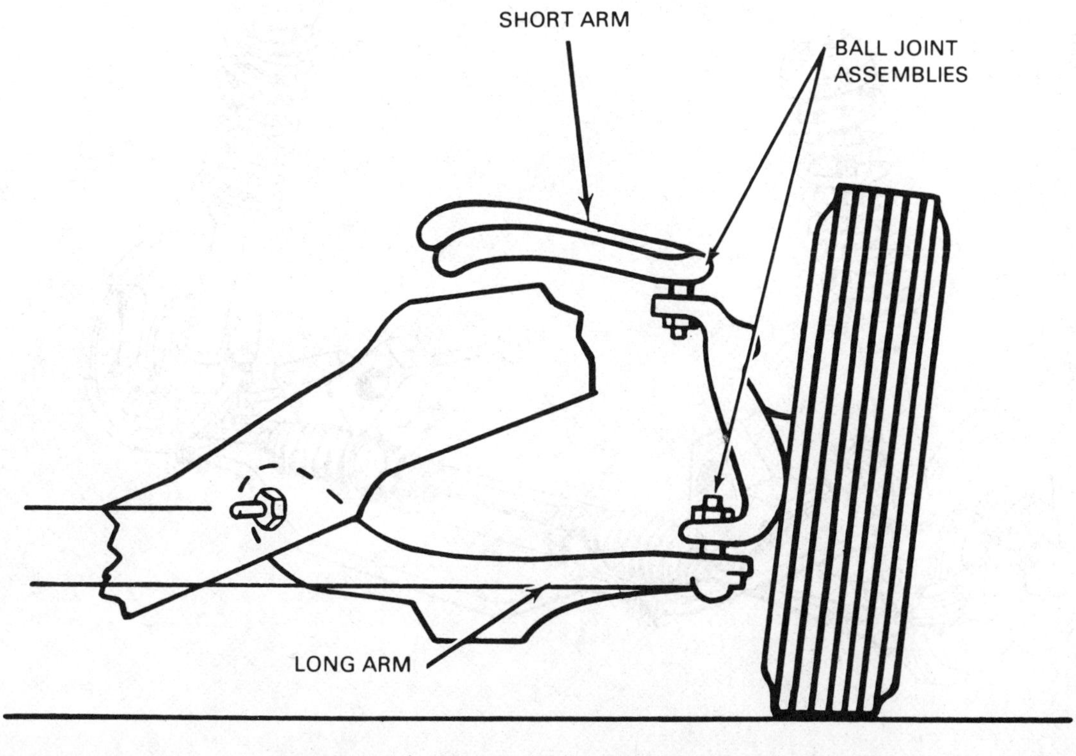

FIGURE II-34 This is a typical independent front suspension short (upper) control arm and long (lower) control arm using ball joints to support steering knuckle. (Courtesy of Chevrolet Motor Division)

As the wheel moves up and down, its vertical position is altered, which in turn changes the distance between the front wheels. The change in distance is called scuff or scrub. Scrub causes tire wear.

With the LASA-type suspension system, the change in distance between front wheels is kept to a minimum during normal driving. This keeps tire wear to a minimum, and is one of the advantages of a LASA type suspension.

As stated above, MacPherson strut suspension systems have only one control arm. This arm connects the frame or crossmember to the ball joint and steering knuckle much the same as the lower control arm on the Type 1 and 2 systems. The strut piston rod (shock absorber) and the coil spring are attached to the body of the vehicle at the upper wheel well. This design does away with the need for an upper control arm and upper ball joint (Fig. II-36).

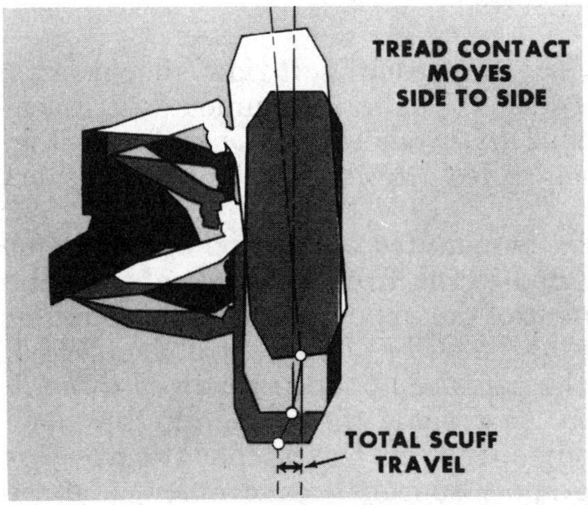

FIGURE II-35 Illustration shows movement of control arms on independent front suspension. (Courtesy of Chrysler Motors Corp.)

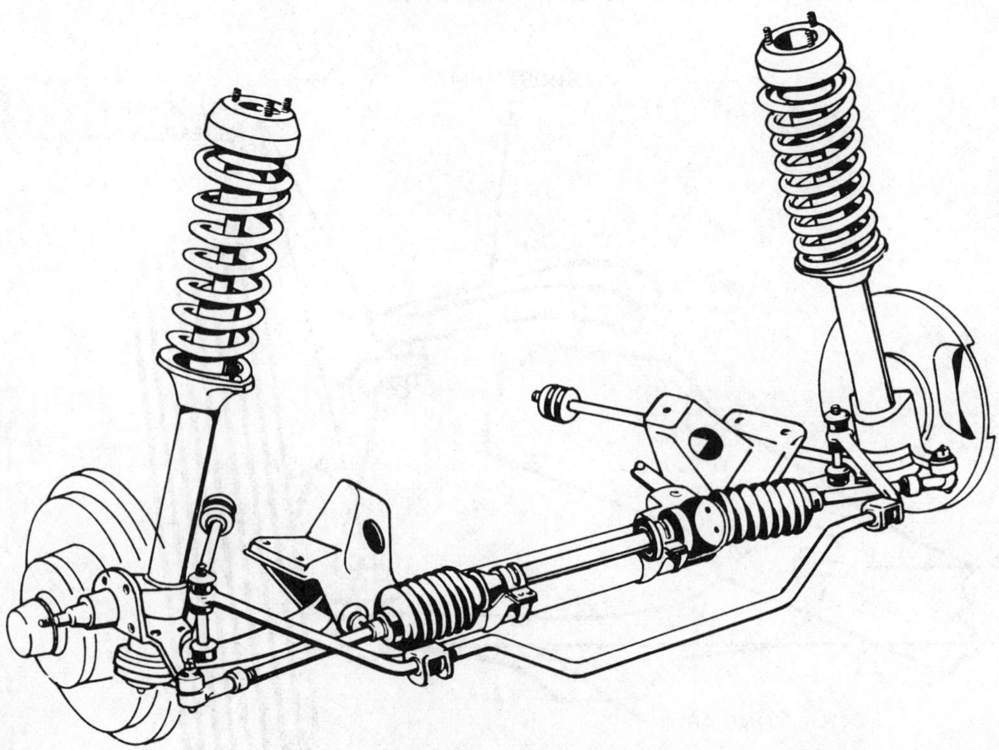

FIGURE II-36 A typical MacPherson strut front suspension

OBJECTIVE II-C1: (manipulative)

Working as a team with one other student and with the aid of Job Report Sheet II-C1, you will remove the control arms and replace the bushings in a minimum of two control arms, either upper or lower, on an instructor assigned automobile. Time recommended: Twice the flat-rate time allowed in Motors or Chilton Flat Rate manuals. This objective may be completed as part of Objective II-F—a complete front end rebuild.

Preparation:

1. Review pages 123 through 125.
2. Study pages 125 through 135.
3. Study manufacturers' or general repair manuals for the automobile to be serviced.
4. Carefully observe instructor's demonstration on bushing removal and replacement.

Materials and Tools Needed:

1. Service (floor) jack and jack (safety) stands
2. Lug or impact wrench and socket
3. Socket set with handles and extension
4. Assorted box/open-end combination wrenches
5. Diagonal pliers
6. Spreader tool (taper breaker) or two large hammers
7. Pneumatic chisel
8. Torque wrench
9. Large safety chain with grab hooks
10. Cotter pins
11. Chassis grease
12. Hand towels or wiping rags
13. Manufacturer's or general repair manual

Procedure:

UPPER CONTROL ARM SERVICE SPRING BETWEEN CONTROL ARMS

Removal:

1. Raise vehicle on a hoist or jack far enough for the wheels to clear the ground.
2. Support the lower control arm with a jack stand at a point as close to the wheel as possible (Fig. II-37).
3. Remove tire and wheel.
4. Separate the upper control arm ball stud from the steering knuckle. Do this by loosening upper ball stud nut one or two turns. Then strike the steering knuckle with a heavy hammer in the area of the taper. After the taper breaks, remove ball stud nut.
5. Remove the two nuts securing the upper control arm cross shaft to the frame bracket. (Fig. II-38).

FIGURE II-37 The photograph shows the proper method of raising a wheel with a jack at the outer edge of the lower control arm. (Courtesy of Moog Automotive, Inc.)

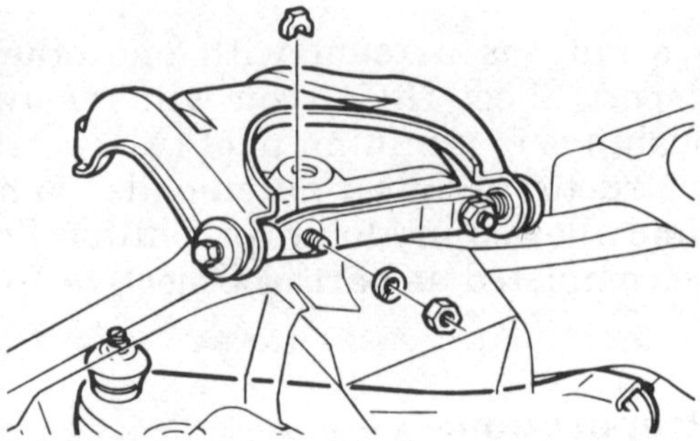

FIGURE II-38 Diagram for removing self-locking nuts and shims on upper control arm. (Courtesy of Chevrolet Motor Division)

6. Remove alignment shims. Tape them together and identify each shim pack as a front bolt or rear bolt shim pack.
7. Remove the upper control arm. It may be necessary to remove the cross-arm shaft bolts before removing the arm (Fig. II-39). Some bolts are splined into the frame and must be driven out with a brass drift or pry bar (Fig. II-40).
8. Install the bushings as described in "Control-Arm Bushing Service" (page 133).

Note: *Do not tighten control arm bushings until car is on the ground.*

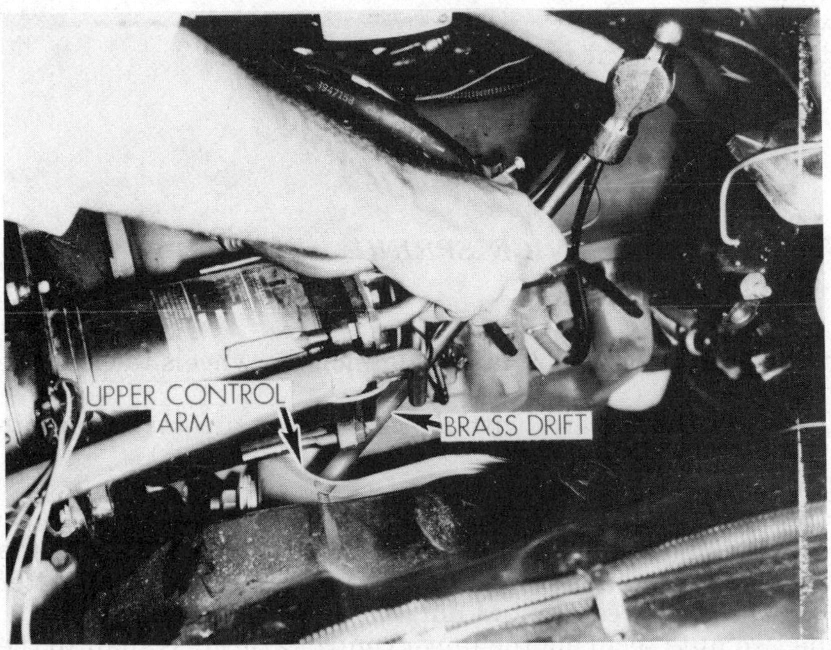

FIGURE II-39 Use a brass drift and hammer to loosen upper control arm cross shaft bolts. (Courtesy of Chevrolet Motor Division)

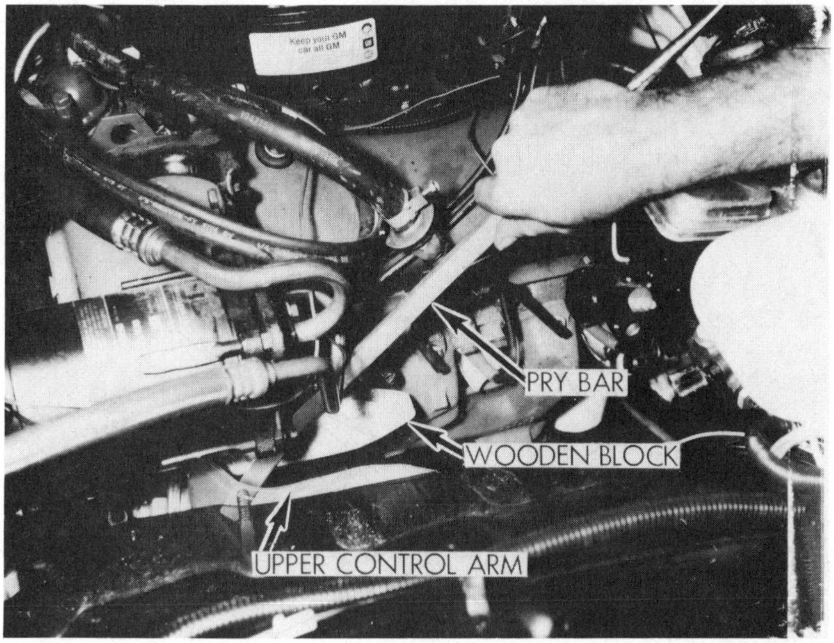

FIGURE II-40 Use a pry bar and wooden block to loosen upper control arm cross shaft bolts. (Courtesy of Chevrolet Motor Division)

Replacement:

1. Place the control arm in position near frame bracket (Fig. II-41).
2. If the bolts have been removed, insert them through frame bracket and cross arm shaft.
3. Use temporary nuts (not self-locking) and tighten cross arm shaft until splined bolts are fully seated in the frame bracket.
4. Remove the nuts and place alignment shim packs in their original positions.
5. Tighten the original self-locking nuts to specifications.

FIGURE II-41 Insert cross arm shaft bolts with upper control arm in position. Non–self-locking nuts should be used on first assembly. (Courtesy of Chevrolet Motor Division)

6. Install the ball stud through the steering knuckle, torque the nut, and install the cotter pin. **Caution:** *Never back off the nut to align the nut slot with the hole in the stud. Always tighten the nut to the next slot so the cotter pin can be installed* (Fig. II-42).
7. Install the tire and wheel and lower vehicle to floor.
8. Front end alignment should be performed to ensure that the proper steering angles are maintained.

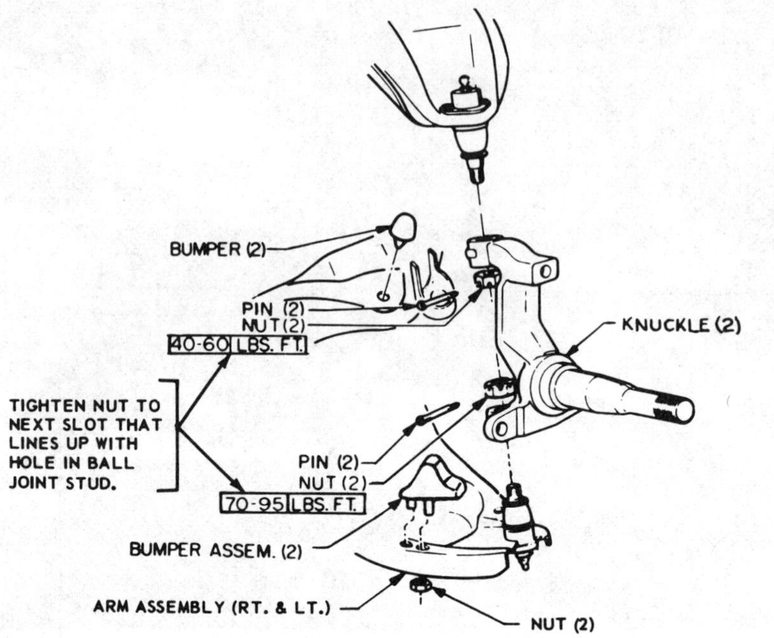

FIGURE II-42 Steering krtuckle and control-arm assembly involves using proper torque settings. (Courtesy of Buick Motor Division)

LOWER CONTROL ARM SERVICE SPRING BETWEEN CONTROL ARMS

Removal:

1. In many front suspension designs it is easier to completely remove the lower control arm for service. This requires removal of the coil spring, which is described in "Coil Spring Service—Spring Between Control Arms" page 153.
2. With the spring removed, loosen the mounting bolts on the inner side of the control arm and remove the arm (Fig. II-43).
3. Install bushings as described in "Control Arm Bushing Service" page 133.

Replacement:

1. Reverse the above procedure.
2. If any new parts were installed, adjust alignment to meet the manufacturer's specifications.

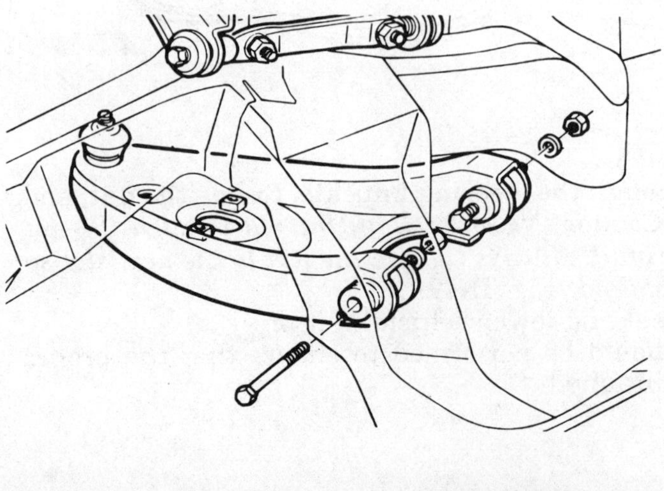

FIGURE II-43 When inserting lower control arm bolts, replace bolts as shown in manufacturers' or repair manuals. (Courtesy of Chevrolet Motor Division)

UPPER CONTROL ARM SERVICE SPRING ABOVE UPPER ARM

Removal:

1. To service the upper control arm, the spring must be removed. Follow specific directions in "Coil Spring Service—Spring Above Upper Control Arm" page 157.
2. With the spring removed, or compressed extract cotter pin and nut from upper ball joint stud.
3. Separate the ball stud from control arm with a spreader tool or strike the steering knuckle in the taper area with a heavy hammer.
4. Remove the control arm mounting bolts and remove the arm.
5. To replace the bushings, see "Control Arm Bushing Service" page 133.

Replacement:

Reverse above procedure but do not tighten the pivot bolt nuts until the car is on the ground. Tightening before this would damage the bushings in a short period of driving. **Note:** *Always use a torque wrench and tighten all bolts to correct torque.*

LOWER CONTROL ARM SERVICE-SPRING ABOVE UPPER CONTROL ARM AND MACPHERSON STRUT

Removal:

1. Raise the front of the car, allowing the wheels to hang free.
2. Remove the wheel and brake drum as a unit from the spindle.
3. Separate the steering arm from the tie rod.
4. Disconnect the stabilizer bar from the control arm, if the vehicle is equipped with a stabilizer bar.
5. Disconnect the strut rod from the control arm.
6. Remove the lower ball joint, cotter pin, and retaining nut.
7. Separate the ball stud from the steering knuckle. Use a separator tool or strike the taper area with a heavy hammer.
8. Remove the inner pivot bolt and remove the lower control arm from the cross member (Fig. II-44).
9. Note and mark the position of the eccentric washers.
10. To replace the bushings, see "Control Arm Bushing Service "page 133.

Installation:

Reverse the above procedure, but do not tighten inner pivot bolt until the car is on the ground. Always use a torque wrench and tighten all bolts to specific torque.

UPPER CONTROL ARM SERVICE–TORSION BAR

Removal:

1. Follow steps 1 through 4 of "Upper Control Arm Service—Spring Between Control Arms" page 125.
2. Note and mark the original position of eccentric washers on the metal.
3. Remove nuts, lock washers, cams, and cam bolts attaching the upper control arm and bushing to the support brackets (Fig. II-45). Lift the control arm up and away from support.

Replacement:

Reverse above procedure.

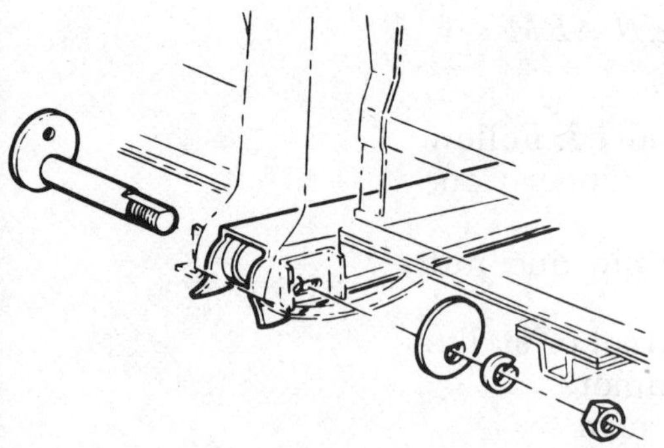

FIGURE II-44 When removing pivot bolts, scribe position of eccentric washers. (Courtesy of Chevrolet Motor Division)

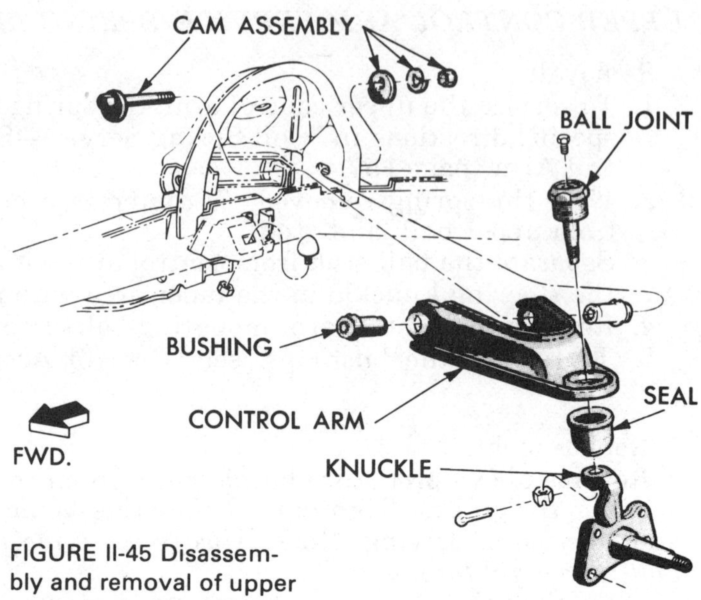

FIGURE II-45 Disassembly and removal of upper control arm from frame bracket. (Courtesy of Chrysler Motors Corp.)

LOWER CONTROL ARM SERVICE–SPRING ABOVE UPPER CONTROL ARM

Removal:

1. Raise the front of the car, allowing the wheels to hang free.
2. Remove the wheel and brake drum as a unit from the spindle.
3. Separate the steering arm from the tie rod.
4. Disconnect the stabilizer bar from the control arm, if the vehicle is equipped with a stabilizer bar.
5. Disconnect the strut rod from the control arm.
6. Remove the lower ball joint, cotter pin, and retaining nut.
7. Separate the ball stud from the steering knuckle. Use a separator tool or strike the taper area with a heavy hammer.
8. Remove the inner pivot bolt and remove the lower control arm from the cross member (Fig. II-44).
9. Note and mark the position of the eccentric washers.
10. To replace the bushings, see "Control Arm Bushing Service."

Installation:
Reverse the above procedure, but do not tighten inner pivot bolt until the car is on the ground. Always use a torque wrench and tighten all bolts to specific torque.

UPPER CONTROL ARM SERVICE–TORSION BAR

Removal:

1. Follow steps 1 through 4 of "Upper Control Arm Service—Spring Between Control Arms" page 125.
2. Note and mark the original position of eccentric washers on the metal.

3. Remove nuts, lock washers, cams, and cam bolts attaching the upper control arm and bushing to the support brackets (Fig. II-45). Lift the control arm up and away from support.

Replacement:
Reverse above procedure.

LOWER CONTROL ARM SERVICE–TORSION BAR

1. To remove the lower control arms on torsion bar suspension cars, it is necessary to first remove the torsion bar. See "Torsion Bar Service."
2. Remove the tie rod end from the steering knuckle (Fig. II-46).
3. Remove the stabilizer bar strap or link attachments from the lower control arm.
4. Separate lower ball joint stud from lower control arm.
5. Remove the strut pin, nut, and bushing from front cross member.
6. Remove the nut, washer, and pin holding the inner end of the lower control arm.
7. Remove the lower control arm and strut as an assembly.

Replacement:
Reverse the above procedure except for the following:
1. Place the rear bushing and retainer on the strut rod and the forward bushing and retainer on the cross member.
2. Install the strut rod.
3. Install the lower control arm.
4. After the vehicle is again resting on the floor, tighten the strut nut at the cross member and install the spring (roll) pin.
5. Adjust the front suspension trim height (Fig. II-47).
6. Adjust the front end alignment as necessary.

FIGURE II-46 A tie rod end puller is used to separate tie rod from steering knuckle arm. (Courtesy of Chevrolet Motor Division)

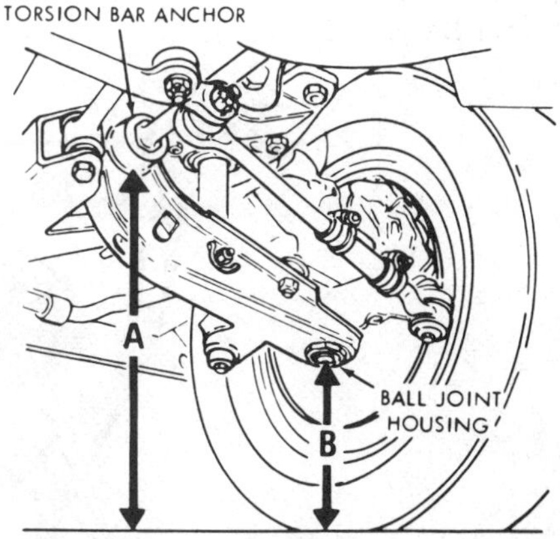

FIGURE II-47 This is a typical method of measuring trim height on torsion-bar front suspension. (Courtesy of Chrysler Motors Corp.)

"A" Dim. – Lowest point of the adjusting blade to floor.
"B" Dim. – Lowest point of the Ball Joint Housing to floor.
"A" Dimension is always greater than "B" dimension. Measurements should always be made on a centerline to control arm.

CONTROL ARM BUSHINGS

The control arm shafts and bushings provide the upper and lower control arms of the front suspension with a pivot point. The shafts and bushings attach the control arms to the vehicle frame in a manner that permits the wheels to move up and down independently (Fig. II-48).

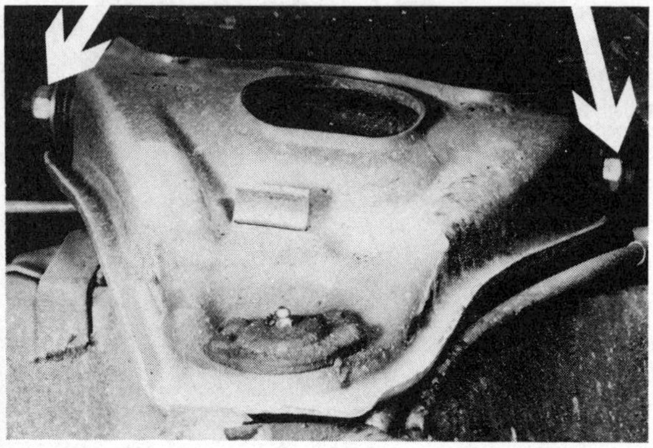

FIGURE II-48 Typical front suspension. The arrows show positions of upper control-arm bushings. (Courtesy of Moog Automotive, Inc.)

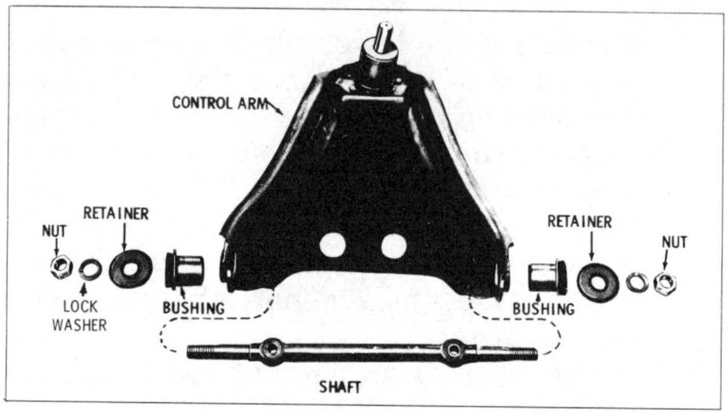

FIGURE II-49 This is an upper control arm assembly with rubber type bushings. (Courtesy of Buick Motor Division)

FIGURE II-50 Upper control arm assembly with metal type bushings. (Courtesy of Ford Customer Service Division)

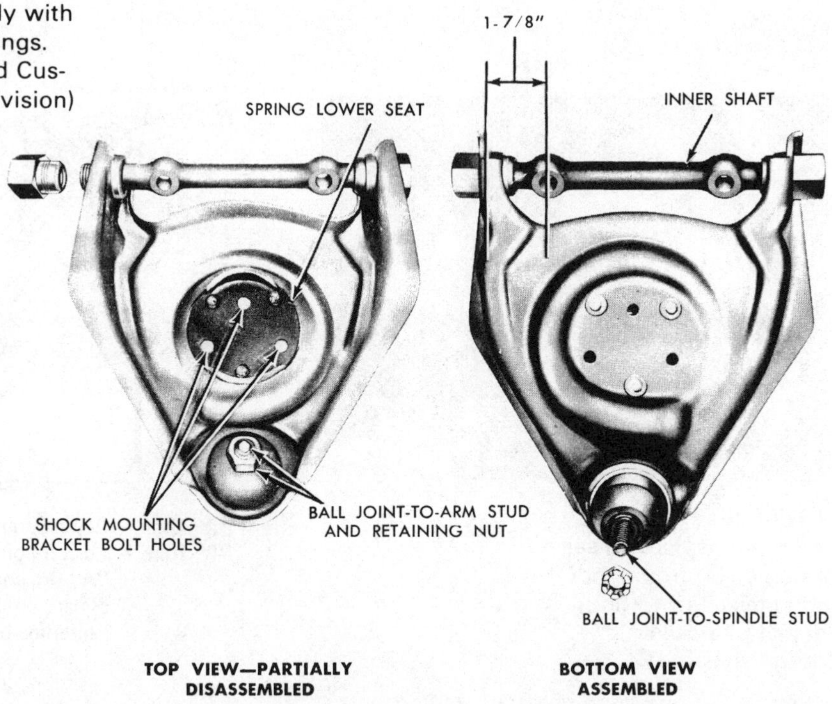

FIGURE II-51 Examples of upper control arm bushings with excessive rubber deterioration.

Control arm bushings are constructed of either rubber or metal. The rubber bushing is more commonly used and has a torsional effect. As the end of the control arm moves up and down, the rubber bushing must be able to twist equally in either direction from the normal load position. Figs. II-49 and II-50 show both the rubber and the metal type bushings.

The metal inner control arm bushing must be manufactured to very close tolerances and be heat treated to allow the regulated movement of the control arm.

Control arm bushings should be replaced when they show signs of:

1. Rubber deterioration (Fig. II-51).
2. Rubber sliding over the metal sleeve and not twisting (usually resulting in a squeak)
3. Excessive wear

CONTROL ARM BUSHING SERVICE

There are several ways to remove and replace control arm bushings. It is often impossible to service these bushings without removing the control arm. See "Control Arm Service" (page 125) for removal procedures. Manufacturers have various special tools to press out old bushings and press in new ones (Fig. II-52). To be able to service most control arm bushings in use today, the following general procedure should apply to the average vehicle. **Note:** *Use only soap and water to lubricate rubber bushings.*

Procedure:

1. Clamp the control arm securely in a vise.

Note: *If the control arm has a cross-shaft, as in Figs. II-49 and II-50, be sure to reinstall it in its original position. To do this, mark one end of the shaft and the control arm so that in assembling the new bushings, the cross-shaft is in its original position when reinstalled.*

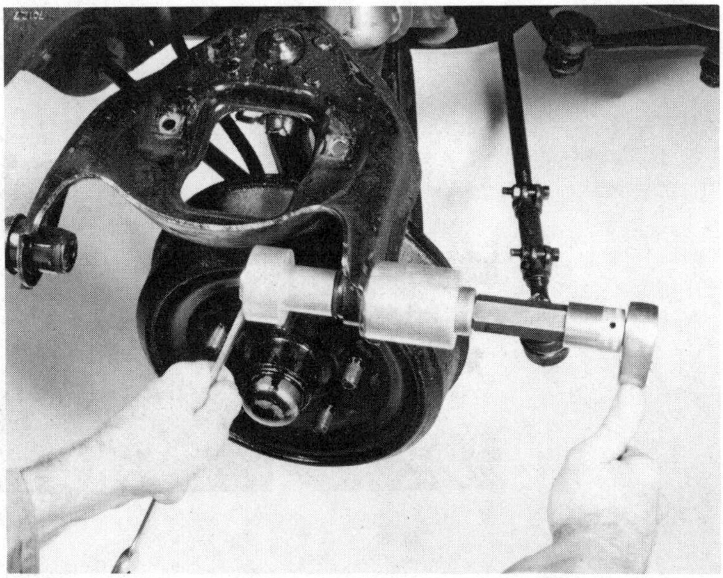

FIGURE II-52 Servicing control arm bushings without removing control arm always requires special tools. (Courtesy of American Motors Corporation)

2. Place the blade of a pneumatic chisel between the control arm and the flange of the metal sleeve of the bushing (Fig. II-53).
3. Operate the chisel, prying upward on the flange of the metal sleeve.
4. Continue to chisel until bushing is free of the control arm. Remove the bushing from the other side.
5. Prepare the openings where the replacement bushings will seat by filing and/or straightening any irregular surfaces.
6. The preferred method of inserting new bushings is with an arbor press or vise.
7. Using a bushing driver, or other firm metal tube or pipe that fits the metal flange of the bushing, press the bushing until seated as the old one (Fig. II-54).
8. If the two openings in the control arm holding the bushing want to collapse during the press-in operation, place a spacer between, to prevent collapse of control arm ends.

FIGURE II-53 Remove control arm bushings with a pneumatic chisel after removal of control arms from the vehicle. (Courtesy of Moog Automotive, Inc.)

FIGURE II-54 Use a bench vice to install control arm bushings. (Courtesy of Moog Automotive, Inc.)

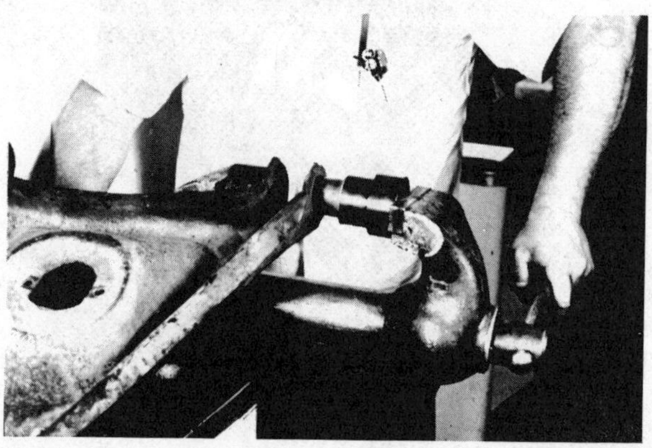

CONTROL ARM BUSHING REPLACEMENT

Student name____________________Date__________

Make of car__________Model__________Year__________

Time started__________Time finished__________Total time__________

Flat rate time ____________________

Specifications:

Ball joint nut torque ____________________ (N-m)

Cross arm shaft nut torque____________________ (N-m)

Pivot bolt nut torque ____________________ (N-m)

Strut Rod torque____________________ (N-m)

Checkup Questions:

1. Did front tires show excessive or unusual wear?____________________
2. Where are self-locking nuts located on the front suspension? ____________________

   ____
3. Why is it necessary to go beyond normal torque on the ball stud nut if the cotter pin hole does not align? ____________________

   ____
4. In what condition did you find the ball joints? ____________________
5. Why must alignment be performed after control arm bushings are replaced?__________

   ____

   ____

DATE COMPLETED____________________INSTRUCTOR'S SIGNATURE ____________________

Student name________________________________Date____________________

Directions: Select the best answer to each question and circle the letter at the left.

1. The lower control arm is known as the:
   I. Short arm
   II. Road contact arm
   A. I only
   B. II only
   C. Both I and II.
   D. Neither I nor II
2. The distance between the front wheels during normal driving conditions remains nearly constant because of:
   I. Control arm design
   II. Ball joint design
   A. I only
   B. II only
   C. Both I and II.
   D. Neither I nor II
3. The major difference(s) in control arm service is prescribed by:
   I. Make of vehicle
   II. Position of spring
   A. I only
   B. II only
   C. Both I and II.
   D. Neither I nor II
4. Common symptom(s) of worn control arm bushings are:
   I. Deteriorated rubber
   II. Squeaking suspension
   A. I only
   B. II only
   C. Both I and II.
   D. Neither I nor II
5. When removing control arm bushings the following procedure(s) is recommended:
   I. Use a torch to burn out old rubber bushings
   II. Use a pneumatic chisel to remove the rubber bushing and its metal sleeve
   A. I only
   B. II only
   C. Both I and II.
   D. Neither I nor II
6. Rubber control arm bushings:
   I. Are the type most commonly used
   II. Assist steering control of the vehicle
   A. I only
   B. II only
   C. Both I and II.
   D. Neither I nor II

7. When tightening ball stud nuts, the procedure(s) that should be used is:
   I. Tighten with a torque wrench
   II. Loosen nut to first cotter pin hole. Insert pin
      A. I only
      B. II only
      C. Both I and II.
      D. Neither I nor II
8. Many cross arm shaft bolts must be removed by:
   I. Prying
   II. Driving out with a brass drift
      A. I only
      B. II only
      C. Both I and II.
      D. Neither I nor II
9. The use of self-locking nuts on the cross arm shaft bolts prevents:
   I. Loss of shims
   II. Loss of eccentric washers
      A. I only
      B. II only
      C. Both I and II.
      D. Neither I nor II
10. Alignment shims, when removed, should be:
   I. Taped together for later replacement
   II. Identified as to front or rear bolt group
      A. I only
      B. II only
      C. Both I and II.
      D. Neither I nor II

DATE COMPLETED________________ INSTRUCTOR'S SIGNATURE ________________

OBJECTIVE II-D
(written test)

After studying the preparation assignment you will be expected to correctly answer a minimum of 14 out of 20 multiple choice questions in a written test without reference materials, on the following front suspension topics:

1. Trim height
2. Leaf springs and mounting parts
3. Coil springs
4. Torsion bars

Preparation:

1. Study pages 139 through 169.
2. Review manufacturer's or general repair manual.

TRIM HEIGHT AND SPRINGS

There are three basic types of springs in use today on the front suspension of American built automobiles and trucks. They are (1) leaf springs, (2) coil springs, and (3) torsion bars (Fig. II-55).

The weight of the automobile can be easily divided into two classifications. One is the sprung weight: all weight carried by the springs. The other is the unsprung weight: all the weight that is not carried by the springs (Fig. II-56). Generally speaking, heavy unsprung weight makes the car handle poorly. The greater the percentage of sprung weight, the better the car will handle.

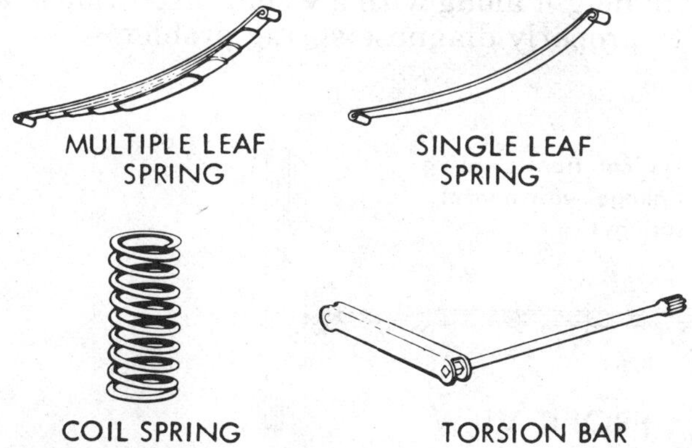

FIGURE II-55 These are basic types of springs used on front suspensions. (Courtesy of Monroe Auto Equipment Company)

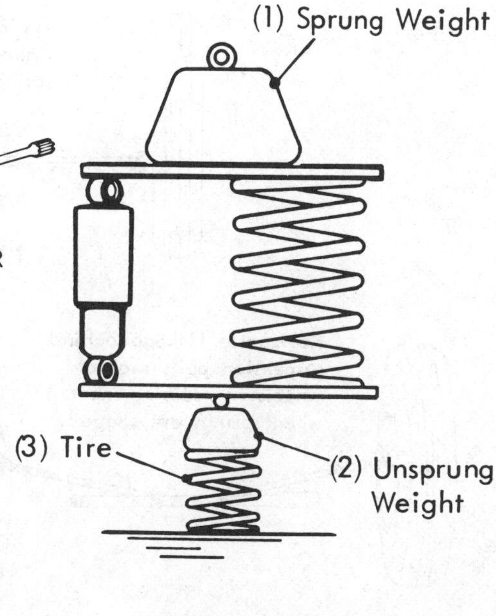

FIGURE II-56 Sprung weight is all the weight carried by the vehicle's springs. The unsprung weight is everything not carried by the springs. This includes the tires, wheels, lower control arms, and differential assembly. (Courtesy of Monroe Auto Equipment Co.)

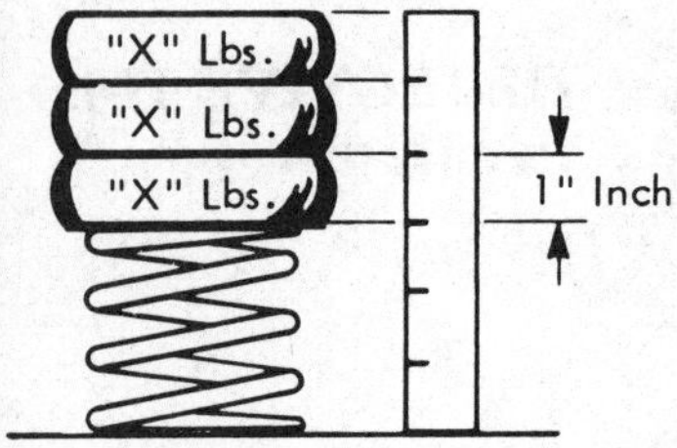

FIGURE II-57 Spring-deflection rates are based on the amount of weight it takes to deflect or compress the spring a distance of one inch. (Courtesy of Monroe Auto Equipment Co.)

Every automobile suspension system is designed to carry a given sprung weight while allowing reasonable up-and-down movement (ride) of the wheels (Fig. II-57). Springs make ride possible, but the strength of springs vary. This strength is expressed as spring deflection rate. Spring deflection rate is determined by the amount of weight needed to deflect or compress a spring through 1 inch of travel. As an example, a 300 pound spring would require 300 pounds to compress or deflect it 1 inch and 600 pounds to compress or deflect it 2 inches. There are wide variations in spring rates and care should be taken to ensure that replacement springs have the proper spring deflection rate. Leaf and coil springs and torsion bars have different rate characteristics. These differences will be discussed individually.

TRIM HEIGHT

Other factors that affect springs are age and the type of service to which the springs were subjected. As a natural occurrence, springs become weak through constant twisting and flexing in normal service. A visible sign of spring age is the sagging front end on automobiles more than five years old. Usually such sagging is on the larger, heavier models.

When a vehicle has been heavily overloaded or accident damaged, permanent distortion of the spring occurs. Listed below are common symptoms of weak or distorted springs:

1. Rapid tire wear
2. Premature failure of other chassis parts
3. Apparent shock absorber malfunction
4. Erratic steering and handling
5. Loss or change of important suspension and steering angles, camber, and toe (Fig. II-58). (Fig. II-59).
6. Reduced ability to control a skid.
7. Faulty headlight aim.

Weak or distorted springs normally cause a change in the trim height of a car. Measurement of vehicle trim height along with a visual inspection of all suspension parts is necessary to properly diagnose spring problems.

FIGURE II-58 This is a typical example of how camber (vertical position of the wheel) changes as a spring becomes weak. (Courtesy of Monroe Auto Equipment Co.)

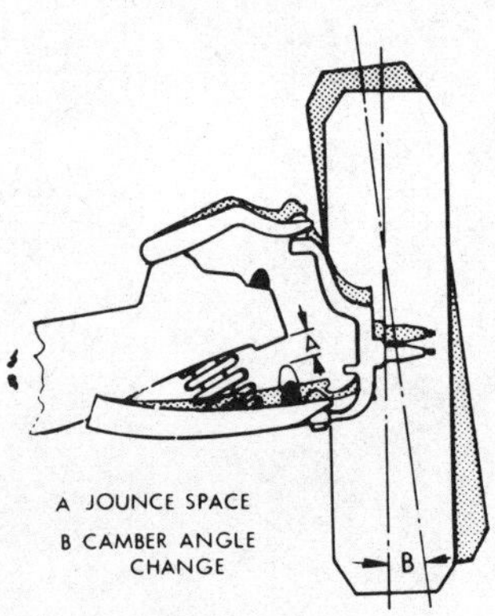

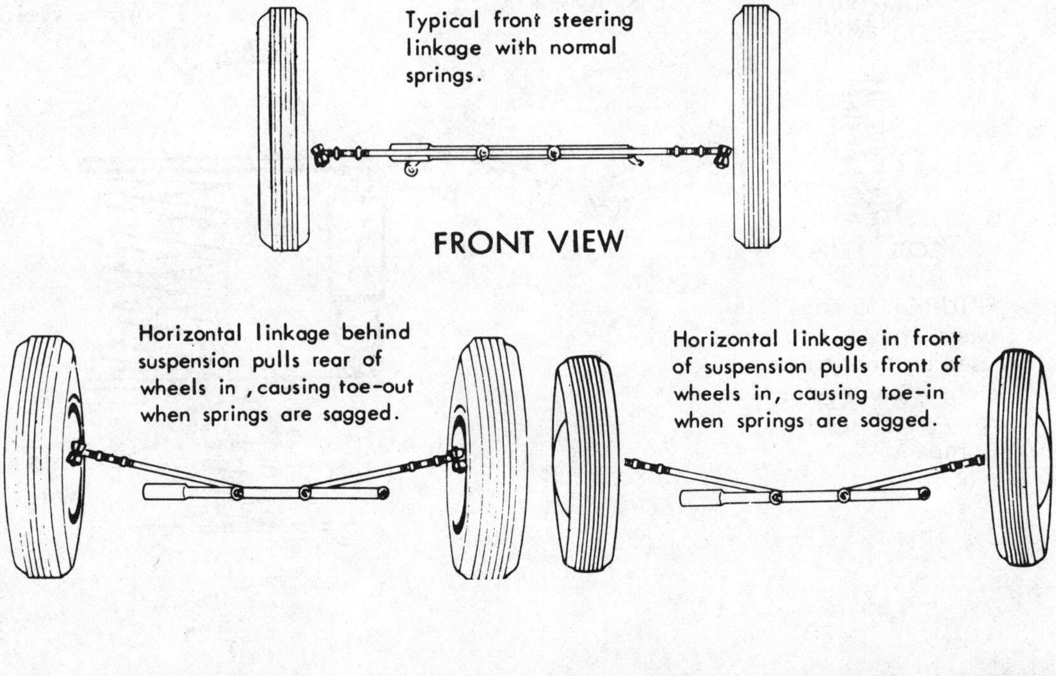

FIGURE II-59 Examples of toe changes as springs sag. (Courtesy of Monroe Auto Equipment Co.)

OBJECTIVE II-D1 (manipulative)

Working alone or with one other student as a team, you will measure and report in writing the trim height condition on two instructor-assigned automobiles. Manufacturer's specifications and recommended procedures must be followed. Use the Job Report Sheet to report the following:
1. Manufacturer's specifications
2. Exact measurements of trim height
3. Recommendations for corrections, if needed

Time recommended: Maximum of 20 minutes per automobile.

Preparation:

1. Review pages 139 through 141.
2. Study pages 141 through 144.
3. Carefully observe instructor's demonstration on measuring trim heights.

Materials and Tools Needed:

1. Steel tape or rule or torsion bar height gauges.
2. Flashlight or trouble light.

TRIM-HEIGHT CHECK PROCEDURE

Fig. II-60 should be studied as an example of how to check for sagging springs when manufacturer's recommended trim height specifications are not available.

1. Measure dimension A (Fig. II-60). The distance between the lower right control arm and the frame should not be more than ¼ inch (6 mm) different from the distance between the lower left control arm and the frame.

FIGURE II-60 Measurement points to be used when no other specifications are available. (Courtesy of Replacement Division of TRW Inc.)

check the left side against the right side

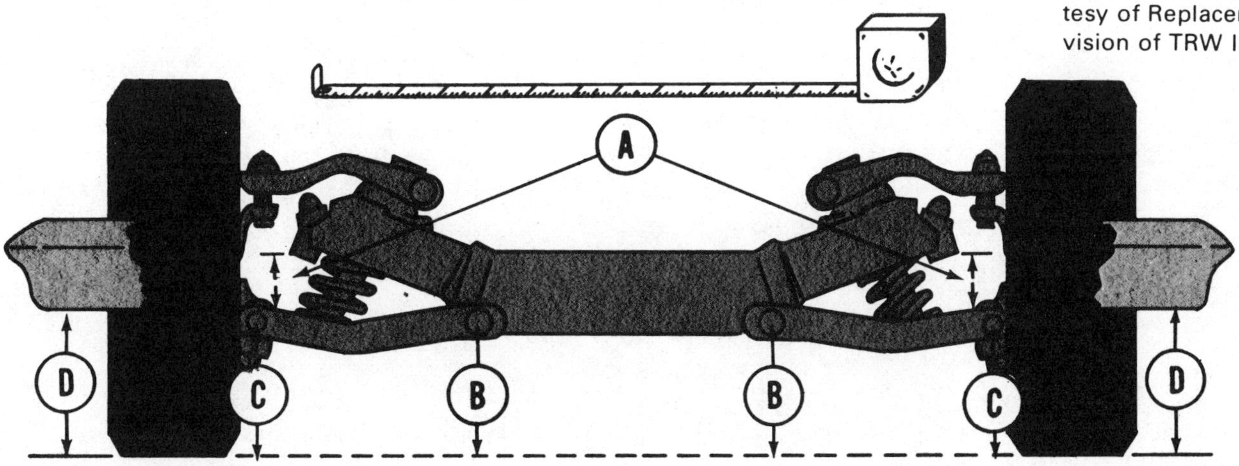

2. Measure dimensions B and C. These are the distances between ground level and the center of the pivot points of the lower control arm. The difference between the measurement of B and C on each side of the automobile should not be more than ¾ inch (19 mm).
3. Make certain that the bumper is not bent or deformed. Then measure the distance at D. The difference between the distance from the lower part of the bumper to ground level should not vary more than ⅜ inch from side to side.

When possible, always use the manufacturer's recommended procedures for measuring or adjusting trim height. An example is shown in Fig. II-61.

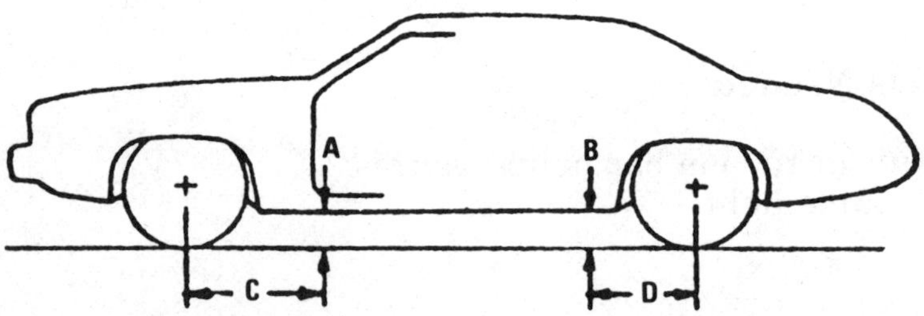

"A" Dim. — Ground to rocker panel at front.
"B" Dim. — Ground to rocker panel at rear.
"C" Dim. — Front wheel centerline to "A".
"D" Dim. — Rear wheel centerline to "B".

FIGURE II-61 Typical trim height adjustment locations and measurements. (Courtesy of Moog Automotive, Inc.)

MODEL	YEAR	BODY STYLE	MEASUREMENT POINTS & SPECIFICATIONS "A" Dim. Specifications	"B" Dim. Specifications	"C" Dim. Specifications	"D" Dim. Specifications
CAMARO	71-72	All	7¾ — 8½	7¼ — 8	36	16
	1973	All	7⅝ — 8⅜	7⅛ — 7⅞	36	20
	74-80	All	7⅞ — 8⅝	7½ — 8¼	36	20

TRIM HEIGHT MEASUREMENT

Student name______________________________Date____________________

Make of car______________Model______________Year____________________

Time started______________Time finished______________Total time______________

Flat rate time ____

Fill in appropriate specifications and procedures before making measurement check. Turn in this worksheet to your instructor when completed.

Specifications:

1. Source of specifications and procedure: ______________________________

   ____

2. Front height: ______________ inches (millimetres) measured between ______________

   and ______________

3. Rear height: ______________ inches (millimetres) measured between ______________

   and ______________

4. Recommended tire size: ______________

5. Recommended tire inflation: Front ____________________________ psi (kPa)

   Rear ____________________________ psi (kPa)

Procedure:

1. Actual measured front height:

   Right____________________Left____________________

2. Actual measured rear height:

   Right____________________Left____________________

Recommendations:

1. List recommendations for correction of all noted defects: ______________________

____

____

____

Checkup Questions:

1. Explain how shock absorber functions are affected by sagging springs:

____

____

2. Were tire sizes the same at all four wheels? ______________________

3. List additional suspension defects noted: ______________________

____

____

____

DATE COMPLETED______________ INSTRUCTOR'S SIGNATURE ______________

LEAF SPRINGS

Leaf springs are not used in most front suspensions today, owing to their large space requirements and their spring rate characteristics.

Leaf springs have a progressive spring rate. This means that after the spring is bent the first inch, the second inch is harder to bend. And the third inch is harder to bend than the second. This characteristic causes a rough ride. Automotive engineers concerned with the smooth riding qualities of cars they design cannot use leaf springs and achieve their goal.

However, this progressive spring rate is used to advantage in small trucks or four wheel drive vehicles where the sprung weight is relatively low. With a light load a soft ride is possible over fairly smooth roads with front leaf springs. With a heavier load or where road surfaces become rough, the leaf stiffens progressively to provide good handling without suspension damage.

The common leaf springs used in modern automotive suspension systems are built from long, flat strips of spring steel. Each strip is called a leaf. When more than one leaf is used, the unit is referred to as a *multi-leaf spring*. Multi-leaf springs generally have four to eight leaves. The one leaf with the round holes at the ends is called the *main leaf*. Other leaves of progressively shorter length are added to the main leaf to build its spring rate (make it harder to bend). A center bolt and clips are used to hold the leaves of the spring assembly together.

When multi-leaf springs are mounted on the front suspension, the rear eye of the main leaf is attached to a shackle which in turn is attached to the frame with a hanger (Fig. II-62). These parts allow a back-and-forth as well as a twisting movement to take place within the eye. As the spring is deflected it must be allowed to lengthen and shorten. The shackle arrangement does this well. The front end of the spring is mounted in a hanger which does not allow any front-to-rear movement. This prevents changes in the axle position which would change the alignment angles and directional stability.

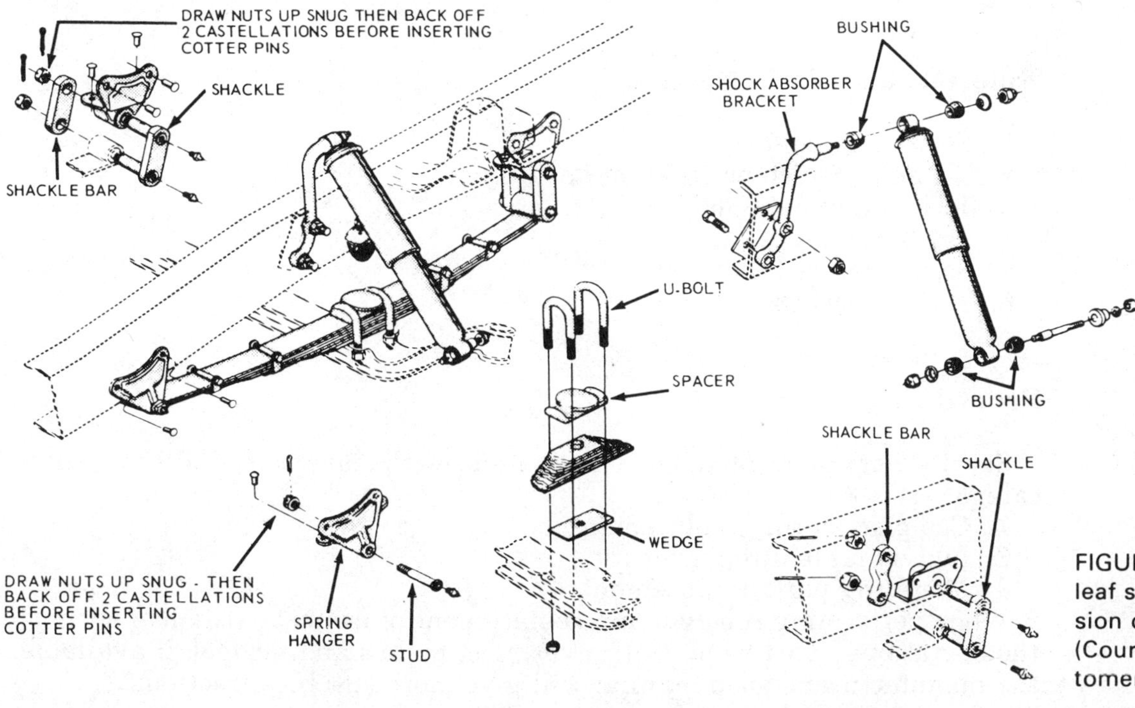

FIGURE II-62 Typical leaf spring front suspension of a light truck. (Courtesy of Ford Customer Service Division)

As stated previously, visual inspection is generally all that is necessary to diagnose leaf spring problems on front or rear suspensions. Things to look for are worn hangers and hanger bolts, worn eye bushings, broken bolts and shackles, and broken leaves. The broken leaves are sometimes more difficult to detect since they may be hidden behind the U bolts or a heavy accumulation of road dirt or grease.

The best job of diagnosis will be done when the car's undercarriage is at eye level (as on a hydraulic lift). A pry bar can be helpful in inspecting the eye bushings. Place the pry bar between the spring eye and the shackle and pull downward. Excessive bushing wear and/or heavy wear on the bolts will be seen as movement between the spring eye and hanger bolts. No wear is indicated by little or no movement. Always work under the best lighting available. A drop light or flashlight offers a good, portable source of light.

Since leaf springs do not lend themselves to the majority of independent front suspensions, they have been discarded for more functional designs such as coil springs and torsion bars.

OBJECTIVE II-D2 (manipulative)

Working as a team with one other student and with the aid of Job Report Sheet II-D2, you will diagnose, remove, repair, and replace a leaf spring on one instructor assigned automobile or light truck. Time recommended: Twice the flat-rate time allowed in Motors or Chilton Flat Rate manuals. This objective may be completed as part of Objective II-F—a complete front end rebuild.

Preparation:

1. Review pages 145 through 146.
2. Study pages 146 through 149.
3. Study manufacturer's or general repair manual for the vehicle to be serviced.

Materials and Tools Needed:

1. Floor (service) jack
2. C-clamp (8-inch or 200 mm opening)
3. Jack (safety) stands
4. Pry bar
5. Socket set with handles
6. Torque wrench

Procedure:

Service work on front suspension systems with leaf springs falls into three categories:

1. Complete spring replacement
2. Individual leaf replacement
3. Mounting parts replacement

When performing removal and replacement of front or rear leaf springs, the operations listed below will cover most makes and models. If available, the manufacturer's shop manual will give more specific directions.

REMOVAL:

1. With a jack under the frame (not under the axle), raise the car until the sprung weight is off the axle and spring. This occurs just before the wheel is lifted off the ground.
2. Remove the nuts on the U bolts which bind the spring and axle together (Fig. II-63).
3. Remove the shackle nuts and bolts.
4. Removal of the shackle nuts and bolts should free the shackle end of the spring and allow the hanger bolt to be removed from the opposite end of the spring.

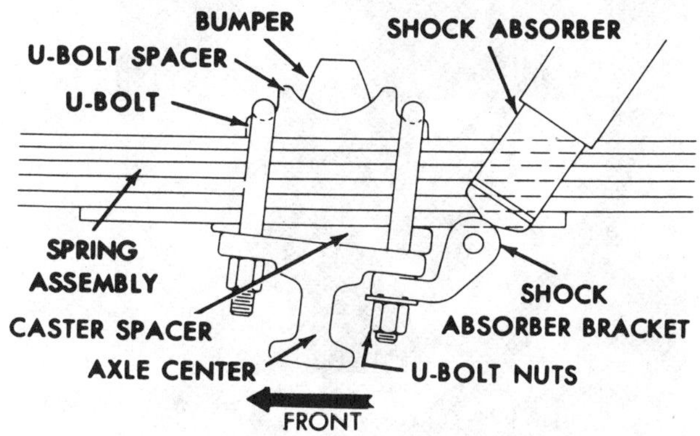

FIGURE II-63 Detailed drawing of leaf-spring front suspension. (Courtesy of Chevrolet Truck Division)

INSTALLATION:

For installation, reverse the procedure above. **Note:** *Do not fully tighten shackle and hanger bolts until the full weight is resting on the car.*

When the entire spring or springs are to be replaced, the technician should be careful to find exact replacements. Incorrect spring replacement will cause trim height irregularities and make correct suspension alignment impossible. The best job will result from replacement of the springs on both sides of the vehicle with new or reconditioned assemblies.

There will be occasions when an automobile owner will want to have an individual leaf or a multi-leaf unit replaced. To replace a single leaf, observe the following procedure:

1. Follow the previous removal procedures until the spring is free of the vehicle.
2. Place the spring on the floor on its edge. Attach a C-clamp to the flat surfaces of the main leaf and bottom leaf close to the center bolt. A bench vise can be substituted for the C-clamp.
3. Remove rebound clips holding the leaves in line.
4. Remove the nut holding the center bolt (Fig. II-64).

FIGURE II-64 Removing center bolt from leaf-spring assembly, clamp the assembly in a vise or use a C-clamp.

5. Slowly release the C-clamp until all the leaves have no tension.
6. Replace the broken leaf in correct order with the proper replacement.
7. Line up the leaves and friction materials by placing the parts on the floor and inserting a long punch or rod into the bolt hole.
8. Attach the C-clamp in the same position as before and tighten until all leaves are compressed (Fig. II-65).
9. Remove the punch or rod and insert the center bolt. Tighten the bolt to approximately 40 ft-lb. (54 N-m).

FIGURE II-65 Drift punch inserted through center-bolt hole aligns leaf springs prior to clamping.

SERVICE:

To service the mounting parts, it is not necessary to remove the spring from the vehicle:

1. With a jack under the frame, raise the car until the sprung weight is off the leaf spring but has not yet caused the spring to lift the axle. Some trial-and-error adjustment of the jack is often necessary to achieve this balance.
2. When the sprung weight is off the spring, remove the damaged or worn parts. Be sure to replace the damaged or worn parts with identical replacements.

Spring eye bushing replacements can be annoying to install if some care is not exercised. It is essential that the eye be completely clean and free of all previous mounting materials. This usually includes steel sleeves, rubber, and rust. Lubricate the new bushings with soapy water just prior to installation and prevent undue distortion to the rubber. Do not use motor oil or rubber lube since it will decompose the rubber.

To bring a new bushing to the correct position in the spring eye, an arbor press, C-clamp, or factory designed tool should be used. Some bushings are two piece assemblies inserted from both sides of the eye.

If only one bushing is to be replaced, the work should be done without removing the entire spring.

JOB REPORT SHEET II-D2

LEAF SPRING SERVICE

Student name________________ Date__________

Make of car__________ Model__________ Year__________

Time started__________ Time finished__________ Total time__________

Flat rate time __________

Specifications:

Refer to manufacturer's or general repair manuals if available.

Checkup Questions:

1. Source of specifications: ________________
2. Torque applied to center bolt: ________________ ft-lb. (N-m)
3. Other suspension defects noted: ________________

DATE COMPLETED__________ INSTRUCTOR'S SIGNATURE __________

COIL SPRINGS

Coil springs replaced leaf springs in front ends mainly because of their constant spring rate and their adaptability to an independent front suspension. When coil springs are compressed, there is very little bending involved. The majority of the spring action comes from the twisting effect of the rod that is bent into a coil. This twisting effect is exactly like that of a torsion bar.

The coil springs are mounted either between the lower control arm and a pocket in the frame or the upper control arm and a pocket built into the unibody construction. See Figs. II-66 and II-67 for examples.

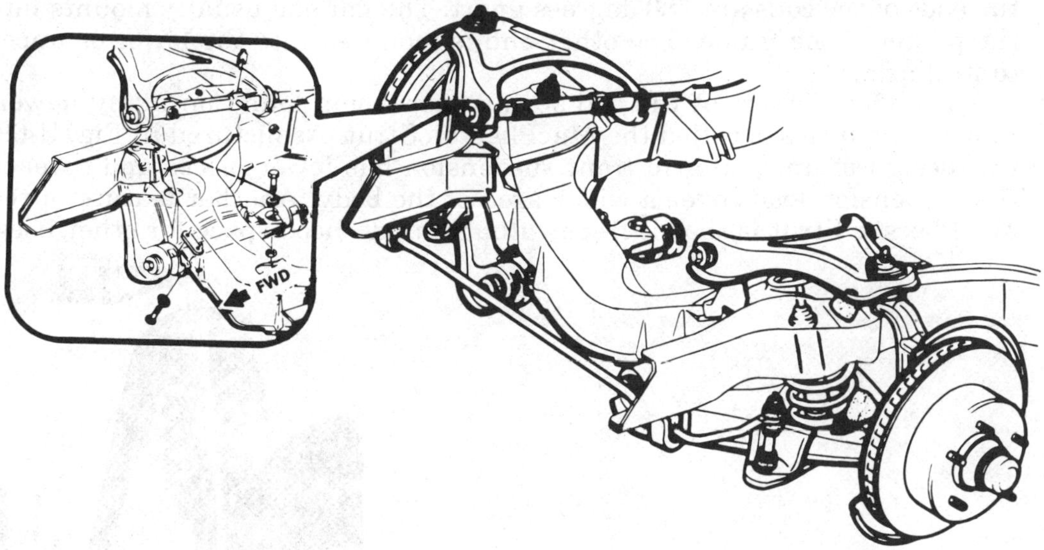

FIGURE II-66 Typical coil spring front suspension with spring between control arms. (Courtesy of Chevrolet Motor Division)

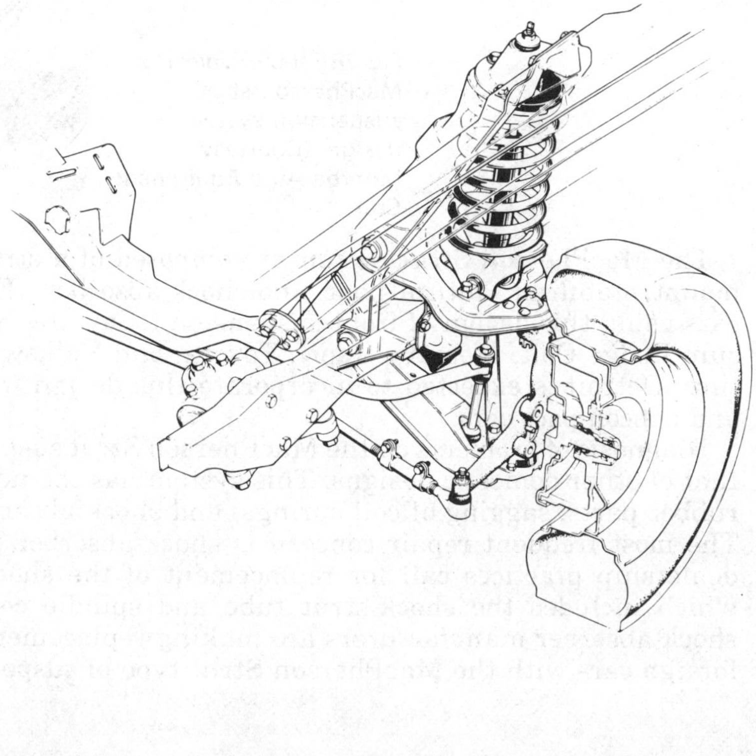

FIGURE II-67 Typical coil spring front suspension with spring above upper control arm. (Courtesy of American Motor Corp.)

FIGURE II-68 Coil spring with flattened upper-end surface to allow use of a shim in adjusting trim height. (Courtesy of Buick Motor Division)

Vehicles vary considerably in their demands for the correct coil spring. The weight of the vehicle, the terrain to be traveled, and the type of service and use of the vehicle are important in the selection of the right coil spring.

The spring rate of a coil spring is determined by the following:

1. Wire size
2. Type of wire and heat treatment
3. Types of ends on the coil
4. Diameter of the coils
5. Number of coils
6. Loaded height

In Fig. II-68, notice that one end of the coil spring is shaped to give it a flat surface, while the other end is cut off square with the wire. Notice also that the ends of the coils are 180 degrees apart. The flat end usually mounts into the pocket of the frame. The other end is mounted into the lower or upper control arm.

A modified version of the coil spring suspension found on many newer, imported vehicles is called the MacPherson Strut. As pictured in Fig. II-69, this design of unequal arm front suspension has fewer pieces and spreads the suspension loads over a wider area of the body. Another feature of the MacPherson Strut is that it frees valuable underhood space for other uses.

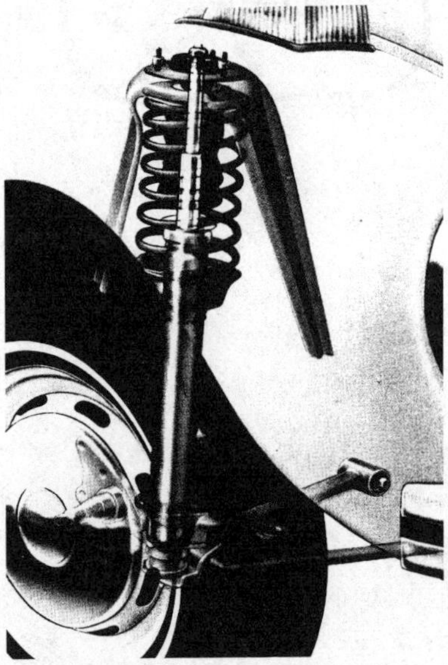

FIGURE II-69 Typical MacPherson strut suspension system design. (Courtesy of Monroe Auto Equipment Co.)

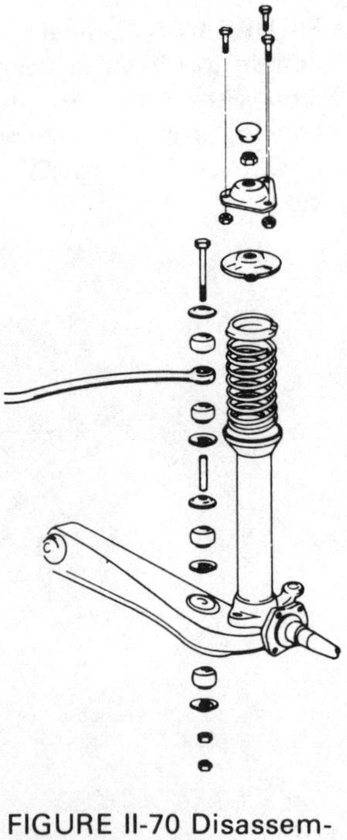

FIGURE II-70 Disassembled view of a MacPherson strut suspension design. No upper control arm is used in this popular import car front suspension system. (Courtesy of Monroe Auto Equipment Co.)

The MacPherson Strut system is composed of a strut, coil spring, upper mount, stabilizer (antiroll) bar, and shock absorber. (Fig. II-70). Among the cars using this design of front suspension today are: Sunbeam, BMW, Datsun, Dodge Colt, Mercury Capri, Toyota, and Volkswagen. In the near future, Detroit is expected to incorporate this design in some of its compact and subcompact cars.

Diagnosis of problems on the MacPherson Strut suspension system follows that of other domestic designs. This system has the normal deterioration of rubber parts, sagging of coil springs, and shock absorber wear and failure. The most frequent repair concern is shock-absorber replacement. Present dealership practices call for replacement of the shock absorber assembly which includes the shock strut tube and spindle components. American shock absorber manufacturers are making replacement cartridges for most foreign cars with the MacPherson Strut type of suspension system.

OBJECTIVE II-D3 (manipulative)

Working with one other student and with the aid of Job Report Sheet II-D3, remove and replace the coil spring of an instructor assigned vehicle with the spring *between* the upper and lower control arms. Objective should be completed in accordance with manufacturer's recommended procedures and specifications. Time recommended: Twice the flat-rate time allowed in Motors or Chilton Flat Rate manuals. This objective may be completed as part of Objective II-F—a complete front-end rebuild.

Preparation:

1. Review pages 151 through 152.
2. Study pages 153 through 155.
3. Study manufacturer's or general repair manual for the vehicle being used.
4. Carefully observe instructor's demonstration of coil spring removal and replacement.
5. Discuss with your instructor all aspects of safety while working with coil springs.

Service work on coil springs is generally a task of removal and replacement. Regardless of the type of coil spring front suspension design, the technician will only be required to change a fatigued or broken spring for a good one.

Materials and Tools Needed:

1. Service (floor) jack
2. Jack (safety) stands
3. Lug or impact wrench and socket
4. Socket set with handles
5. Assorted box/open end combination wrenches
6. Large safety chain with grab hooks
7. Cotter pins
8. Coil spring tools (Moog T-469C, T-476, T-490, and T-493)
9. Manufacturer's or general repair manual.

The next consideration is determining whether the spring is above or below the upper control arm.

COIL SPRING SERVICE—SPRING BETWEEN CONTROL ARMS

In the case where the spring is below the upper control arm, the following procedure may be used:

1. Raise the front of the vehicle and support it with jack stands.
2. Remove the wheel (Fig. II-71).

FIGURE II-71 Use an impact wrench to remove front wheel lug nuts. Notice whether threads are right-hand or left-hand.

3. Remove the shock absorber (Fig. II-72).
4. Place a service jack under the lower control arm in a position where the jack handle and its controls are perpendicular to the side of the car. Raise the lower control arm with the jack so the spring is loaded.
5. Disconnect the stabilizer bar at the lower control arm.
6. Install the chain through the lower control arm opening, through the spring and around the jack axle. Leave 6 to 8 inches (150–200 mm) of slack.
7. Remove cotter pin and loosen nut on lower ball joint approximately three turns. Do not remove the nut.
8. With a heavy hammer, strike the lower mounting end of the steering knuckle to release the tapered end of the ball joint. If this does not work, use a spreader tool to separate the knuckle and ball joint.
9. When ball joint and knuckle release, remove the ball joint nut and carefully lower jack to its lowest position (Fig. II-73).
10. Remove the chain.
11. The spring can now be removed.
12. Reverse this procedure to install and lower vehicle to floor.

Note: *If available, use coil-spring tools such as Moog's T-469C, T-476, or T-490 or T-493. During reassembly, place the end of the coil spring in its correct position. Do not tighten rubber parts until the vehicle is on the floor.*

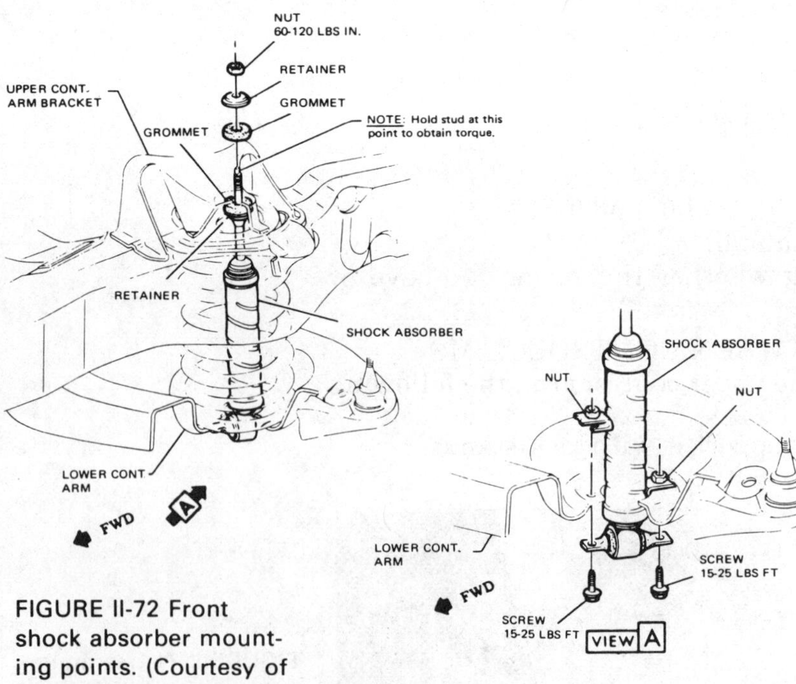

FIGURE II-72 Front shock absorber mounting points. (Courtesy of Buick Motor Division)

FIGURE II-73 Lower the jack to allow the lower control arm to release spring tension. Notice how a chain is used to protect against injury if the spring releases unexpectedly. (Courtesy of Chevrolet Motor Division)

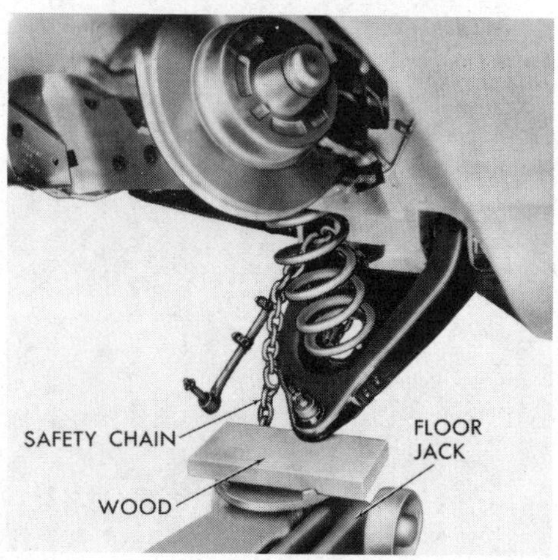

JOB REPORT SHEET II-D3

COIL-SPRING SERVICE—SPRING BETWEEN CONTROL ARMS

Student name____________________ Date____________

Make of car____________ Model____________ Year____________

Time started____________ Time finished____________ Total time____________

Flat-rate time____________________

Specifications:

Refer to manufacturer's or general repair manuals.

Checkup Questions:

1. What was the reason for using a chain around the axle of the jack?____________

2. What precautions should be observed when tightening the ball nut stud?____________

3. What other suspension parts should be replaced?____________

   Why?____________

4. In what condition was the spring?____________

DATE COMPLETED____________ INSTRUCTOR'S SIGNATURE____________

OBJECTIVE II-D4 (manipulative)

Working with one other student and with the aid of Job Report Sheet II-D4, remove and replace the coil spring of an instructor assigned vehicle with the spring *above* the upper control arm. Objective should be completed in accordance with manufacturer's recommended procedures and specifications. Time recommended: Twice the flat-rate time allowed in Motors or Chilton Flat Rate manuals. This objective may be completed as part of Objective II-F—a complete front end rebuild.

Preparation:

1. Review pages 151 through 152.
2. Study pages 157 through 159.
3. Study the manufacturer's or general repair manual for the vehicle being used.
4. Carefully observe instructor's demonstration on coil spring removal and replacement.
5. Discuss with your instructor all aspects of safety while working with coil springs.

Materials and Tools Needed: See Objective II-D3, page 153.

COIL SPRING SERVICE—SPRING ABOVE UPPER CONTROL ARM

To remove and replace coil springs on unitized body cars where the spring is above the upper control arm, the following can serve as a general procedure. A coil spring compressing tool is required.

1. Raise the car with a floor jack or a hoist.
2. Lower the car, placing the jack stands under the lower control arms (Fig. II-74).

Note: *On 1962–1965 Fairlanes and on 1960–1966 Thunderbirds, put the jack stand under the frame, then jack up one side at a time to permit straightening of the spring. Next raise the lower control arms slightly with the floor jack until bowed springs are straight.*

3. Lower the car further until some of its weight is on the lower control arms.
4. Remove the shock absorber and bracket assembly (Fig. II-75).
5. Remove the upper strike-out bumper.
6. Raise the car until the lower control arm is free.

FIGURE II-74 Use a hoist and jack (safety) stands to load the lower control arm. (Courtesy of Moog Automotive, Inc.)

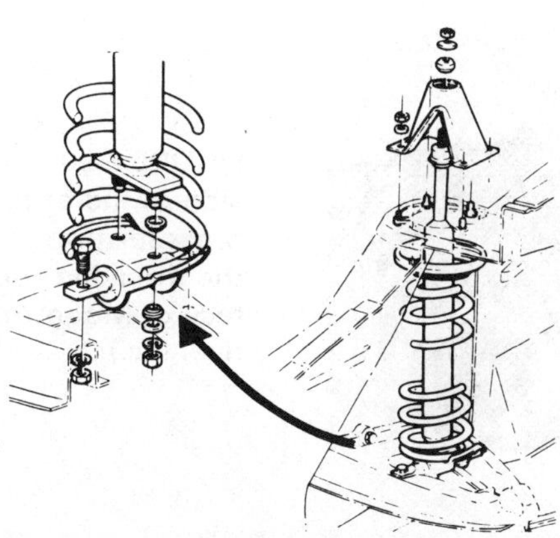

FIGURE II-75 Remove the shock absorber and bracket assembly where the spring is above the upper control arm. (Courtesy of Chevrolet Motor Division)

7. Insert compressing tool plates as close to both ends of the spring as possible (Fig. II-76). **Note:** *Be certain the rod's screwheads are lubed.*
8. Insert the threaded rod from the top to engage both plates. Be sure the rod is locked in position on the lower plate (Fig. II-77).
9. Hold the lower plate and turn the threaded rod compressing the spring. Compress it to a height that will just allow its removal (Fig. II-78).
10. Measure the compressed spring. Note the location of plates by counting the number of coils from both ends in order to reposition the plates properly in the new spring (Fig. II-79).
11. Support the old spring and release its tension by turning the thread rod while holding the lower plate (Fig. II-80).
12. Reverse the above procedure for installation of the new spring.

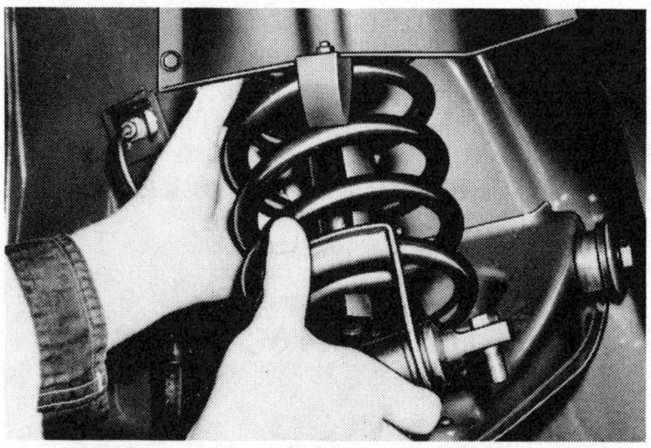

FIGURE II-76 Insert spring compressor tool plates as close to the spring ends as possible. (Courtesy of Chevrolet Motor Division)

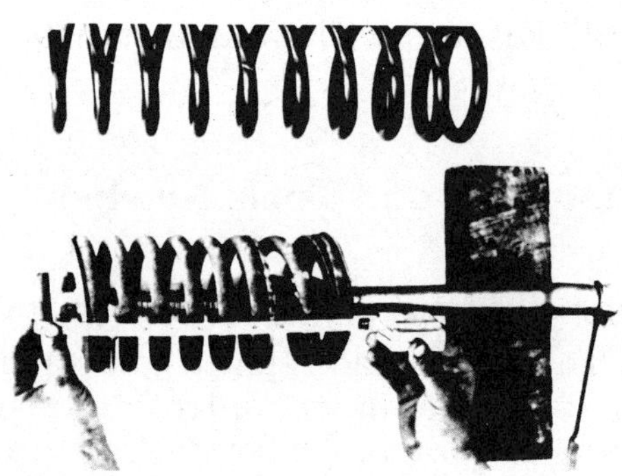

FIGURE II-79 Measure the length of the compressed spring. (Courtesy of Moog Automotive, Inc.)

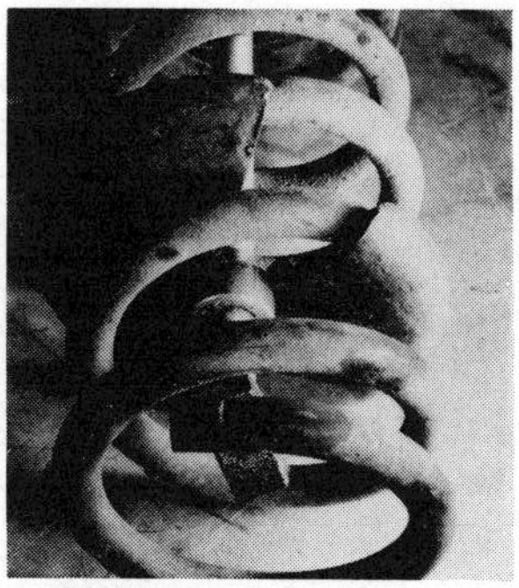

FIGURE II-77 The circle shows correct position for a spring compressor plate with threaded rod locked in place.

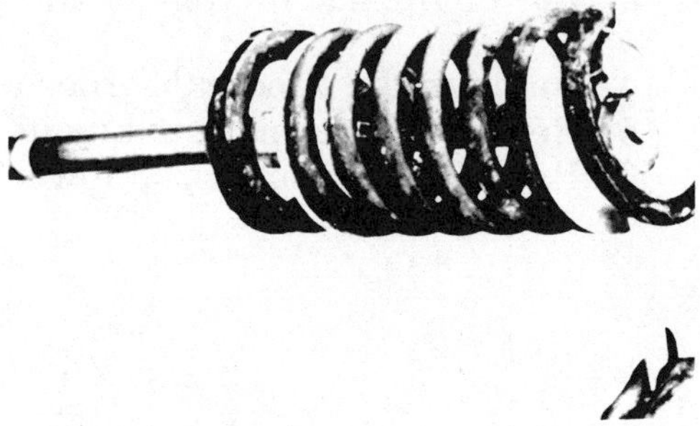

FIGURE II-80 Release tension of the original compressed spring by turning the threaded rod. (Courtesy of Moog Automotive, Inc.)

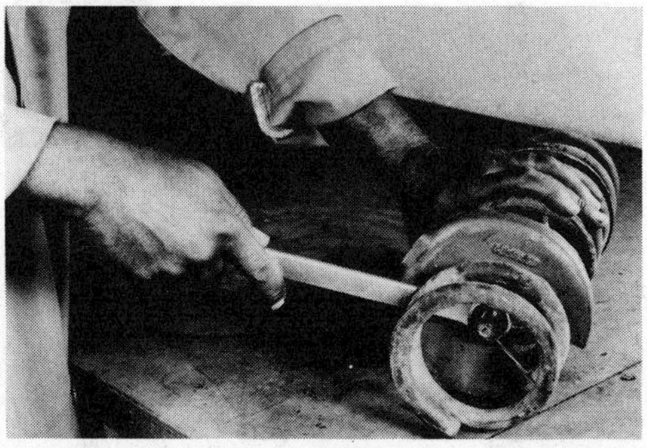

FIGURE II-78 A wrench is used to turn the threaded rod and compress the spring.

COIL SPRING SERVICE—SPRING ABOVE UPPER CONTROL ARM

Student name____________________ Date __________

Make of car__________ Model__________ Year__________

Time started__________ Time finished__________ Total time__________

Flat-rate time ____________________

Checkup Questions:

1. Why was it necessary to use the spring compressor?____________________

2. Did the tires show any abnormal wear pattern? __________

   If so, what are possible causes? ____________________

3. In what condition was spring?____________________

4. List other suspension defects noted: ____________________

DATE COMPLETED__________ INSTRUCTOR'S SIGNATURE __________

OBJECTIVE II-D5 (manipulative)

Working with one other student and with the aid of Job Report Sheet II-D5, remove and replace the coil spring of an instructor-assigned vehicle with MacPherson strut suspension. Objective should be completed in accordance with manufacturer's recommended procedures and specifications. Time recommended: Twice the flat-rate time allowed in Motors or Chilton Flat Rate manuals. This objective may be completed as part of Objective II-F—a complete front end rebuild.

Preparation:

1. Review pages 157 through 159.
2. Study pages 161 through 163.
3. Study the manufacturer's or general repair manual for the vehicle being used.
4. Carefully observe instructor's demonstration on coil spring removal and replacement.
5. Discuss with your instructor all aspects of safety while working with coil springs.

Materials and Tools Needed:

1. Service (floor) jack or hydraulic lift
2. Jack (safety) stands
3. Lug or impact wrench and socket
4. Socket set with handles
5. Assorted box/open end combination wrenches
6. Large link chain with grab hooks
7. Cotter pins
8. Coil spring tools
9. Manufacturer's or general repair manual

The next consideration is determining whether the spring is above or below the upper control arm.

COIL SPRING SERVICE—MACPHERSON STRUT

The following directions are for service involving the disassembly of the entire strut. The directions include replacing the coil spring and shock absorber.

1. Place the vehicle on a frame-type hoist but do not lift it. Mark the position of the strut upper bearing plate (Fig. II-81).

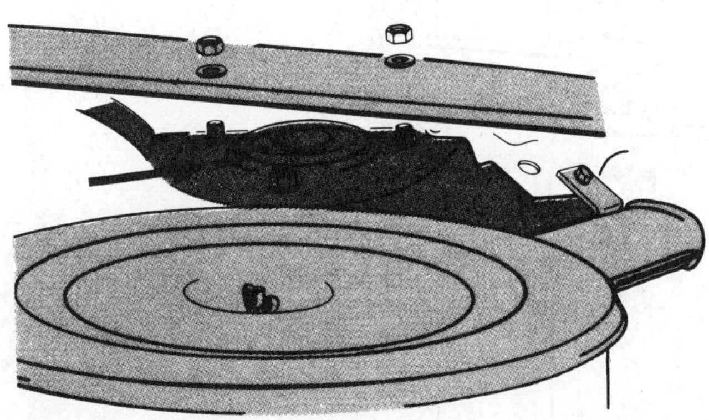

FIGURE II-81 Mark the upper plate. This is necessary for reassembly in the correct position. (Courtesy of Monroe Auto Equipment Co.)

2. Remove two of the three upper bearing plate nuts.
3. Loosen the third nut, but do not remove it.
4. Remove the rod nut cap.
5. Lift the vehicle to convenient work level and mark the lug stud nearest the tire valve stem.
6. Remove the wheel.
7. Disconnect the brake hose where necessary (Fig. II-82). Plug or cover fluid holes. **Caution:** Disconnect the speedometer cable at the wheel if necessary (an example would be the Volkswagen).
8. Remove the lower strut attachment bolts which allow separation of the strut from the lower control arm. Use a pry bar if necessary.
9. Remove the third (and final) retaining nut from upper bearing plate and remove the strut from the vehicle.

Note: *Do not let the strut strike the rubber axle boot on front wheel drive vehicles.*

10. Firmly clamp the strut assembly in a vise. Use the lower end mounting plate for clamping. *Do not clamp the strut tube.*
11. Compress the coil spring until it is unloaded from spring seats (Fig. II-83).
12. Remove the upper bearing dust cap to expose the piston rod nut.
13. Remove the upper bearing plate, upper spring seat, coil spring, and rubber dust boot (if used).
14. Replace all worn or defective parts in reverse order from above. Make sure the spring is fully seated and bearing plate is positioned according to your original mark. Bleed brakes and refill the system with brake fluid recommended by vehicle's manufacturer.

Note: *Alignment should be checked if any new parts other than shock absorbers were replaced.*

FIGURE II-83 Use the coil spring compressors to compress the spring enough to remove upper spring seat. (Courtesy of Monroe Auto Equipment Co.)

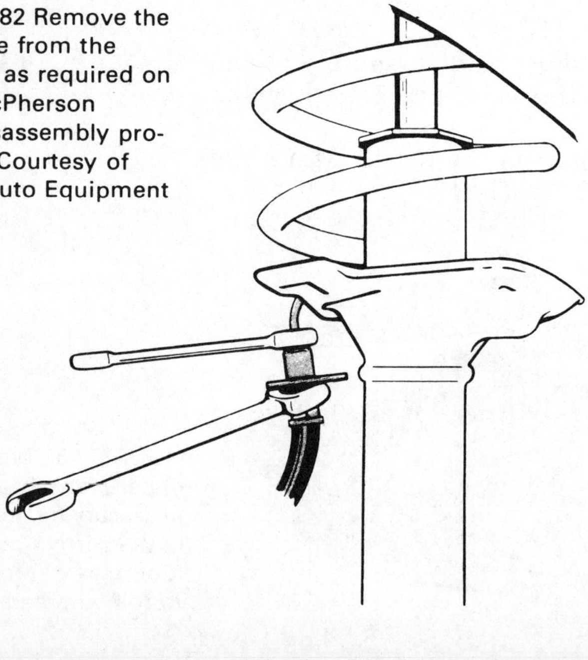

FIGURE II-82 Remove the brake hose from the brake line as required on some MacPherson design disassembly procedures. (Courtesy of Monroe Auto Equipment Co.)

JOB REPORT SHEET II-D5

COIL-SPRING SERVICE—MAC PHERSON STRUT

Student name ______________________ Date ______________

Make of car ____________ Model ____________ Year ____________

Time started ____________ Time finished ____________ Total time ____________

Flat-rate time ______________________

Specifications:

Refer to manufacturer's or general repair manuals.

Checkup Questions:

1. What was the reason for marking the wheel lug nut and the upper bearing plate? ____________

____

____

2. What precautions should be observed when using the spring compressor? ____________

____

____

3. What suspension parts should be replaced? ____________

____

Why? ____________

____

4. In what condition was the spring? ____________

____

DATE COMPLETED ____________ INSTRUCTOR'S SIGNATURE ____________

TORSION BARS

Torsion bars have replaced coil springs in some cars and light trucks (Fig. II-84). In most cases, they are used only on the front suspension. However, there have been some cases where the rear suspension has been supported with torsion bars. The 1955 Packard was probably the first automobile to use torsion bars and they were used on all four wheels.

The torsion bars' ability to hold the weight of the vehicle in suspension is obtained by putting a twist in the bars.

Torsion bars are not interchangeable from left to right. They are stamped at one end with an L or an odd number (left) and an R or an even number (right). This marking indicates the side of the car where the bar should be used (Fig. II-85).

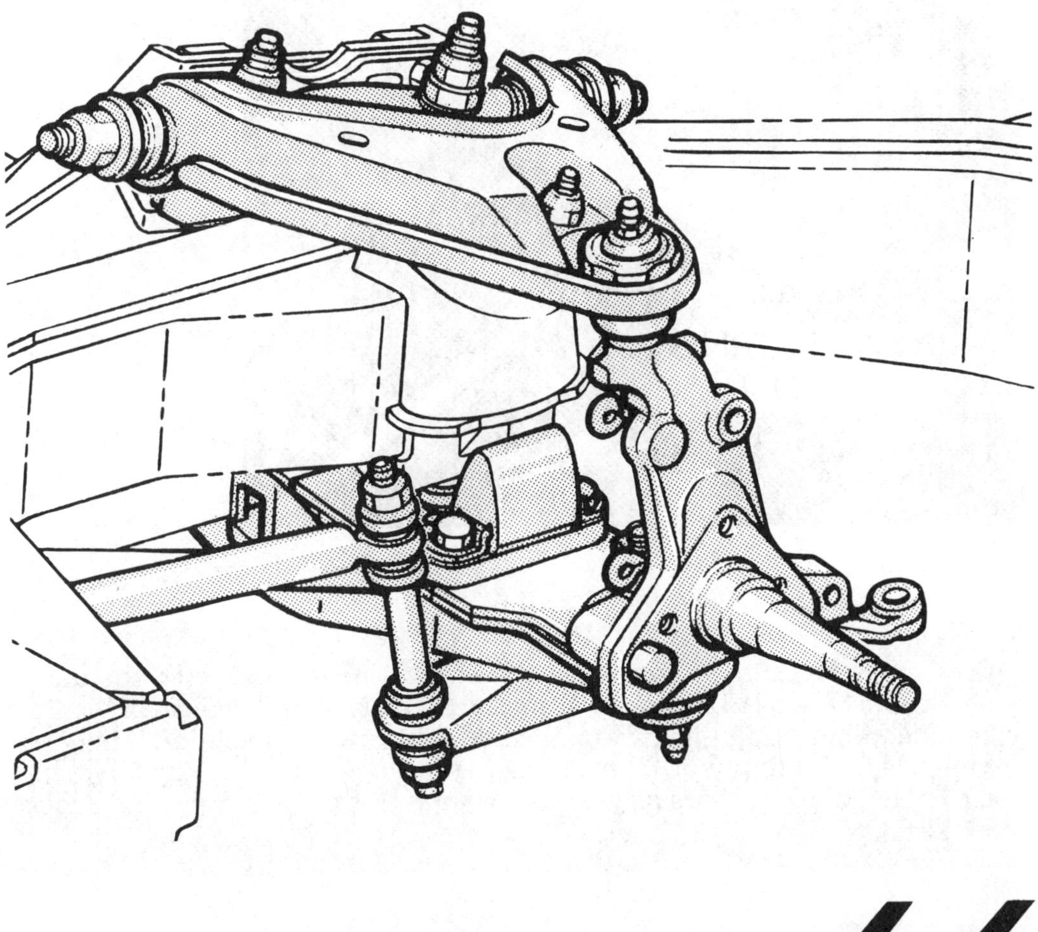

FIGURE II-84 Typical torsion-bar front suspension, left-front view. (Courtesy of Chrysler Motors Corp.)

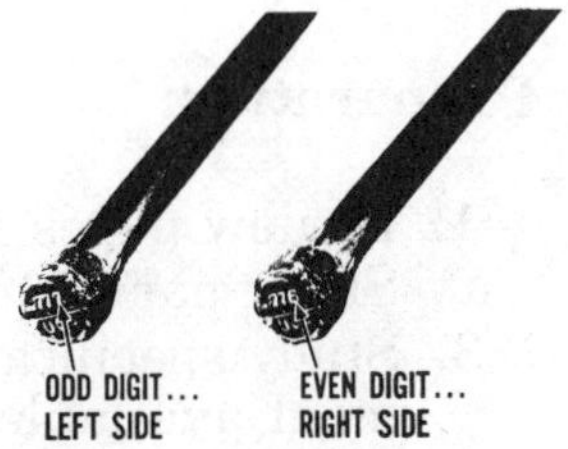

FIGURE II-85 Identification marking of torsion bars on Chrysler Corporation cars. (Courtesy of Chrysler Motors Corp.)

Torsion bars are designed with hexagonal shaped ends that fit into matching sockets. This allows for a secure mounting at each end.

The front end of each torsion bar fits into the lower control arm through a socket in the swivel. The swivel mounts into the inner control arm bushing along with the pivot shaft for the lower control arm.

The rear end of each torsion bar is mounted in a socket on a cross member that is bolted to a section of the frame.

Twist is applied to the torsion bar by an adjusting bolt that is mounted through the lower control arm and into the swivel. When this adjusting bolt is turned clockwise, the swivel and lower control arm are drawn together (Fig. II-86). This puts the twist or spring effect into the torsion bar. The more twist given the torsion bar, the higher the car will ride.

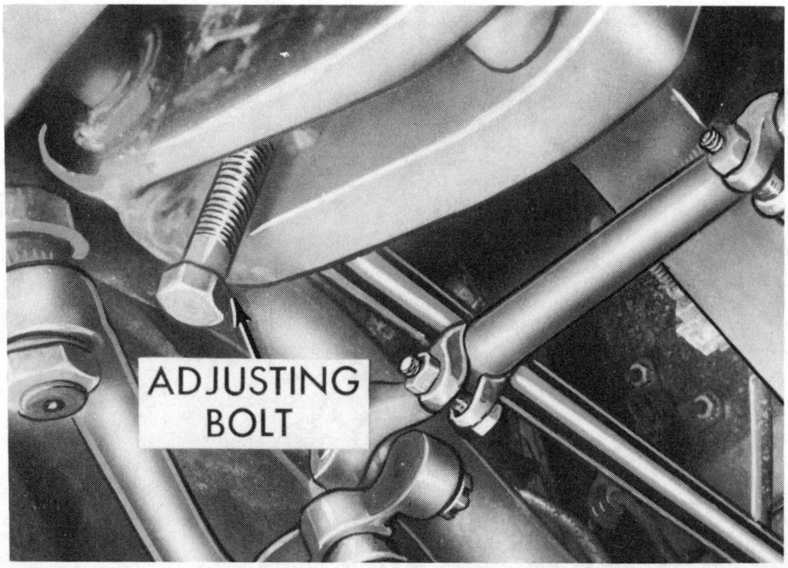

FIGURE II-86 The adjusting bolt puts the spring effect in the torsion bar and adjusts trim height. (Courtesy of Chrysler Motors Corp.)

OBJECTIVE II-D6 (manipulative)

Working alone or as a team with one other student and with the aid of Job Report Sheet II-D6, you will measure, adjust, and report in writing the torsion bar height of two instructor assigned automobiles. Time recommended: Maximum of 20 minutes per automobile. This objective may be completed with others as part of a complete front end rebuild—Objective II-F.

Preparation:

1. Review pages 165 through 166.
2. Study pages 166 through 169.
3. Study specifications and procedures for specific automobile and equipment available.
4. Carefully observe instructor's demonstration on torsion bar height measurements and adjustments.

Materials and Tools Needed:
1. Tape measure or torsion-bar-height gauges
2. ½-inch drive-socket set with handles
3. Tire-pressure gauge

The torsion bar, unlike the coil and leaf spring, can be adjusted to control vehicle height. In many of today's cars, this is done by measuring the distance between the lowest point of the lower ball joint to the floor and from the lowest point of the front torsion bar anchor to the floor (Fig. II-87). The difference between these two measurements should be equal to the specification, which in this case is 2⅛ inches (54 mm).

Many front end alignment equipment manufacturers make and sell gauges that allow you to set the torsion bar specification in the gauge and then adjust the torsion bar until the level bubble in the gauge is centered (Fig. II-88).

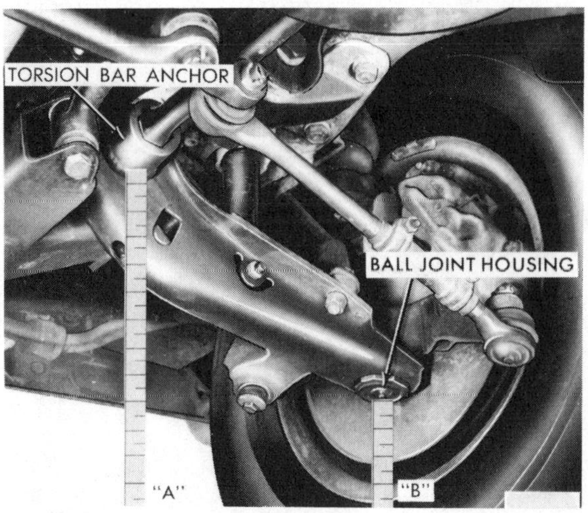

FIGURE II-87 Positions of measurement recommended by Chrysler Corporation. View is of right-front wheel as seen from rear to front. (Courtesy of Chrysler Motors Corp.)

FIGURE II-88 Torsion-bar height gauge in position on right front of a typical Chrysler Corporation automobile. (Courtesy of Bear Manufacturing Co.)

Procedure:

1. Check tire sizes of all four wheels to be certain they are the same size. Do not adjust height unless they are equal.
2. Inflate all tires to recommended pressure.
3. With vehicle on alignment rack or other level surface, measure or attach gauges to lower control arms.
4. Lubricate adjusting bolt with light oil.
5. Adjust to specifications.
6. Be sure to jounce the vehicle as adjustments are made.
7. Remove gauges and return them to their proper place.

TORSION-BAR-HEIGHT ADJUSTMENT

Student name____________________ Date________________

Make of car____________ Model____________ Year____________

Time started____________ Time finished____________ Total time____________

Flat-rate time ____________

Specification:

Refer to manufacturer's or general repair manuals for specifications.

Checkup Questions:

1. What was the source of specifications and procedures?____________________

____

2. List recommended tire pressures: Front____________psi.

   Rear____________psi.

3. What was the height as first measured: ____________________

4. Any suspension damage visable?____________________

   If so, name parts: ____________________

____

____

DATE COMPLETED____________ INSTRUCTOR'S SIGNATURE ____________________

Working with one other student and with the aid of Job Report Sheet II-D7, remove and replace two torsion bars on an instructor assigned vehicle. Objective should be completed in accordance with manufacturer's recommended procedures and specifications. Time recommended: Twice the flat-rate time allowed in Motors or Chilton Flat Rate manuals. This objective may be completed as part of Objective II-F—a complete front end rebuild.

OBJECTIVE II-D7 (manipulative)

Preparation:

1. Review pages 165 through 169.
2. Study pages 171 through 173.
3. Study manufacturer's or general repair manuals for the vehicle being used.
4. Carefully observe instructor's demonstration on torsion bar removal and replacement.
5. Discuss with your instructor all aspects of safety while working with torsion bars.

Procedure:

Torsion bars are usually replaced individually. The only other service work is lubricating the sockets which secure the ends of the torsion bars (Fig. II-89).

Torsion bars are usually removed in the following manner:

a. Raise the front of the vehicle off the floor far enough for the wheels to clear.
b. Place jack (safety) stands under the frame slightly behind the lower control arm.
c. Remove the lower shock absorber attaching bolt and swing the shock absorber out of the way.

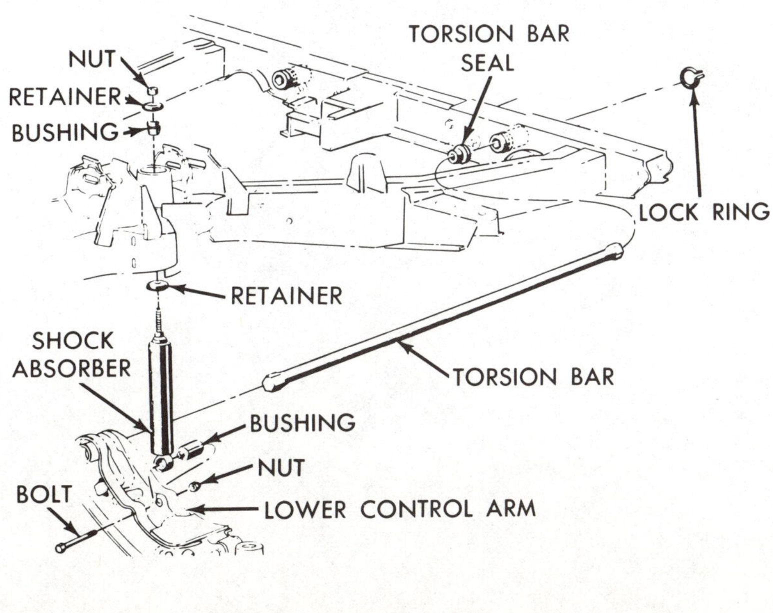

FIGURE II-89 Disassembled view of typical torsion-bar front suspension. (Courtesy of Chrysler Motors Corp.)

d. Loosen the torsion bar anchor until it is completely released.
e. Remove the torsion bar and rear hanger.
f. For installation, reverse this procedure.

Note: *Torsion bar ends are marked R or even numbers (right) and L or odd numbers (left) for the side of the vehicle in which they are to be installed.*

JOB REPORT SHEET II-D7

TORSION BAR SERVICE

Student name________________________________ Date _______________________

Make of car______________ Model______________ Year_______________________

Time started______________ Time finished______________ Total time____________

Flat-rate time ____

Check-up Questions:

1. Did the torsion bar show any signs of damage?_________________________________
2. Were the ends of the torsion bars lubricated? _________________________________
3. Which end was marked R or L or numbered? _________________________________

DATE COMPLETED__________________ INSTRUCTOR'S SIGNATURE __________________________

TEST II-D

Student name____Date_____________________________

Directions: Select the best answer to each question and circle the correct letter at the left.

1. The trim height of a vehicle should be maintained to prevent:
   I. Rapid tire wear
   II. Erratic steering and handling
   A. I only
   B. II only
   C. Both I and II.
   D. Neither I nor II
2. Trim height should not vary more than _____ from side to side, measured at the frame and lower control arm:
   I. ½ inch
   II. ¼ inch
   A. I only
   B. II only
   C. Both I and II.
   D. Neither I nor II
3. The cause of trim height loss is:
   I. Sagging springs
   II. Accident damage
   A. I only
   B. II only
   C. Both I and II.
   D. Neither I nor II
4. The amount of unsprung weight in an automobile should be kept:
   I. As low as possible
   II. As high as possible
   A. I only
   B. II only
   C. Both I and II.
   D. Neither I nor II
5. The deflection rate of a spring refers to the:
   I. Amount of flex before it distorts
   II. Number of vibrations per minute
   A. I only
   B. II only
   C. Both I and II.
   D. Neither I nor II
6. Which of the following affect coil spring rate:
   I. Wire size
   II. Types of ends on the coil
   A. I only
   B. II only
   C. Both I and II.
   D. Neither I nor II

7. A spring that deflects 3 inches with 300 pounds is called:
   I. 300-pound spring
   II. 100-pound spring
      A. I only
      B. II only
      C. Both I and II.
      D. Neither I nor II
8. Which statement(s) is true?
   I. Springs should be replaced in pairs.
   II. Springs should be replaced individually only as they need it.
      A. I only
      B. II only
      C. Both I and II.
      D. Neither I nor II
9. Rubber bushings are used in spring eyes to:
   I. Allow more spring movement
   II. Dampen noise and vibration
      A. I only
      B. II only
      C. Both I and II.
      D. Neither I nor II
10. Which part of a leaf spring should *not* be tightened until the weight of the vehicle rests on the ground?
   I. Center bolt
   II. Hanger bolt
      A. I only
      B. II only
      C. Both I and II.
      D. Neither I nor II
11. The shackle end of the front leaf spring is positioned toward the:
   I. Front of the car
   II. Shock absorber
      A. I only
      B. II only
      C. Both I and II.
      D. Neither I nor II
12. Leaf springs on front suspension systems have been replaced by:
   I. Torsion bars
   II. Coil springs
      A. I only
      B. II only
      C. Both I and II.
      D. Neither I nor II
13. To permit a leaf spring to lengthen as it is loaded, one end is fastened to a:
   I. Hanger
   II. Shackle
      A. I only
      B. II only
      C. Both I and II.
      D. Neither I nor II
14. Which of the following is(are) used to keep multi leaf springs assembled?
   I. Center bolt
   II. Clips
      A. I only
      B. II only
      C. Both I and II.
      D. Neither I nor II

15. Leaf springs have a progressive deflection rate. This means they:
    I. Get softer as they are loaded
    II. Will hold more weight
        A. I only
        B. II only
        C. Both I and II.
        D. Neither I nor II
16. What tool(s) can be used to separate the ball joint taper from the steering knuckle?
    I. Tie-rod fork
    II. Hammer
        A. I only
        B. II only
        C. Both I and II.
        D. Neither I nor II
17. A spring compressor is normally required for spring removal on which of the following types of coil spring front suspensions?
    I. Coil spring between control arms
    II. Coil spring above upper control arm
        A. I only
        B. II only
        C. Both I and II.
        D. Neither I nor II
18. Which of the following must be removed before extracting a coil spring?
    I. Tie rod
    II. Steering knuckle
        A. I only
        B. II only
        C. Both I and II.
        D. Neither I nor II
19. When a torsion bar has been replaced, which of the following operations is(are) necessary:
    I. Alignment of front suspension
    II. Adjustment of trim height
        A. I only
        B. II only
        C. Both I and II.
        D. Neither I nor II
20. Advantage(s) of torsion-bar suspension is:
    I. Tension adjustment is provided
    II. Lubrication is eliminated
        A. I only
        B. II only
        C. Both I and II.
        D. Neither I nor II

DATE COMPLETED____________________INSTRUCTOR'S SIGNATURE ____________________

OBJECTIVE II-E (written test)

After studying the preparation assignment you will be expected to correctly answer a minimum of seven out of ten multiple choice questions in a written test without reference materials on the following topics:
1. Stabilizer bars and their bushings
2. Strut rods and their bushings

Preparation:

1. Study pages 179 through 185.

STABILIZER BAR AND BUSHINGS

The stabilizer bar is a steel bar that is mounted to the frame or a cross member with a rubber bushing (Fig. II-90). The ends of the stabilizer bar are extended out over the lower control arm. The stabilizer bar is then connected to each lower control arm by a stabilizer link that consists of a bolt and two sets of rubber bushings and steel washers.

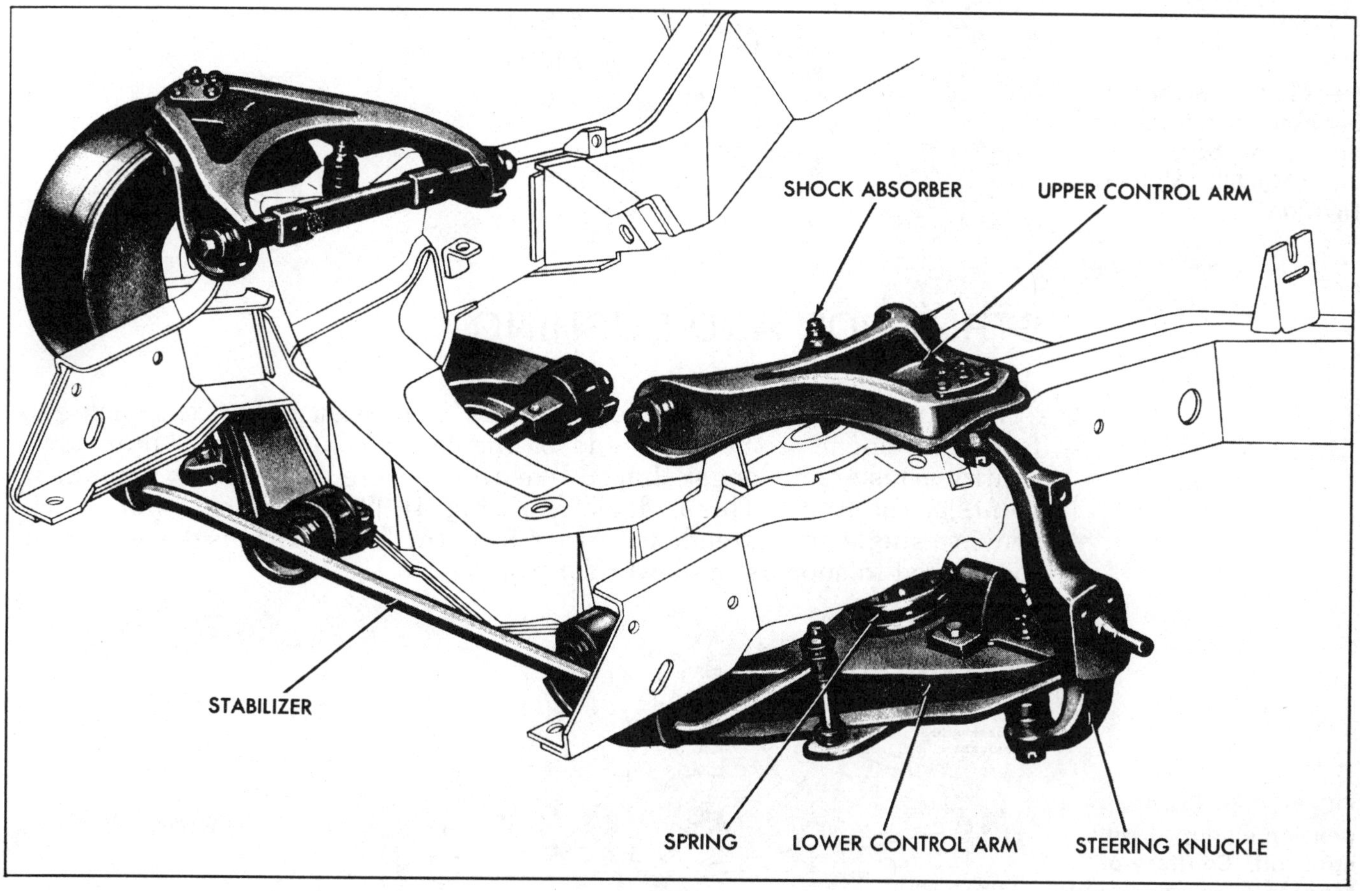

FIGURE II-90 Front suspension equipped with stabilizer bar. (Courtesy of Chevrolet Motor Division)

The stabilizer bar is named for the function it performs in the suspension system: stabilizing the action of each side of the independent front suspension (Fig. II-91). When the entire front end encounters a bump, both springs are deflected the same amount. This causes the stabilizer arm to move inside its mounting bushings with no effect. When only one spring deflects, the stabilizer bar is moved on one end only, causing a twisting effect. This twist makes the stabilizer bar a miniature torsion bar that connects one side of the car to the other. This miniature torsion bar acts as a resistance spring and helps to dampen any pitching effect on the car because of the heavy coil spring action.

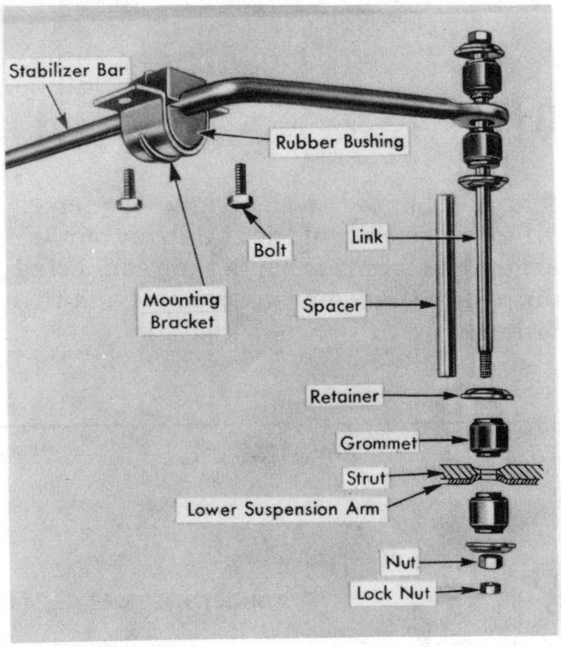

FIGURE II-91 Stabilizer-bar assembly. Note position of link (bolt). (Courtesy of Cadillac Motor Car Division)

STRUT ROD AND BUSHING

The strut rod is a rod used to eliminate, or at least reduce the back-and-forth movement of the outer end of the lower control arm. Many lower control arms are no longer shaped like an A, where there are two mounting points at the frame. This is depicted in Fig. II-92, where the lower control arm is a single arm extended outward from the frame. In this design a strut rod is used to support the lower control arm.

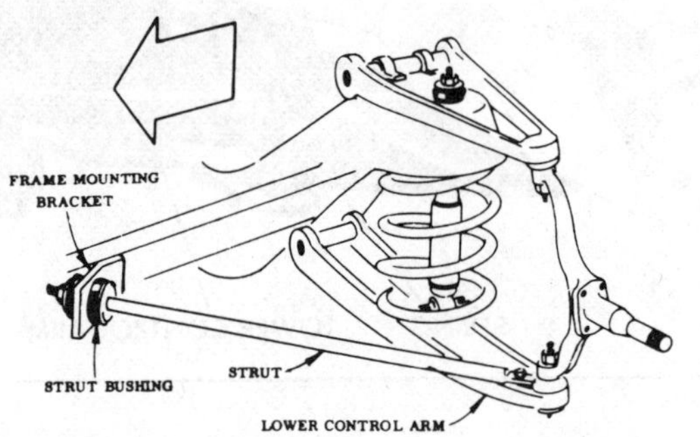

FIGURE II-92 Front suspension equipped with strut rod. (Courtesy of Monroe Auto Equipment Co.)

The lower control arm end of the strut rod (the rear) is bolted solidly to the lower control arm. The front end of the strut rod is mounted through a rubber bushing to the frame cross member. This dampens road shock and prevents excessive front-to-rear movement of the end of the control arm.

Strut rod bushings suffer severe stress when the front wheels strike irregularities in the road or during heavy periods of braking. Consequently, the bushings need frequent replacing (Fig. II-93). Visual inspection is usually adequate to detect worn bushings. If the customer complains of a loud clunk when braking, give the strut rod and its bushing careful attention.

FIGURE II-93 The white circle shows the location of the rubber strut-rod bushing. (Courtesy of Moog Automotive, Inc.)

OBJECTIVE II-E1 (manipulative)

Working alone or as a team with one other student and with the aid of Job Report Sheet II-E1, you will remove and replace the stabilizer bushings or strut rod bushing on one instructor assigned vehicle. Time recommended: Twice the flat-rate time allowed in Motors or Chilton Flat Rate manuals. This objective may be completed as part of Objective II-F—a complete front end rebuild.

Preparation:

1. Review pages 179 through 181.
2. Study pages 181 through 185.
3. Consult instructor for individual problems.

STABILIZER BAR SERVICE

Aside from deterioration of rubber bushings through normal wear, stabilizer bar units seldom need other service. A severe collision often causes enough damage to the bar and its bushings to require their replacement. Visual inspection of the stabilizer bar unit is usually adequate to discover such damage.

Materials and Tools Needed:

1. Hoist or service (floor) jack and jack (safety) stands.
2. Socket set with handles.
3. Diagonal pliers or hammer and correct size pin punch.
4. Torque wrench.

Procedure:

When replacing stabilizer bushings, the following can serve as a general procedure:

1. Raise the front end of the vehicle on a lift or jack and place on jack (safety) stands.
2. Disconnect the stabilizer bar from lower control arms, then remove the frame clamps (brackets) (Fig. II-94).
3. Remove the stabilizer bar.
4. Reverse the procedure to install the bar with new bushings (Fig. II-95).

Note: *Center the stabilizer bar on the frame before final torquing any of the mounting bolts.*

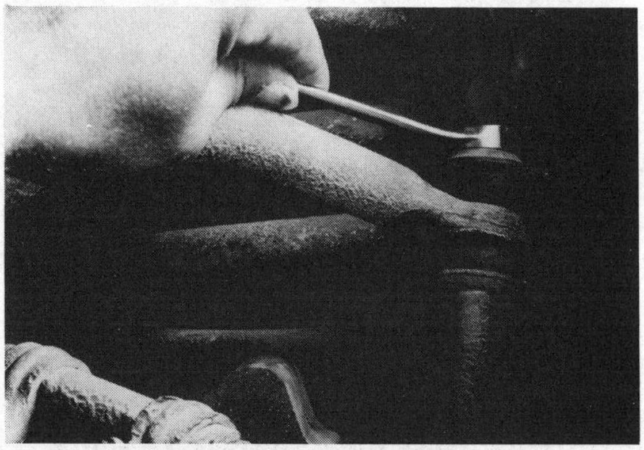

FIGURE II-94 Remove the stabilizer link after loosening frame brackets and disconnecting the stabilizer bar from the lower control arms.

FIGURE II-95 Install the stabilizer with new bushings and link.

STRUT ROD BUSHING SERVICE

Procedures may vary from vehicle to vehicle, but the following can serve as a typical service for strut rod bushings:

1. With the car on the floor or on a twin post (front-rear) lift, remove the cotter pin and loosen the front nut of the strut rod (Fig. II-96). Note the approximate position of the nut on the rod.

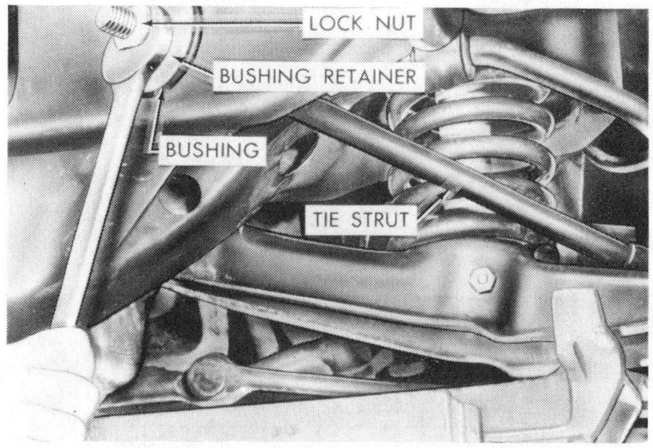

FIGURE II-96 Loosen the nut on forward end of the strut rod. (Courtesy of Moog Automotive, Inc.)

2. Remove mounting bolts at the lower control arm. Turn the front wheel for clearance and remove the strut rod from the rear side (Fig. II-97).
3. Remove old bushings and install replacements (Fig. II-98).

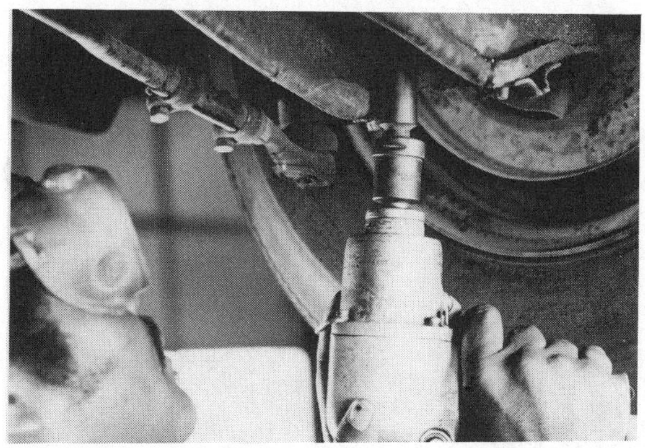

FIGURE II-97 Remove mounting bolts on the lower control arm. Note use of safety glasses by technician.

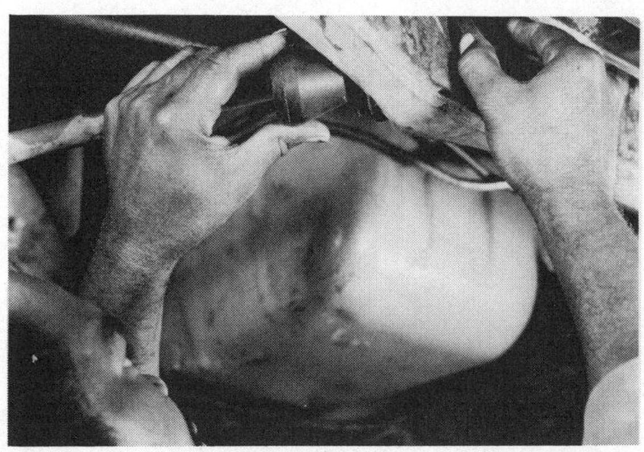

FIGURE II-98 Installing new strut rod bushings.

4. Reverse the procedure to install the strut rod. **Note:** *Do not torque the front bolts until the rear bar bolts are tightened.* Turn the front nuts to their original positions, as some front suspensions lengthen or shorten the strut rod to change caster settings (Fig. II-99).

Since front end alignment should follow bushing replacement, adjust the strut rod to approximate the correct position.

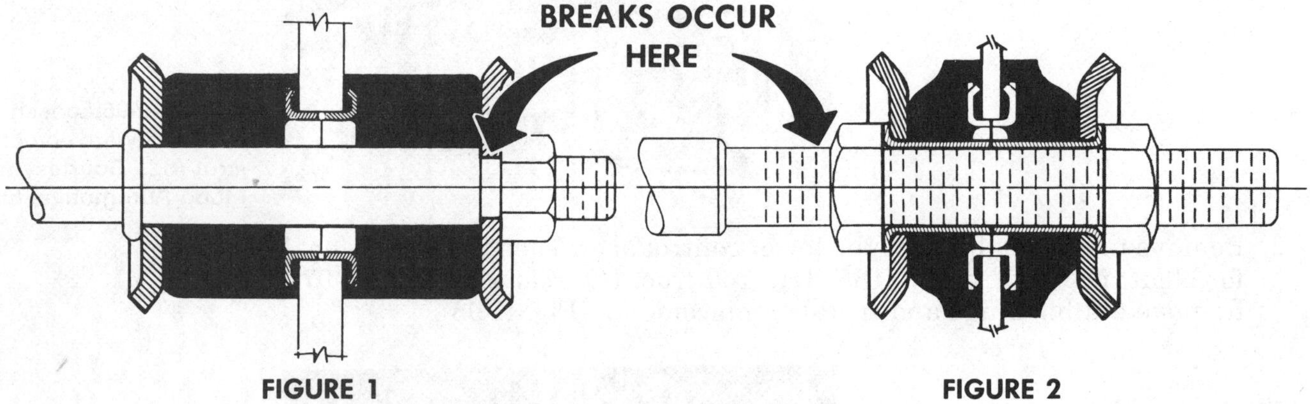

FIGURE II-99 The figure on the left shows a double-shouldered strut rod. The figure on the right shows a threaded strut rod the length of which can be adjusted to provide caster setting.

JOB REPORT SHEET II-E1

STABILIZER AND STRUT ROD BUSHING SERVICE

Student name ______________________ Date ______________

Make of car ____________ Model ____________ Year ____________

Time started ____________ Time finished ____________ Total time ____________

Flat-rate time ______________________________

Specifications:

1. Torque required for stabilizer bolts (bar to control arm)? ____________ ft-lb. (N-m)
2. Torque required for stabilizer bar clamps? ____________ ft-lb. (N-m)
3. Strut-rod bolt torque (rod to control arm)? ____________ ft-lb. (N-m)
4. Strut-rod front-nut torque (rod to frame)? ____________ ft-lb. (N-m)

Check-up Questions:

1. Was the stabilizer bar centered before torquing clamps? ____________

   ____

2. What rubber parts showed greatest wear? ____________

   ____________ Why? ____________

   ____

3. Which way do the strut rods move during braking? ____________
4. Is the strut rod on this vehicle used to change caster setting? ____________

   ____

DATE COMPLETED ____________ INSTRUCTOR'S SIGNATURE ____________

TEST II-E

Student name ______________________ Date ______________

Directions: Select the best answer to each question and circle the correct letter at the left.

1. Which of the following statement(s) is true?
   I. Stabilizer bars help prevent body roll and pitch
   II. Stabilizer bars have no effect on trim height
      A. I only
      B. II only
      C. Both I and II.
      D. Neither I nor II
2. Stabilizer bars affect:
   1. One side at a time
   II. Both sides at the same time
      A. I only
      B. II only
      C. Both I and II.
      D. Neither I nor II
3. Stabilizer bars are commonly mounted to:
   I. Frame or crossmember
   II. Control arms
      A. I only
      B. II only
      C. Both I and II.
      D. Neither I nor II
4. Stabilizer bushings that need replacement can be detected by:
   I. Turning steering wheel side to side and listening for noise
   II. Bouncing front end and watching for dampening movement
      A. I only
      B. II only
      C. Both I and II.
      D. Neither I nor II
5. If a stabilizer bar is bent, it should be:
   I. Heated with a torch and bent to original shape
   II. Replaced
      A. I only
      B. II only
      C. Both I and II.
      D. Neither I nor II
6. The strut rod controls the:
   I. Inner end of upper control arm
   II. Outer end of stabilizer shaft
      A. I only
      B. II only
      C. Both I and II.
      D. Neither I nor II

7. Strut rod bushings are usually mounted:
   I. On the lower control arm
   II. On the upper control arm
      A. I only
      B. II only
      C. Both I and II.
      D. Neither I nor II
8. Rubber bushings are used between strut rods and the frame end mount to:
   I. Absorb road shock on control arms
   II. Allow up-and-down movement of control arms
      A. I only
      B. II only
      C. Both I and II.
      D. Neither I nor II
9. The importance of tightening the strut-rod nut lies in the fact that:
   I. Some front suspension alignment angles are controlled by the strut rod nut position
   II. Proper cushioning effect can only be achieved through proper tension on strut rod nut
      A. I only
      B. II only
      C. Both I and II.
      D. Neither I nor II
10. The strut rod nut should be:
   I. Torqued to specifications
   II. Tightened only after the weight of the vehicle is on its tires
      A. I only
      B. II only
      C. Both I and II.
      D. Neither I nor II

DATE COMPLETED ______________________ INSTRUCTOR'S SIGNATURE ______________________

Working alone or as a team with one other student you will completely rebuild the front suspension of an instructor assigned automobile. This objective requires the completion of at least three of the following jobs in conjunction with the removal and replacement of coil spring or torsion bars. Each job will be done in accordance with the manufacturer's specifications and recommended procedures. Time recommended: twice the flat-rate time allowed in the Motors or Chilton Flat Rate manual.

1. Replace the upper and/or lower ball joint.
2. Replace the upper or lower inner control arm bushings.
3. Remove and replace the torsion bar (coil spring replacement may be substituted in assignment).
4. Replace stabilizer bushings.
5. Replace strut rod bushings.

OBJECTIVE II-F1 (manipulative)

Preparation:

1. Review all material for Objectives II-A, II-B, II-C, II-D, and II-E.
2. Study suspension rebuild section of general, or manufacturer's repair manual for the automobile assigned.
3. Carefully observe instructor's demonstration on coil spring removal.
4. Carefully observe instructor's demonstration on bushing removal and replacement.

JOB REPORT SHEET II-F1

COMPLETE FRONT END REBUILD

Student name__________ Date__________

Make of car__________ Model__________ Year__________

Time started__________ Time finished__________ Total time__________

Flat-rate time __________

This Job Report Sheet is a summary of all objectives in Unit II, "Suspension Systems." As your work progresses on a front end rebuild, complete each one of the appropriate Job Report Sheets and turn them in to your instructor for credit. Turn in this Job Report Sheet when rebuild is completed.

	Job Report Sheet Number	*Date Turned In*		*Job Report Sheet Number*	*Date Turned In*
1.			7.		
2.			8.		
3.			9.		
4.			10.		
5.			11.		
6.					

REAR SUSPENSION

After studying the preparation assignment you will be expected to correctly answer a minimum of 14 out of 20 multiple-choice questions on the following topics:

1. Axle housings
2. Control arms
3. Coil and leaf springs
4. Tracking

OBJECTIVE II-G (written test)

Preparation:

1. Study pages 191 through 203.

The rear suspension consists of the components that support the rear portion of the automobile. The combination of rear and front suspension allows the vehicle to be driven in a safe and comfortable manner.

Leaf spring and coil spring are the two rear suspension systems commonly used in today's automobiles. In many imported cars, the rear suspension is an independent suspension with coil springs.

LEAF SPRING SUSPENSION

The major components of a typical leaf spring design suspension are shown in Fig. II-100. The components are:

1. Live axle (solid axle housing with driving mechanism).
2. Two leaf springs, longitudinally mounted (front to rear)
3. Leaf spring mounts (shackles and hangers)
4. Shock absorbers (two)

Leaf spring suspension requires little service but usually gives a harsher ride than the coil spring type.

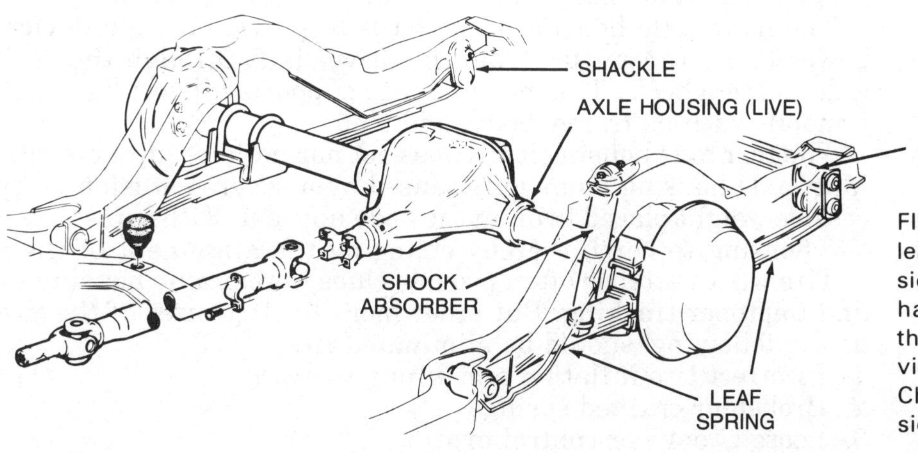

FIGURE II-100 Typical leaf spring rear suspension. The shock absorber has been removed from the right side for ease of viewing. (Courtesy of Chevrolet Motor Division)

COIL SPRING SUSPENSION

The major components of this type of suspension are:
1. Axle housing (live axle)
2. Two coil springs
3. Control arms (one to four)
4. Track bar (not shown)
5. Shock absorbers (two)

Coil spring rear suspension provides a soft ride with average handling qualities (Fig. II-101). It is currently the more popular rear suspension design.

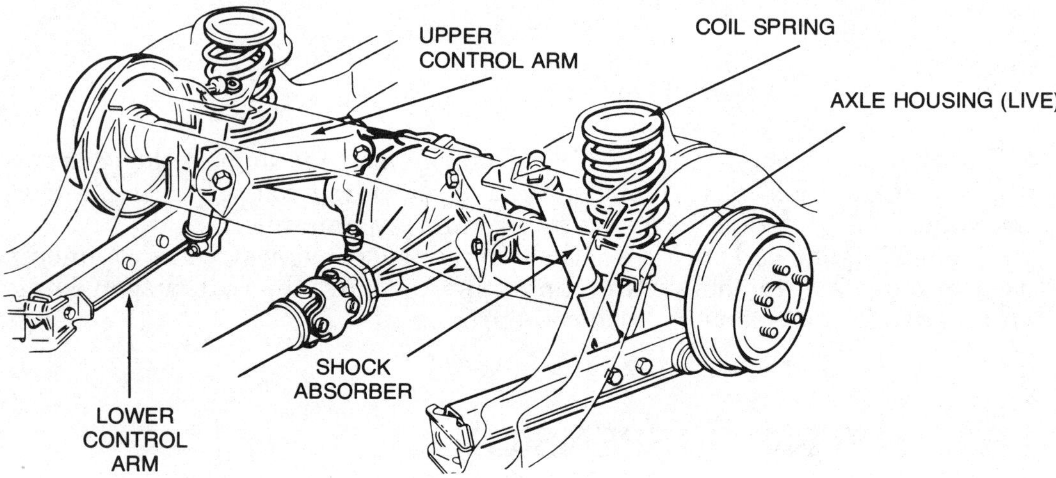

FIGURE II-101 Typical coil spring rear suspension. (Courtesy of Chevrolet Motor Division)

AXLE HOUSINGS

The basic design of most domestic automobile rear axle housings has not changed in the last 50 years. Exceptions are rear axle housing designs of Corvette and Corvair. These cars have a so-called swing or independent rear suspension, as do many imported vehicles (Fig. II-102).

The main axle housing function is to provide a rigid device that allows power to be transmitted from the drive shaft through the differential and axle to the wheels. This housing also supports all sprung weight of the automobile such as frame, body, and load.

The rear axle housing itself does not normally require major service work. Most axle housing damage is caused by a severe collision or by having excessive weight placed in or on the body above it. Either instance would cause the housing to bend, thereby changing the alignment of the rear wheels.

Tire wear patterns often provide clues to rear axle housing misalignment and improper tracking. But other more likely causes of the tire wear, such as the following, should be eliminated first:
1. Incorrect tire inflation and wrong tire size
2. Broken or cracked springs
3. Loose U bolts or control arm
4. Possible collision damage

In most cases, a careful visual inspection will reveal something other than the rear-axle-housing misalignment is causing tire wear. Review Objective I-B1, of this text for information about tire-wear patterns.

In the rear axle, as in the front suspension, the running alignment should be of greatest concern. Ideally, when moving, all four car wheels should be straight up and down and roll along in a straight line. Since we usually cannot check running alignment, we must compensate in our static alignment (as the car rests on the alignment rack).

There are several forces acting upon the rear axle while it is being driven that are not apparent while the automobile is sitting on the alignment rack. This is why a zero camber (wheels perpendicular to the road), zero toe setting (wheels parallel to each other) is undesirable on an empty car. The load forces change with the weight of passengers and baggage, trailer towing, or even fuel in the gas tank.

A car going over a dip in the road or being braked experiences a load force change on the rear axle. Such changes affect the camber or vertical position of the rear wheels. To compensate for load forces, some positive camber usually is desirable (less than ¼-degree outward tilt) when one is checking alignment on the rack (when the automobile is at rest).

Another force affecting suspension occurs during acceleration. Normally the wheels of the rear axle seek to move slightly inward at the front of the tire, causing toe-in. However, rolling deflection (the forces affecting the wheels in motion) has a tendency to force the rear wheels to toe out. Thus, a slight amount of toe-in (1/32 inch or .79 mm) may be encountered during rear alignment checks. But this toe-in disappears when the vehicle is in motion. For specific information concerning rear axle alignment, refer to Unit IV.

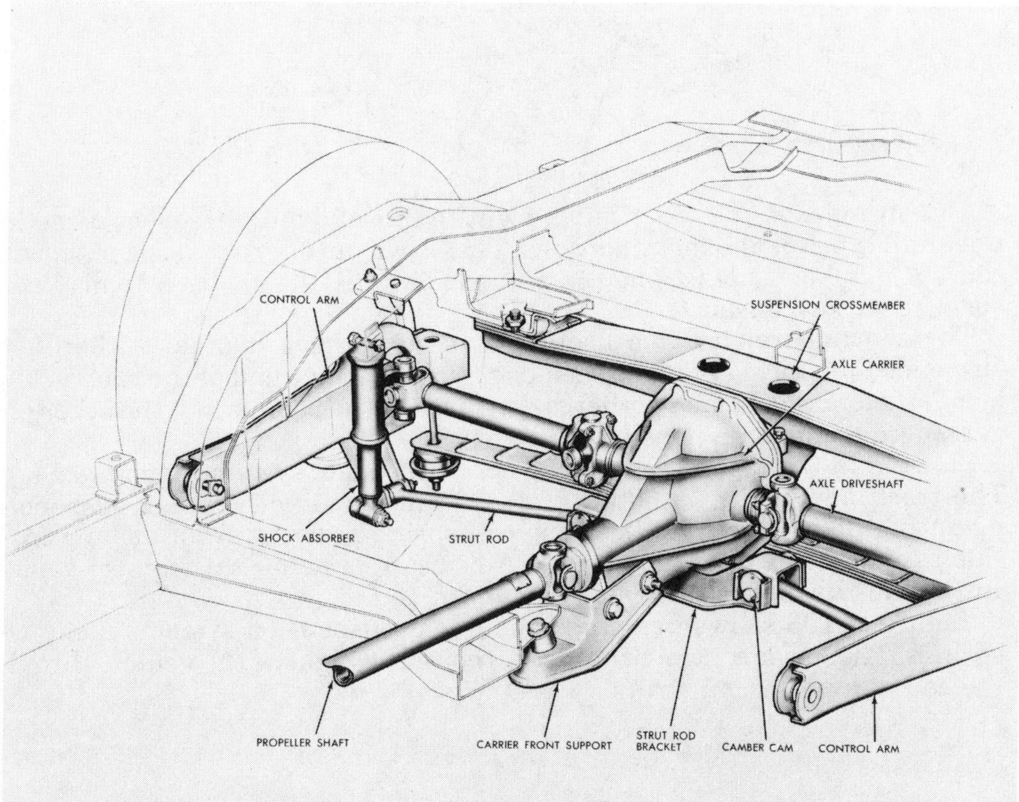

FIGURE II-102 Independent rear suspension. Movement of rear wheels is controlled by transverse leaf spring, strut rods, and arms. (Courtesy of Chevrolet Motor Division)

CONTROL ARM

Every axle housing, during acceleration periods, has a tendency to twist upward at the front. This is known as axle housing *wind-up* or *wrap-up*.

With a leaf spring rear suspension, as shown in Fig. II-103, wind-up movement is effectively controlled by the U bolts holding the housing firmly in its intended position with the spring. The axle housing has a flat steel pad located between the housing and the spring. This mounting pad has an aligning hole that corresponds to a pin in the leaf spring known as the *center bolt*. The pin passes completely through the spring and serves to hold the leaves in place as well as to keep the axle housing from shifting from its intended position. Because of the inherent strength of the spring, it controls not only the wind-up problem but also the side sway. This is the reason why most leaf spring rear suspensions do not require control arms.

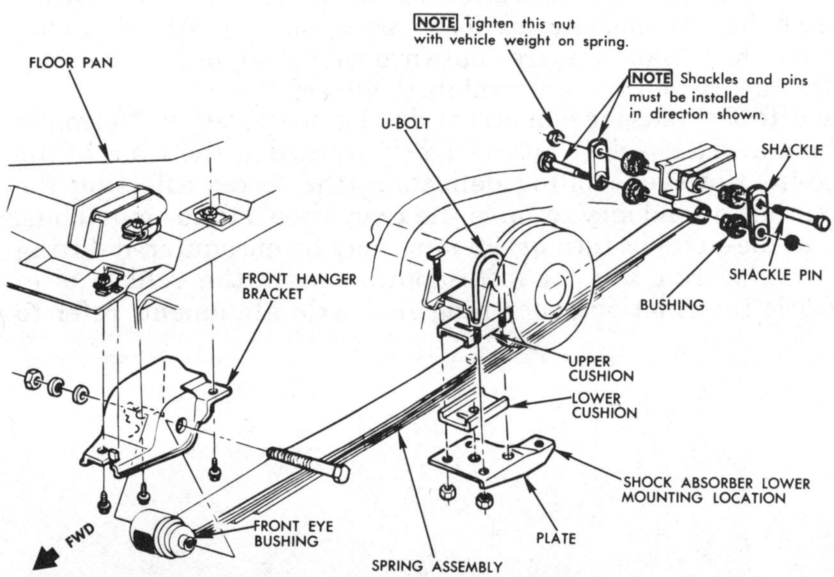

FIGURE II-103 Typical leaf spring rear suspension, left side. Read notes. (Courtesy of Chevrolet Motor Division)

Coil springs, on the other hand, want to flex and move in every direction under all types of driving conditions. This is why every rear suspension with coil springs has at least one control arm. Fig. II-104 shows a typical coil-spring rear suspension.

The control arm itself is usually a piece of steel channel with rubber suspended mounting holes at each end. Normally one end of the control arm is mounted to the frame or other strong area of the body, while the other end is secured to the axle housing (Fig. II-105).

Sharing in the task of controlling the suspension system is the *track bar*. The track bar, which is nothing more than a rod with rubber suspended mounting holes, serves to keep the coil spring rear end from moving from side to side. *Sway bar* is a common American name for this control device. In Britain this bar is known as a *panhard rod*.

Figure II-106 shows two versions of rear suspension systems using coil springs. The lower control arms carry the rear weight of the vehicle through the coil springs or coil struts.

FIGURE II-104 Typical coil spring rear suspension designed with upper and lower control arms. (Courtesy of Chevrolet Motor Division)

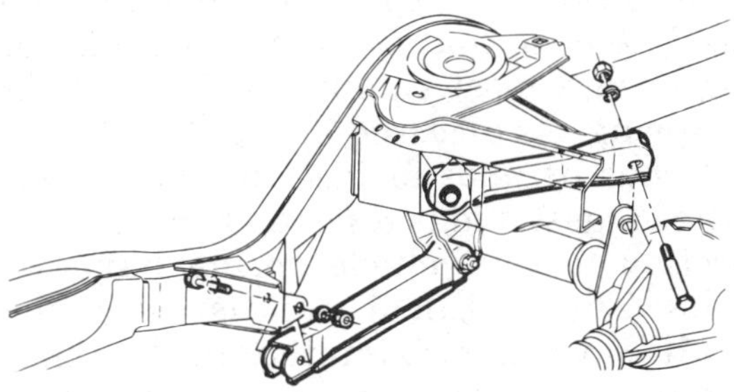

FIGURE II-105 Upper control arm is positioned diagonally to reduce both side sway and axle housing wind-up (acceleration) and wind-down (braking). Lower control arm allows up-and-down movement while maintaining average wheelbase distance. (Courtesy of Chevrolet Motor Division)

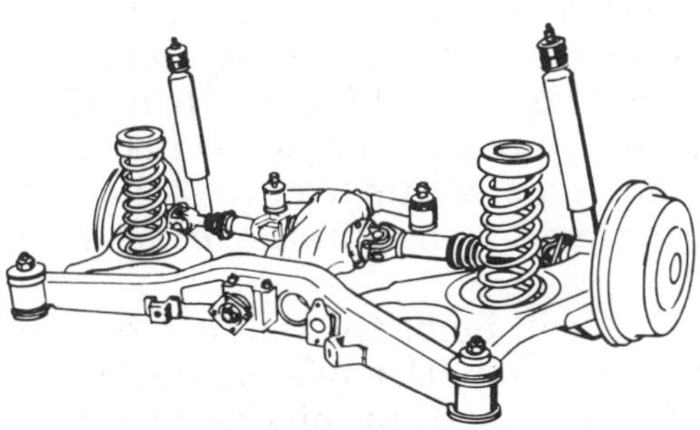

FIGURE II-106A Independent rear suspension has suspension arms (lower control arms) to isolate each wheel and to carry the vehicle weight.

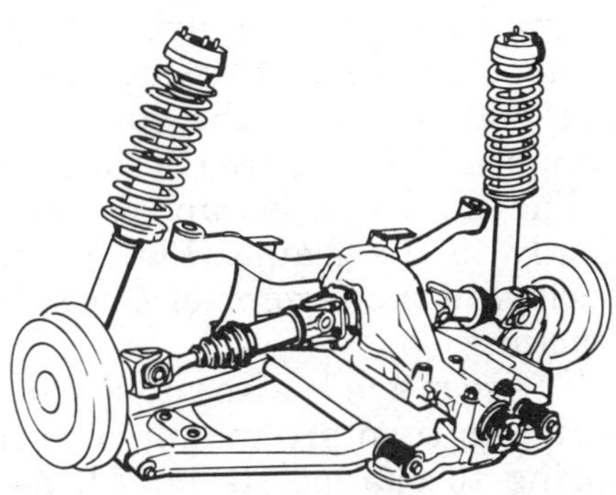

FIGURE II-106B MacPherson strut rear independent suspension uses a MacPherson strut attached to a lower control arm (transverse link) to isolate each wheel.

Pinpointing control arm or track bar problems is a matter of diagnosis through visual inspection. If the rubber is badly deteriorated or has fallen out of the mounts, remove the arms from the car and extract and replace the rubber bushings. In cases of damage to the metal arm, it would be best to replace the entire arm.

Minor bending of a control arm to restore alignment is possible but *not recommended.* Specifications for dimension of control arms are not listed in repair or shop manuals. And some control arms and many track bars are curved as part of their original design. Before attempting to do any straightening, you must have an undamaged part for comparison.

REAR AXLE SPRINGS

The springs designed for solid axle rear suspension are of two general types: (1) leaf spring and (2) coil spring. Both the leaf spring and the coil spring are widely used in rear suspension systems.

Since the springs themselves are basically the same as those of the front suspension, the operation and characteristics of each type are discussed in the section on front suspension springs under the headings "Leaf Springs" and "Coil Springs." Refer to that portion of the text for further information.

The leaf spring design most commonly used at the present time is the semi-elliptic leaf spring. As pictured in Fig. II-107, it consists of one or more leaves and usually has its ends formed into circles known as *eyes*. The leaf with the eyes is the longest and is referred to as the *main leaf.* This main leaf allows for connection to the vehicle spring hanger by means of bolts or shackles and bolts. The multileaf unit is assembled with a center bolt passing through all leaves approximately in the center. This bolt holds the leaves together while its head provides an alignment pin for correct positioning to the rear axle. Spring clips are often used to help keep the leaves aligned and hold any friction material between the leaves in position.

Most leaf springs are attached to the rear section of the car at three places. The ends (eyes) of the spring are attached by bolts through shackles and hangers. The mid area of the spring is held to the solid rear axle by U-bolts.

The front eye is rubber mounted in a hanger that is riveted or bolted firmly to the frame (Fig. II-108). The front mount controls the forward-backward movement of the spring and axle to maintain correct rear-axle tracking.

The rear eye is also rubber mounted. A bolt holds the rear eye to shackles, which in turn are mounted to the rear hanger. Use of shackles allows the spring to change its length as deflection (bending) of the spring occurs. Change of length takes place any time the rear wheel changes its position relative to the body.

A vehicle having leaf springs will not usually have control arms or track (sway) bars. The spring itself has enough built-in strength to resist axle wind up and/or side sway. Some manufacturers offer heavy-duty leaf springs as optional equipment when the vehicle will be more heavily loaded or used in trailer towing.

For service of leaf springs, refer to the section on front suspension leaf springs pages 146 through 148.

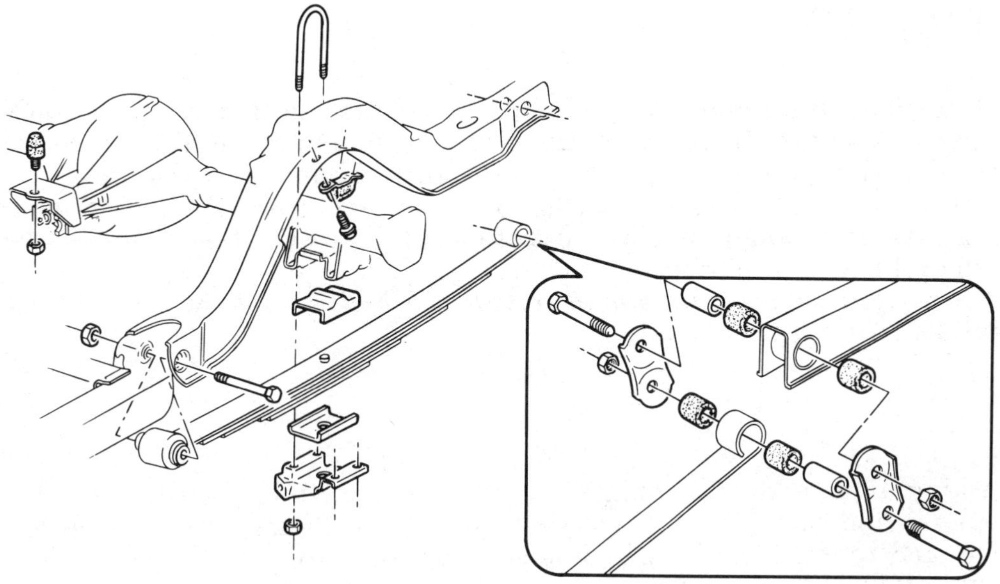

FIGURE II-107 Semi-elliptic leaf spring rear suspension. Center bolt holds leaves together and helps maintain axle alignment. (Courtesy of Chevrolet Motor Division)

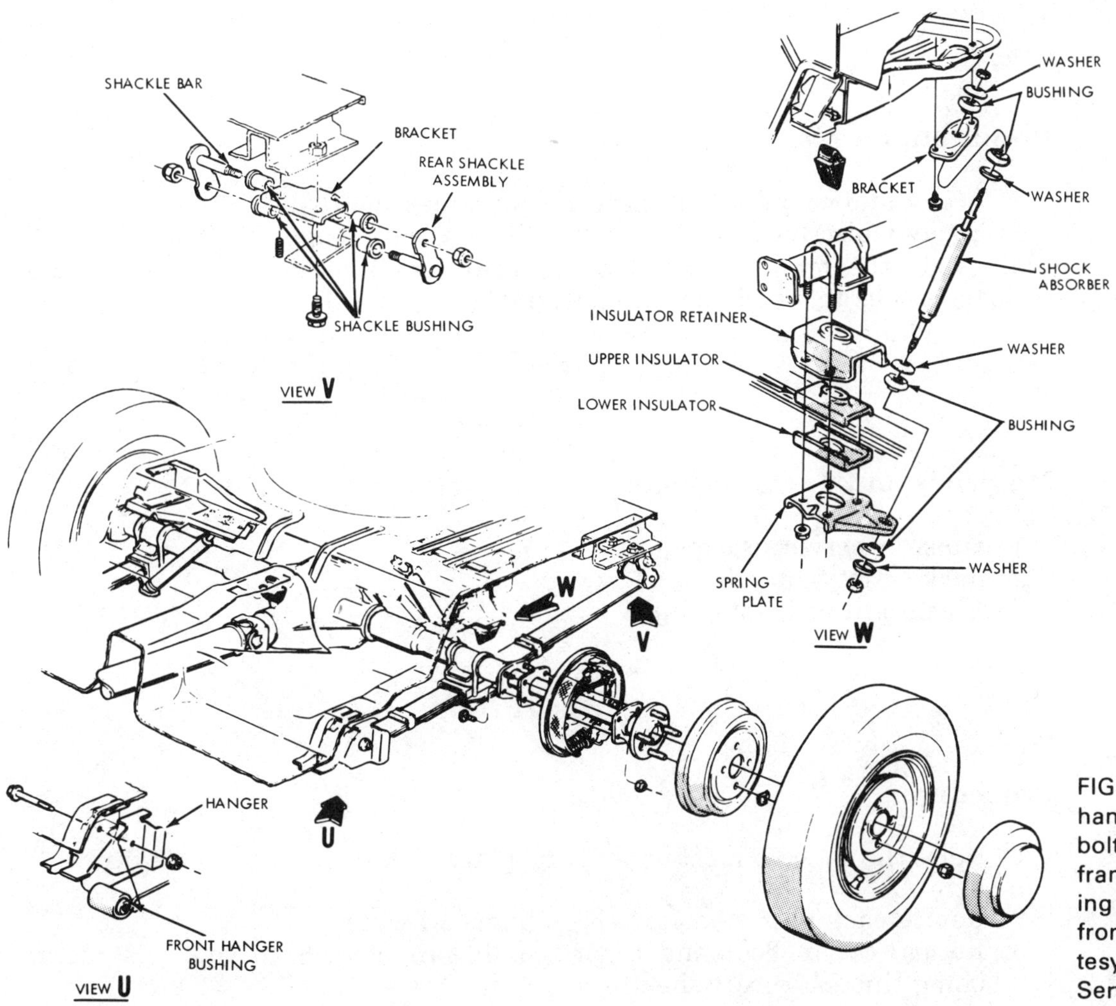

FIGURE II-108 The front hanger is riveted or bolted solidly to the frame. Note view *V* showing shackle bolts entering from both sides. (Courtesy of Ford Customer Service Division)

TRACKING

A fundamental requirement of a vehicle chassis is that it must "track" properly. A vehicle tracks when the rear wheels follow the front wheels in a true parallel position when driven on a straight and level roadway.

A vehicle out of track will not handle properly, especially in emergency situations. Handling problems most often relate to oversteer or understeer as the vehicle is cornering.

Every effort should be made to correct a vehicle out of track before it is returned to its owner.

OBJECTIVE: II-G1 (manipulative)

Working alone and with the aid of reference material, visually inspect, diagnose, and report in writing to your instructor the rear tire wear patterns and conditions of the rear suspension of two automobiles. Time recommended: 20 minutes.

Preparation:

1. Study and complete all material for Objective II-B.
2. Study Objective I-B1, "Abnormal Tire Wear," pages 37 through 43.
3. Study the manufacturer's or general repair manual for specific automobiles to which you are assigned.

Materials and Tools Needed:

1. Hoist or service (floor) jack
2. Jack (safety) stands
3. Flashlight or trouble light

Procedure:

1. Raise the car on a hoist or jack and place jack stands under the frame or axle.
2. Rotate each rear wheel, noting abnormal wear.
3. Return car to floor and report conditions of tires to your instructor using this job report sheet.

JOB REPORT SHEET II-G1

REAR SUSPENSION INSPECTION — ASSIGNMENT 1

Student name ______________________________ Date ____________________

Make of car ______________ Model ______________ Year ____________________

Time started ______________ Time finished ______________ Total time ____________

Flat-rate time ____

Specifications:

Refer to illustrations of tire wear patterns in Unit I of text.

Checkup Questions:

1. Did the vehicle have the same tire size on each side? ______________ If not, how would this affect wear patterns? ____

____

2. Was there excessive wheel run-out (wobble)? ______________________________

3. List any suspension damage or excessive wear: ______________________________

____

____

4. Make a sketch of the tire wear patterns or consult directly with your instructor.

LEFT REAR

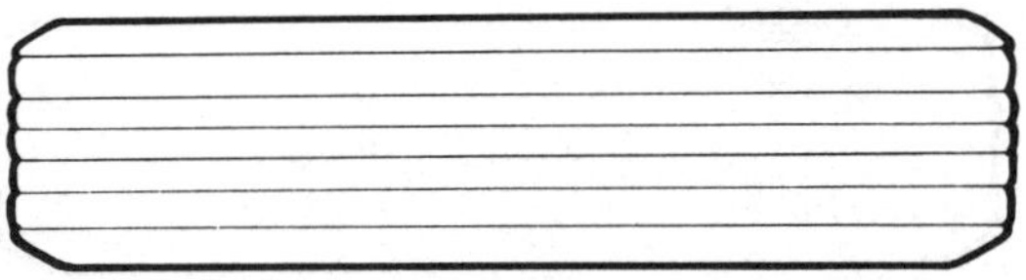

DIRECTION ← OF TRAVEL

RIGHT REAR

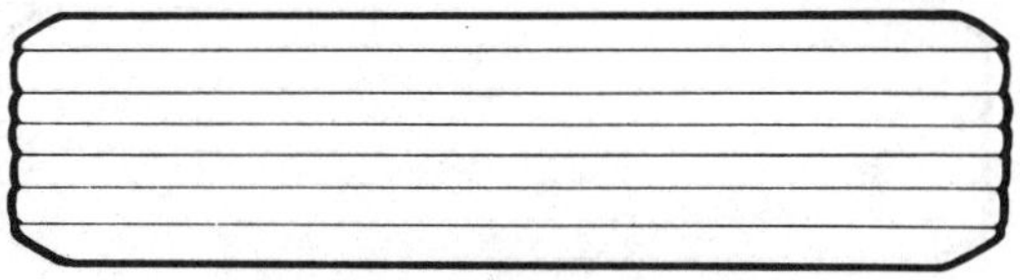

DIRECTION ← OF TRAVEL

DATE COMPLETED ______________ INSTRUCTOR'S SIGNATURE ______________________

JOB REPORT SHEET II-G1

REAR SUSPENSION INSPECTION

ASSIGNMENT 2

Student name____________________ Date__________

Make of car__________ Model__________ Year__________

Time started__________ Time finished__________ Total time__________

Flat-rate time ____________________

Specifications:

Refer to illustrations of tire wear patterns in Unit I of text.

Checkup Questions:

1. Did the vehicle have the same tire size on each side?__________ If not, how would this affect wear patterns? ____________________

2. Was there excessive wheel run-out (wobble)? __________

3. List any suspension damage or excessive wear: ____________________

4. Make a sketch of the tire wear patterns or consult directly with your instructor.

LEFT REAR

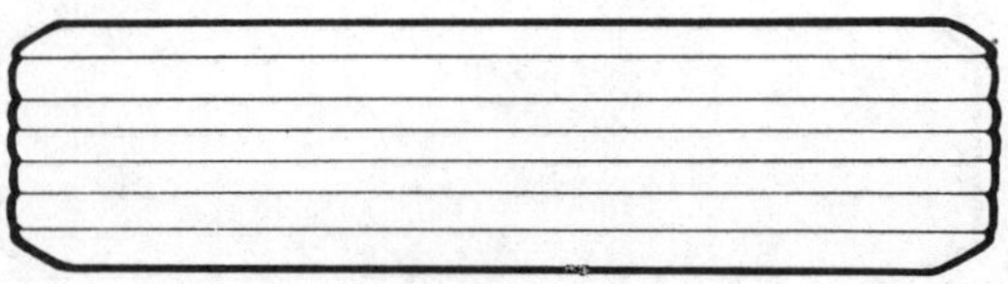

DIRECTION ← OF TRAVEL

RIGHT REAR

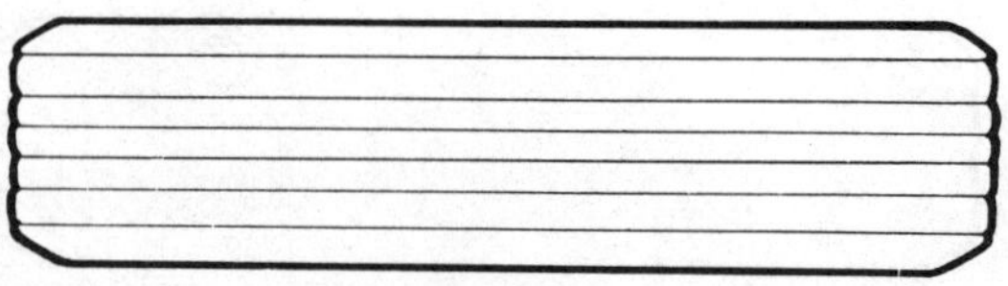

DIRECTION ← OF TRAVEL

DATE COMPLETED__________ INSTRUCTOR'S SIGNATURE __________

Working as a team with one other student and with the aid of Job Report Sheet II-G2, visually inspect and measure the wheelbase of an instructor assigned vehicle, checking for differences in length between the front and rear wheels. Time recommended: 20 minutes.

OBJECTIVE: II-G2 (manipulative)

Preparation:

1. Study pages 201 through 203.
2. Locate wheel base specifications for car being measured.
3. Carefully observe instructor's demonstration on the use of the tracking gauge.

Materials and Tools Needed:

1. Tracking gauge
2. Alignment rack or level surface

Procedure:

When rear axle misalignment is suspected, two checking operations may be required.

First, measuring the wheelbase with a tracking gauge should show if the differential is incorrectly positioned to the frame and body.

Second, placing the vehicle on an alignment rack with gauging devices attached should reveal axle housing misalignment, as discussed earlier in this section. To make axle housing checks, refer to the section, "Rear Axle Alignment," in Unit IV.

To gauge the tracking of a vehicle, it should be within specifications of front-end alignment.

Follow the instructions listed below:

1. Secure the tracking gauge and wheelbase specifications for the vehicle being measured.
2. Set the length of the gauge to approximate wheelbase length.
3. Place gauge on the wheels at hub height with the single pointer at the front and double pointers at the rear (Fig. II-109).

FIGURE II-109 Proper position of tracking gauge with pointers set.

4. Adjust the front pointer so it contacts the tire and rim at the rear of the wheel.
5. Adjust the two rear pointers so they contact the tire at both the front and rear of the rim at axle height.
6. Tighten all thumb screws so the gauge is rigid.
7. Recheck settings to be certain all pointers are making contact in the proper position.
8. Move the gauge to the other side of the vehicle and place it in the same relative position. The pointers should make contact at all three places.
   Tolerance: ⅛ inch (3.17 mm) or less in length for automobiles
   ¼ inch (6.35 mm) or less in length for trucks
9. If the gauge shows more than the allowable tolerance, tracking misalignment is indicated. Consult your instructor for further action. (See Fig. II-110, II-111, II-112 and II-113 for typical tracking problems.)

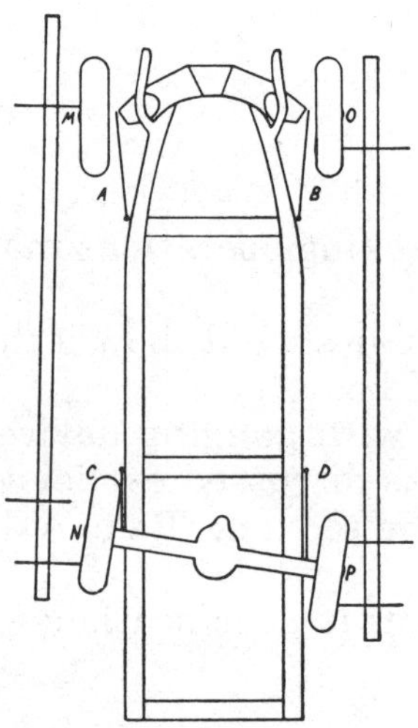

FIGURE II-110 Swung rear end in which (1) *A* and *B* are equal, (2) *C* is shorter than *D*, (3) one pointer does not contact rear wheel, (4) *M* to *N* is shorter than *O* to *P*. (Courtesy of Bear Manufacturing Co.)

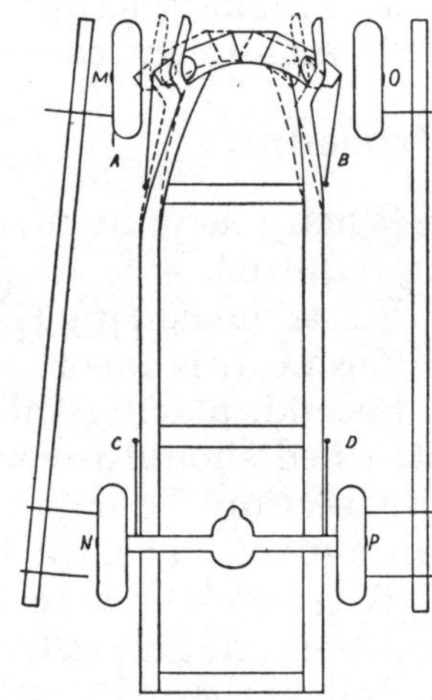

FIGURE II-111 Swayed frame in which (1) *A* and *B* are equal, (2) *C* and *D* are equal, (3) one pointer does not contact rear wheel, (4) *M* to *N* is longer than *O* to *P*. (Courtesy of Bear Manufacturing Co.)

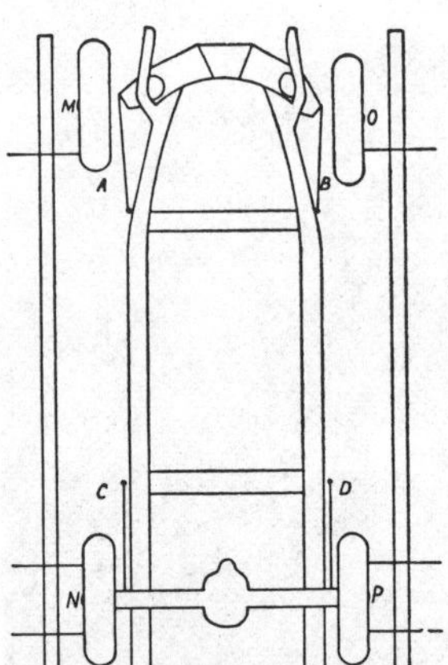

FIGURE II-112 Knee back on right side in which (1) *A* is longer than *B*, (2) *C* and *D* are equal, (3) pointers OK at rear wheel, (4) *M* to *N* is longer than *O* to *P*. (Courtesy of Bear Manufacturing Co.)

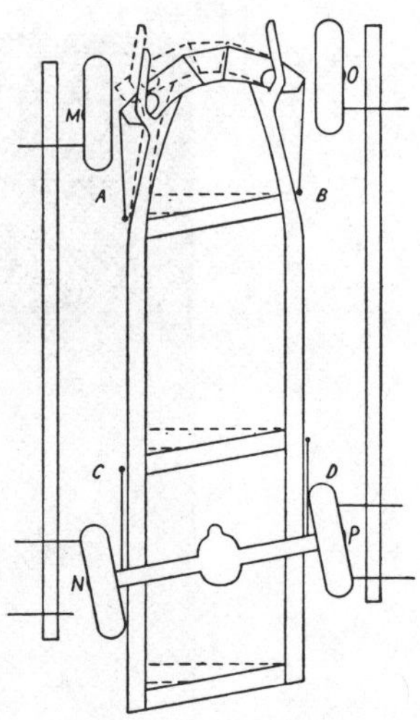

FIGURE II-113 Diamond frame in which (1) *A* and *B* are equal, (2) *C* and *D* are equal, (3) one pointer does not contact rear wheel, (4) *M* to *N* and *O* to *P* are equal. (Courtesy of Bear Manufacturing Co.)

JOB REPORT SHEET II-G2

TRACKING MEASUREMENT

Student name________________________ Date________________

Make of car______________ Model______________ Year________________

Time started______________ Time finished______________ Total time____________

Flat-rate time ____

Specifications:

Wheelbase of vehicle being checked is______________inches.

Checkup Questions:

1. Make sketch of tracking condition of vehicle checked.

Left (circle one) *Right (circle one)*

Length { Longer / Shorter / Same }   Length { Longer / Shorter / Same }

2. How would larger or smaller than normal tire size change the wheelbase? ______________

____

3. Would a bent wheel affect the wheelbase length? ______________

Why or Why not?____

____

DATE COMPLETED______________ INSTRUCTOR'S SIGNATURE ______________

TEST II-G

Name______________________________Date __________________________

Directions: Select the best answer to each question and circle the correct letter at the left.

1. Control arm(s) are used on rear suspensions with coil springs to:
   I. Maintain trim height
   II. Prevent axle housing wind-up
   A. I only
   B. II only
   C. Both I and II.
   D. Neither I nor II
2. Rear control arms have rubber bushing pivots to:
   I. Reduce noise and vibration
   II. Provide some torsional resistance
   A. I only
   B. II only
   C. Both I and II.
   D. Neither I nor II
3. Springs for rear suspension systems have:
   I. The same characteristics as front springs
   II. Generally higher deflection rates
   A. I only
   B. II only
   C. Both I and II.
   D. Neither I nor II
4. Rear suspensions with both upper and lower control arms usually have:
   I. Coil springs
   II. Shock absorbers
   A. I only
   B. II only
   C. Both I and II.
   D. Neither I nor II
5. Which of the following forces cause a deflection of the rear suspension?
   I. Acceleration
   II. Overloading
   A. I only
   B. II only
   C. Both I and II.
   D. Neither I nor II
6. The vertical position (camber) of the rear wheels can be changed slightly:
   I. When the vehicle goes through a dip
   II. When the vehicles brakes are heavily applied
   A. I only
   B. II only
   C. Both I and II.
   D. Neither I nor II

7. Rear tire wear patterns can provide clues to possible:
   I. Bent axles
   II. Collision damage
      A. I only
      B. II only
      C. Both I and II.
      D. Neither I nor II
8. Which of the following statements about rear springs are true?
   I. Coil springs provide a softer ride.
   II. Leaf springs provide greater loading capacities.
      A. I only
      B. II only
      C. Both I and II.
      D. Neither I nor II
9. The rear axle housing is a rigid device designed to:
   I. Allow application of power to the rear wheels
   II. Support the spring weight of the rear of the vehicle
      A. I only
      B. II only
      C. Both I and II.
      D. Neither I nor II
10. On leaf spring assemblies, the function of spring clips is to:
    I. Hold material between leaves
    II. Maintain leaf alignment
       A. I only
       B. II only
       C. Both I and II.
       D. Neither I nor II
11. Leaf spring rear suspension systems have which of the following?
    I. Control arms
    II. Track bar
       A. I only
       B. II only
       C. Both I and II.
       D. Neither I nor II
12. The eye of a spring is:
    I. The smallest diameter coil
    II. The center bolt that holds all leaves together
       A. I only
       B. II only
       C. Both I and II.
       D. Neither I nor II
13. The function of the track bar is to:
    I. Prevent axle housing wind-up
    II. Prevent side-to-side body movement
       A. I only
       B. II only
       C. Both I and II.
       D. Neither I nor II
14. A track bar is known by several other names, such as:
    I. Sway bar
    II. Panhard rod
       A. I only
       B. II only
       C. Both I and II.
       D. Neither I nor II

15. Track bars are found on coil spring rear suspensions because:
    I. Coil springs want to flex from front to rear
    II. They reduce road shock
        A. I only
        B. II only
        C. Both I and II.
        D. Neither I nor II
16. A vehicle with improper tracking will:
    I. Not handle properly in emergency situations
    II. Show abnormal tire wear
        A. I only
        B. II only
        C. Both I and II.
        D. Neither I nor II
17. Tracking refers to the requirement of the rear suspension where the:
    I. Rear axle must be parallel to the road
    II. Rear tires must follow the pattern of the front tires
        A. I only
        B. II only
        C. Both I and II.
        D. Neither I nor II
18. Tracking gauges measure:
    I. Wheel base differences from side to side
    II. Tire tracks
        A. I only
        B. II only
        C. Both I and II.
        D. Neither I nor II
19. For tracking gauge measurements to be accurate:
    I. The vehicle must be resting on a level surface
    II. The gauge must be held at hub height on both sides
        A. I only
        B. II only
        C. Both I and II.
        D. Neither I nor II
20. The allowable tolerance of vehicle tracking is:
    I. ⅛ inch (3.17 mm) for automobiles
    II. ¼ inch (6.35 mm) for trucks
        A. I only
        B. II only
        C. Both I and II.
        D. Neither I nor II

DATE COMPLETED ______________________ INSTRUCTOR'S SIGNATURE ______________________

After studying the following preparation material, you will be expected to correctly answer a minimum of seven out of ten multiple choice questions in a written test without reference materials on the following topics:
1. Shock absorber theory
2. Shock absorber testing, diagnosis
3. Removal and replacement procedures

OBJECTIVE II-H (written test)

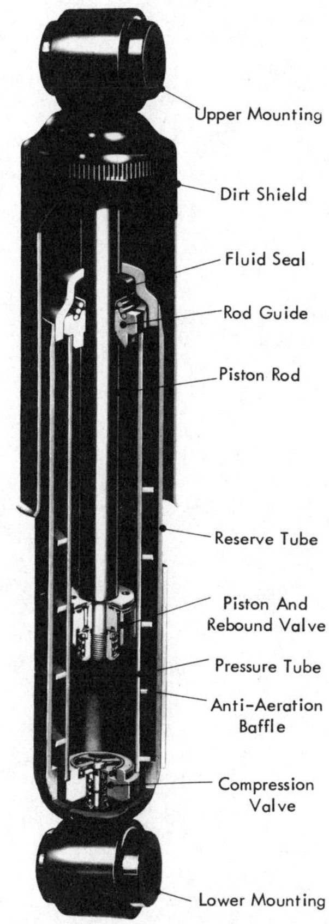

FIGURE II-114 This cutaway shows typical shock absorber construction. (Courtesy of Monroe Auto Equipment Co.)

Preparation:

Study pages 209 through 219.

SHOCK ABSORBERS

In the simplest of terms, the shock absorber is a hydraulic unit that acts to dampen the spring oscillation (action) and retard body movement (Fig. II-114). The most basic physical principle in the operation of a shock absorber is the conversion of the energy of movement into heat. By slowing down the body and spring movement, the oil and metallic parts of the shock absorber are heated. This converted heat energy is then dispelled into the passing air.

In reality the shock absorber is misnamed, for it is the springs and tires that absorb most of the road shock as the car is being driven (Fig. II-115). The British have a better name for shock absorbers. They call them *dampeners,* which is actually what shock absorbers do: dampen the spring oscillations.

As pointed out earlier in the section on springs, the material of a spring is a very elastic type of steel with both good and bad habits. As a bump or hole in the road is struck by the wheel, the spring reacts by moving in the opposite direction to the body. The spring and body move toward or apart from one another, depending on what has been struck. In effect, the spring stores the energy from the bump. To release this energy, the spring reverses the direction of movement. In the process of oscillating back and forth, the energy generated by the bump is discharged. This action is desirable for what we call *ride.* The more flexibility the spring has, the softer and springier the ride is, and vice versa.

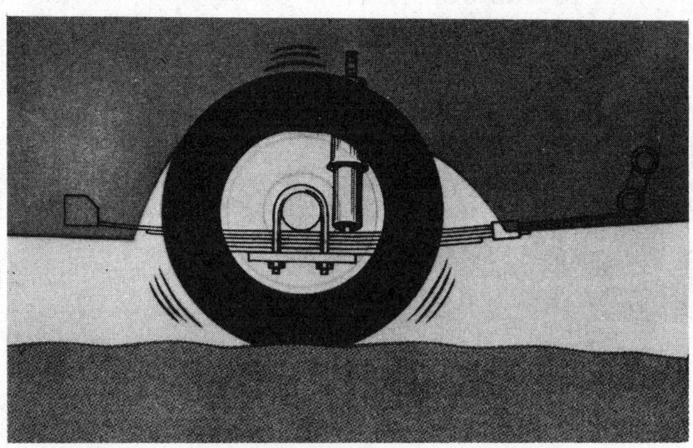

FIGURE II-115 Tires and springs absorb most of the road shock, as shown in this drawing. (Courtesy of Chrysler Motors Corp.)

The undesirable action of a spring is its tendency to spend this stored energy by going beyond its starting positions in decreasing amounts of movement. Such spring oscillation is apparent on a car with worn or defective shock absorbers. Even though the car is not moving, by pushing heavily on one corner of the body, the mechanic can make the suspension and body move downward, which is called *jounce* or *compression* (Fig. II-115). The body and suspension then reverse this direction, called *rebound* or *extension*, past its original position. The springs continue to oscillate in decreasing amounts unless dampened by good shock absorbers, as seen in Figure II-116. Shock absorbers draw off the energy the springs store when they absorb the initial jolt.

The sole responsibility of a car's shock absorbers is to reduce spring oscillation. Shock absorbers other than the spring or air assisted types do not support any of the vehicle's weight (Fig. II-117). A worn or leaky shock absorber will not cause a car to sag or lean. Since the springs (leaf, coil, or torsion bar) support the weight of the car, none of the shock absorbers do any work until the car is in motion with the wheels moving up and down.

The design of most shock absorbers in use today is called *aircraft* or *telescopic* design, since it was originally designed for aircraft landing gear.

FIGURE II-116 As a bump in the road is struck, the spring and body move toward one another in an action called *jounce*. When the spring and body move away from each other, this action is called *rebound*. Good shock absorbers dampen this undesirable spring action. (Courtesy of Chrysler Motors Corp.)

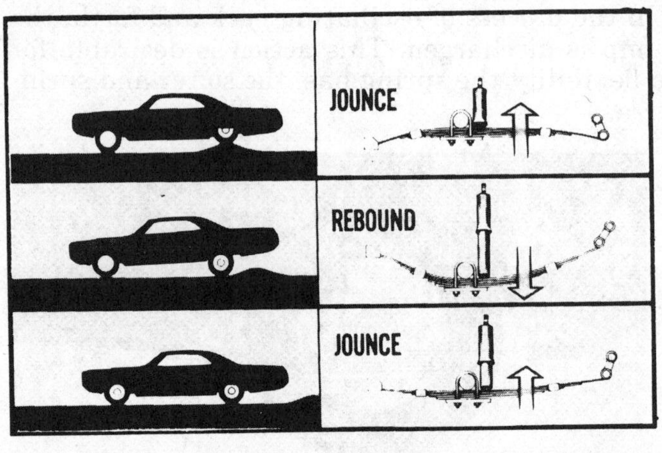

FIGURE II-117 Air lift or air ride shock absorbers are used to regulate rear suspension heights under varied loading conditions. (Courtesy of Cadillac Motor Car Division)

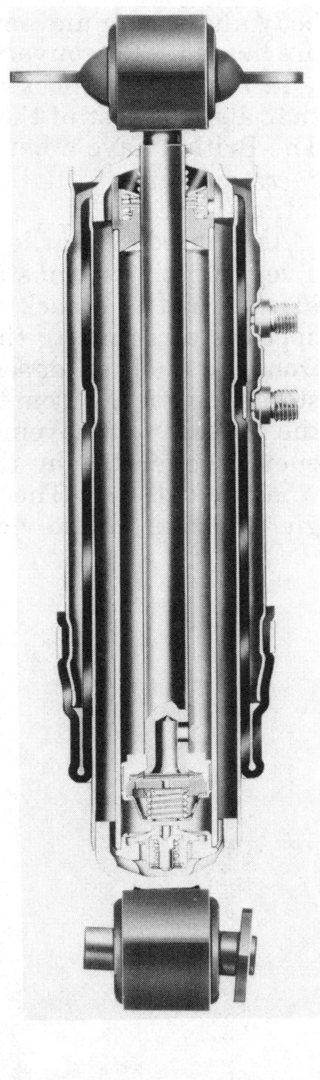

One end of the shock absorber has a mount that is attached to the shock absorber rod and piston. The piston moves up and down in a cylinder filled with oil. The flow of the oil is regulated by small valves. These valves control the amount of oil that passes through the piston device during rebound actions. The end of the shock also contains an oil seal with wiping edges, as shown in Fig. II-118. Its purpose is to retain the oil (which provides lubrication to the rod) in the grooves. A trace of oil around the rod seal is normal, so do not replace the shock because of small amounts of oil leakage. Rear shocks normally have a dirt or stone shield that protects the shaft from physical damage from flying objects beneath the car.

The opposite end of the shock absorber has the cylinder and other mounts. It contains a valve or valves to assist in controlling the oil movement through the piston during the jounce or compression movements. The entire dampening action of the shock is a result of movement of the oil through specific-sized openings within a confined space.

The shock absorber in the MacPherson strut suspension system (front or rear) performs the same function as one mounted separately from the spring. The MacPherson design requires a shock with greater dampening ability. Thus, larger shock absorbers are used (Fig. II-119).

One of the important facts about shock absorbers a technician should understand is that a great many factors are considered in the design of a shock for one specific make and model of car. Never attempt to install shock absorbers that are not specified for that particular car model. Methods of diagnosing shock-absorber failure vary with different authorities.

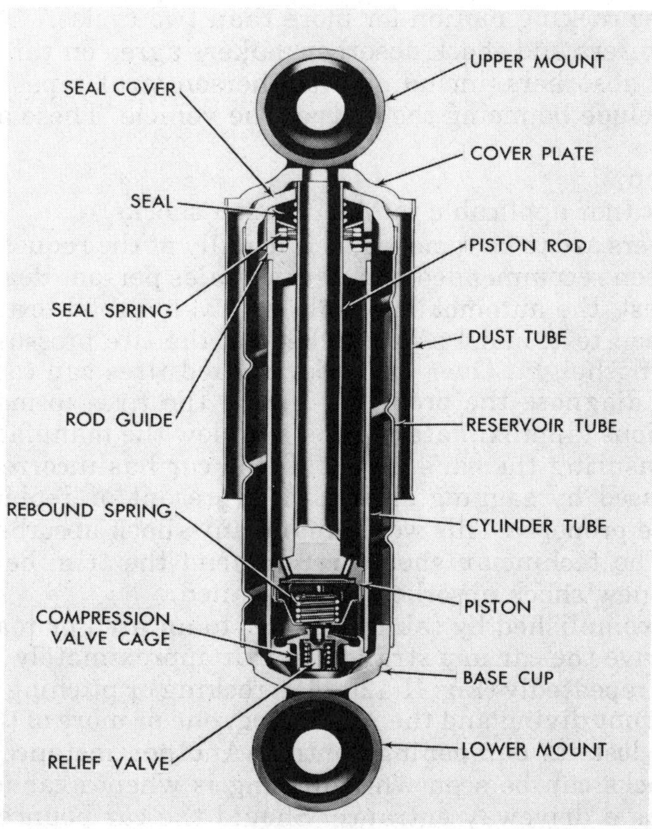

FIGURE II-118 Shock-absorbers contain oil seals with wiping edges. (Courtesy of Chevrolet Motor Division)

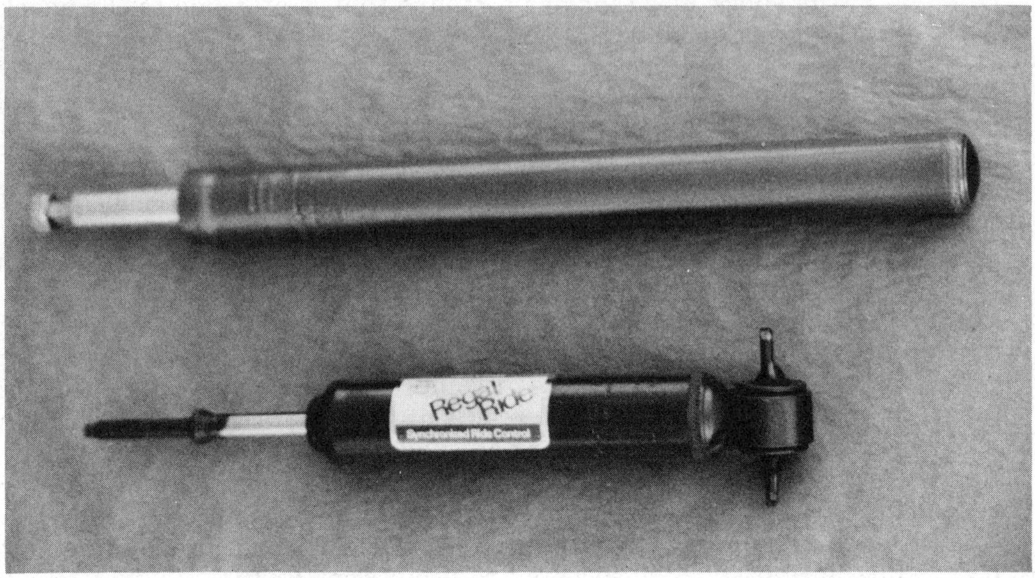

FIGURE II-119 Comparing the MacPherson strut with a regular shock.

The federal government (National Highway Safety Administration) says the following in Section 570.8 (i) of its inspection manual: "Examine shock absorbers for oil leaking from within, then with vehicle on a level surface, push down on one end of vehicle and release. Note number of cycles of free-rocking motion. Repeat procedure at other end of vehicle. The vehicle shall not continue free rocking motion for more than two cycles."

Most car manufacturers and shock absorber makers agree on three steps of diagnosis of shock absorbers, including MacPherson strut types. Notice that none of these include bouncing the ends of the vehicle. These are:

1. Road test
2. Visual inspection
3. Operational test (not applicable to MacPherson shocks)

When shock absorbers are to be removed, it is usually at the request of the car owner or it has been recommended by a service sales person. Regardless of the source of request, the automobile should be given a road test.

Just prior to the road test, make a quick check of the tire pressures and the car's trim or riding height. Over or underinflated tires can cause the driver to improperly diagnose the problem. Inflate the tires to mânufacturer's recommendations (approximately 28 psi). Follow the manufacturer's procedures when measuring the car's height. If the car has incorrect trim height, normally caused by sagging springs, the present or replacement shocks cannot operate properly. This would make any shock absorber diagnosis meaningless. The technician should recommend the trim height be corrected before any new shock absorbers are installed.

FIGURE II-120 Action of car as brakes are being applied in the road test. (Courtesy of Monroe Auto Equipment Co.)

The road test is accomplished by taking the car to an area or road with smooth pavement. Drive the car in a straight line at approximately 10 mph and apply the brakes repeatedly (Fig. II-120). If a rocking or pitching motion is felt with the car's front diving and the rear rising, one or more of the car's shock absorbers has lost its dampening control. Another instance where poor or worn-out shocks can be seen while driving is when a car is going through a dip such as a driveway entrance. Should the car bounce excessively and continue to oscillate up and down for several cycles after the brakes have been applied rapidly, poor shock condition is evident. Most often, wear and defects in the internal parts can not be seen during visual inspection. This is the major reason for the road test.

Visual inspection is the second step in proper shock diagnosis (Fig. II-121). With the car on a lift or placed on high jack (safety) stands, look specifically at each shock absorber with a droplight or flashlight. Some of the more common reasons shocks fail can be easily seen as loss of fluid (oil), broken or worn mountings, worn bushings, body damage, and bent or rusted shafts.

Fluid loss normally takes place at the top seal. The loss may be slight and will have no effect on the shock absorber operation. On the other hand, excessive oil loss will completely disable a shock absorber.

A slight oil loss will have a small amount of dust covering the oil and be confined to a small area nearest the shock absorber rod seal. In comparison, a heavy oil loss will actually look wet and oil will have run down the body of the shock in large amounts. As pointed out earlier, rear shocks having a dirt shield have to be extended fully to reveal slight oil loss. When rear shocks experience heavy oil loss, it can be seen running down the body past the shield in any position (Fig. II-122).

Further visual diagnosis should be directed at the mountings of the shocks. Broken or lost bolts and nuts or frame and/or car body damage may be responsible for ineffective shock absorber function (Fig. II-123).

The bushings at both ends of most shock absorbers are made of high quality rubber to assist in the dampening effect and to reduce road noise. The most frequent bushing problem occurs when the rubber deteriorates and is lost in small bits until it is completely gone (Fig. II-124). Although the bushings can be replaced separately, most often the shocks are also worn out. Before replacing bushings separately, the technician should make sure the shock absorbers are still good.

Should the body or tube of the shock absorber become heavily dented or the shaft or rod have a bend, the unit and its bushings should be replaced (Figs. II-125 and 126).

The last major diagnosis for shock absorber condition is the physical testing of each unit in question. If the road test revealed a poor shock or shocks,

FIGURE II-121 Visual inspection of shock absorber. (Courtesy of Monroe Auto Equipment Co.)

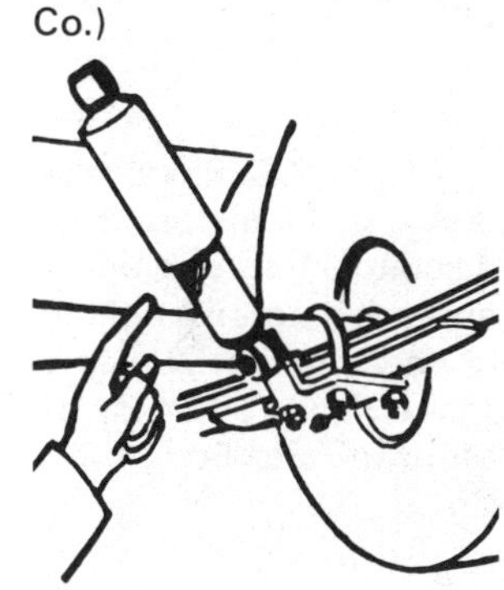

FIGURE II-122 Excessive oil leakage from shock-absorber-seal area. (Courtesy of Monroe Auto Equipment Co.)

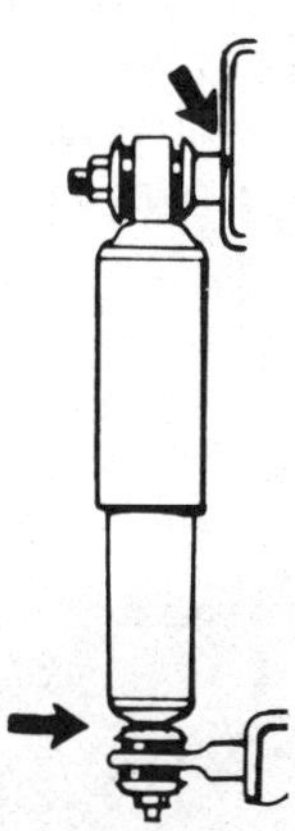

FIGURE II-123 Areas to inspect visually where mounting parts and brackets usually malfunction. (Courtesy of Monroe Auto Equipment Co.)

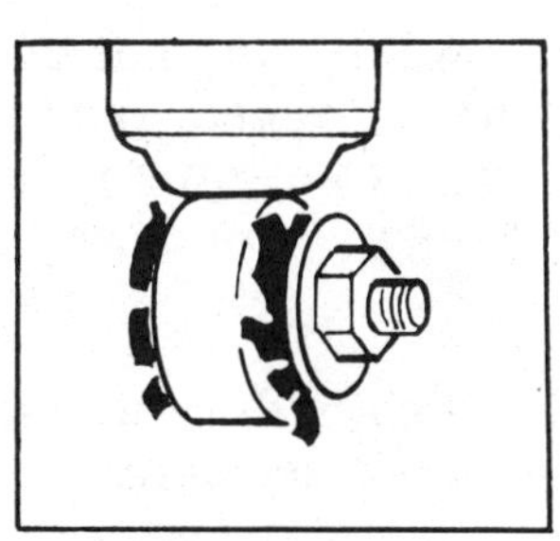

FIGURE II-124 Typical rubber-bushing deterioration. Rubber often looks shredded. (Courtesy of Monroe Auto Equipment Co.)

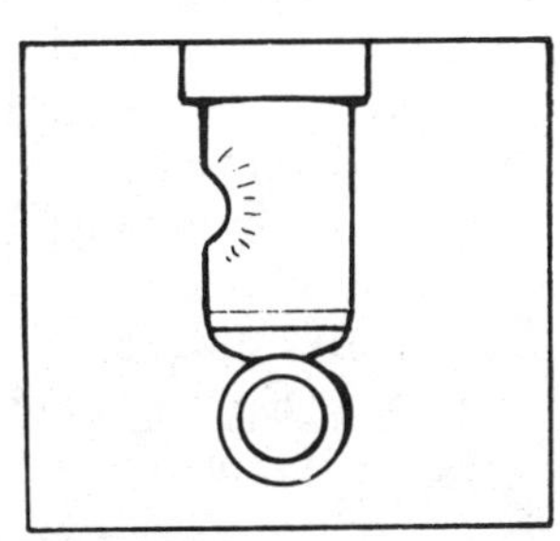

FIGURE II-125 Body damage to the shock absorber is cause for replacement. (Courtesy of Monroe Auto Equipment Co.)

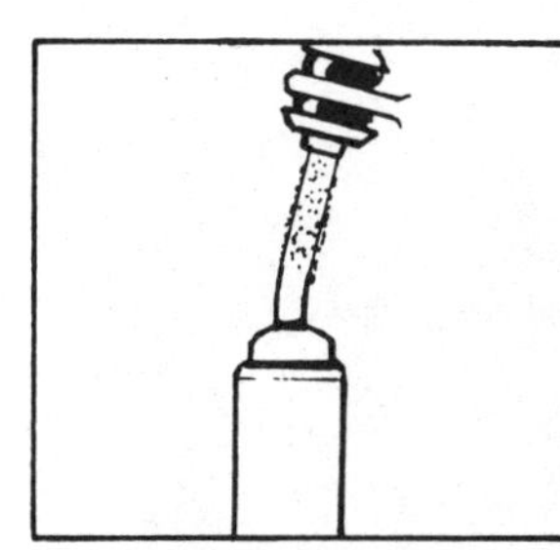

FIGURE II-126 A rusted or bent piston rod will not allow a shock absorber to function properly. (Courtesy of Monroe Auto Equipment Co.)

yet there were no visual problems, the next step is to disconnect the lower end of each suspected unit (Fig. II-127). Take hold of this lower end and work it up and down. A good shock will show resistance to this movement. Little or no resistance generally indicates the shock is worn or has completely failed. If there is any doubt about the resistance to up-and-down movement, compare it to a new unit for that same vehicle and position.

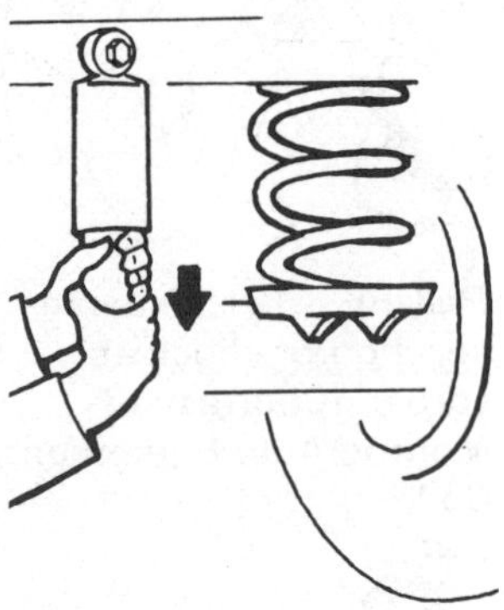

FIGURE II-127 Physically test a shock absorber by disconnecting one end and pushing and pulling to check for resistance of movement. (Courtesy of Monroe Auto Equipment Co.)

If a skip in resistance is felt in the mid-travel position, this can indicate air in the oil. Depending on the brand of shock, this unit may need replacing. Before condemning the shock, attempt to remove (purge) the air as outlined in the installation procedure.

In some applications, disconnecting the lower end of the shock absorber and testing it is nearly impossible. Therefore, remove it completely from the vehicle in order to check it properly. See installation procedure for directions.

The selection of replacement shocks may or may not be a decision the technician makes or assists in making. However, a basic knowledge of factors concerning the selection of replacement units is important to understand.

The choice of new units is generally between original equipment replacement or commonly called heavy duty shock absorbers.

Vehicle requirements should be determined before making a decision on replacement equipment:

1. Are the springs in good condition?
2. What is present trim height?
3. Over what types of roads is the car usually operated?
4. Is the car driven at average or high speeds frequently?
5. Are heavier than normal loads carried?

MacPherson strut shock absorbers are serviced in one of three ways:

1. Replacing the entire shock-strut.
2. Rebuilding the original unit with replacement parts.
3. Replacing the internal shock parts with an after-market cartridge (Fig. II-128).

The following are typical examples of symptoms, conditions of driving, and what would be a good recommendation:

1. Light sedan with large engine, automatic transmission, and air conditioning. Front springs have sagged, rear springs OK. Car is usually operated over normal roads at normal speeds. Driver complains of excessive front end bottoming and poor riding qualities. Front tires have "spotty" wear.

Recommendations:

Front: Spring assisted or heavy duty units
Rear: Original equipment replacements

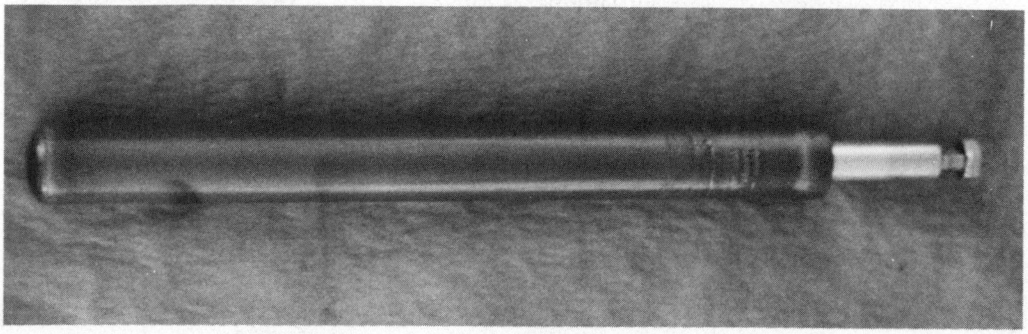

FIGURE II-128 MacPherson strut cartridge

2. Station wagon, frequently overloaded, tail drags when going into or coming out of driveways. Driver complains of instability in cross winds and poor handling. Vehicle is frequently operated over secondary roads at fairly high speeds.
   Recommendations:
   Front: Oversize heavy duty units
   Rear: Spring assisted or air assisted units.
3. Medium sized car whose driver complains of poor handling and rear wheel hop. All tires have usual wear. Car is operated on rough roads at higher than normal speeds. Owner wants a better handling car.
   Recommendations:
   Front: Spring assisted units.
   Rear: Spring assisted units.
4. Large car, tires show average wear, one shock found leaking. Car is operated on average roads at medium speeds. Owner prefers average to firm ride.
   Recommendation:
   Front: Original or heavy duty units
   Rear: Original or heavy duty units

OBJECTIVE II-H1 (manipulative)

Working alone and with the aid of reference material and Job Report Sheet II-H1, visually inspect, diagnose, and report to your instructor the condition of the front and rear shock absorbers of two automobiles. Following the report, remove and replace two front and two rear shock absorbers.

Time recommended: Twice the flat-rate time allowed in the Motors or Chilton Flat Rate manual.

Preparation:

1. Study pages 215 through 219.
2. Study manufacturer's or general repair manual for specific automobiles assigned.

Shock absorber service is a procedure of removal and replacement. Regardless of the make, model, and position of the shock absorber, follow the directions enclosed with the new shock absorber unit.

The following is a list of general precautions related to the placement of shocks.

1. Use adequate lighting.
2. Do not "hang" the rear axle and wheels on cars with coil springs or air springs. This means do not use a frame contact hoist on cars with those suspension systems. Use an axle or wheel-contact type of hoist.
3. Compare the new units to be installed with the present ones to see if they are roughly the same. Also, check to see that all new mounting parts are included.
4. Do not damage mounting pins or studs that have to be used to mount the new shocks. Use penetrating oil or a nut splitter to remove stubborn nuts.
5. Follow the directions printed on the package or separate instruction sheet.

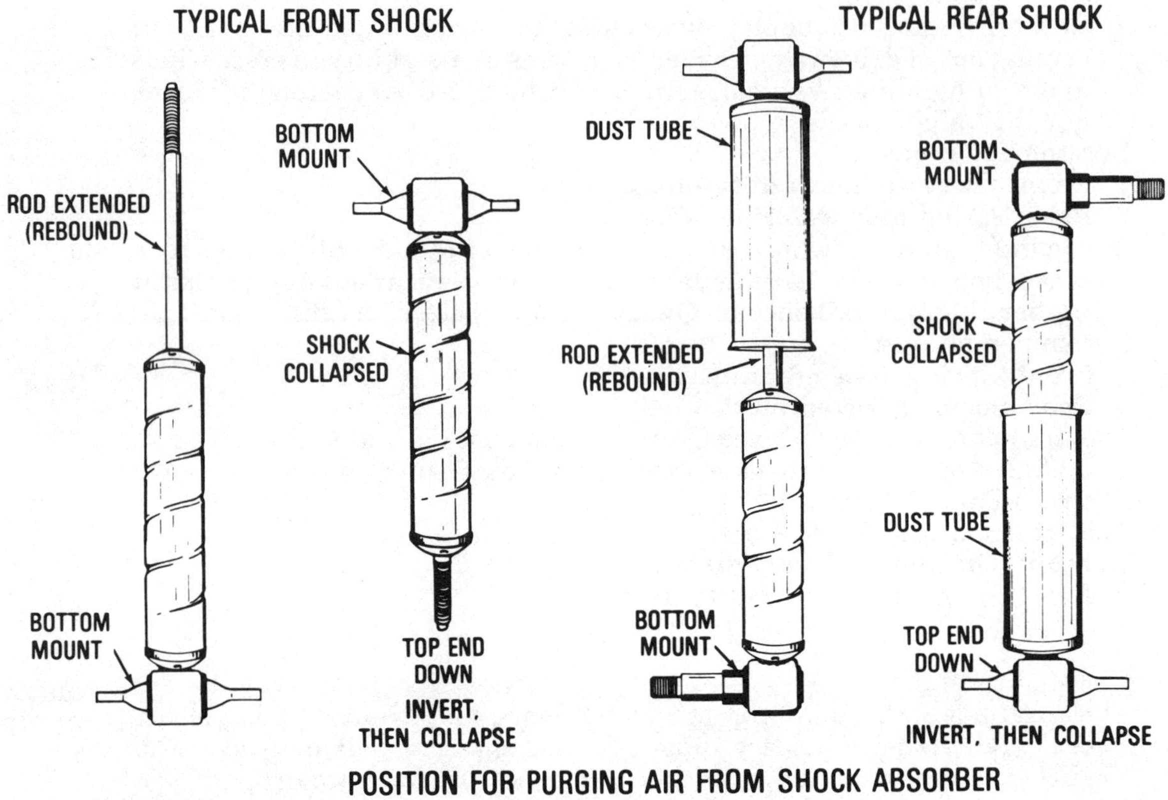

FIGURE II-129 Positions and movements in purging a shock absorber of air. (Courtesy of Chevrolet Motor Division)

6. Purge the shock to prevent it from being installed in an aerated (air in the oil) condition. Do this by holding it upright (normally installed position) and extend it all the way (Fig. II-129). Turn the shock upside down and compress it until it is completely collapsed. Extend and compress the new unit four or five times. Remember, extend right side up, compress upside down.
7. Do not grip the polished piston rod with any tool during installation. This will cut the rod seal and leakage will take place shortly.
8. One of the most important factors in shock absorber installation is to tighten the mounting bolts and brackets only after the weight of the vehicle is on the tires (Fig. II-130). If this is not done, the mounts are not centered, and the rubber in the bushings is compressed in the wrong position. This will cause the bushing rubber to deteriorate very rapidly.
9. Do not use any lubricant on the rubber parts other than soapy water.
10. Do not overtighten stem mountings. Rubber cushions should not bulge outside of metal washers.
11. After tightening the mounts with the car on the ground, use a flashlight and check to see that the new shocks are clear of any other parts on the chassis. These might be brake lines, tail pipes, mufflers, or others.
12. Last but not least, check tire pressure and inflate tires to manufacturer's recommendations.

Replacing the MacPherson strut shock absorber may or may not include removing the strut unit from the vehicle. If the strut unit is removed, this is the procedure to follow, in accord with Coil Spring Service Objective II-D5.

1. After the spring and spring plate are removed, use a spanner wrench to unscrew the top retaining ring from the shock tube (Fig. II-131).
2. Remove the shock rod and all other parts from the tube.
3. Drain the oil and save it. Check inside the tube for grit or debris.
4. Replace the internal seals on the rod and piston (Fig. II-132) *or* insert a new replacement cartridge.
5. If new internal parts are used, pour the specified amount of new oil into the cartridge. In other cases, pour the old oil (⅛ cup or 0.03 ml.) back into the cartridge.
6. Install the piston rod with new parts *or* a new cartridge with its centering collar, if required.
7. Install the locking ring and tighten it securely.
8. Purge the rebuilt shock or new cartridge of air (Fig. II-129).
9. Install the spring on the strut and replace it in the vehicle as outlined in Objective II-D5.

Note: *Be certain to bleed the brakes if the brake line was disconnected.*

FIGURE II-130 Tighten shock-absorber mounts with the weight of the vehicle on its tires. (Courtesy of Chevrolet Motor Division)

FIGURE II-131 Using wrench to remove top retainer ring.

FIGURE II-132 Replacing seal

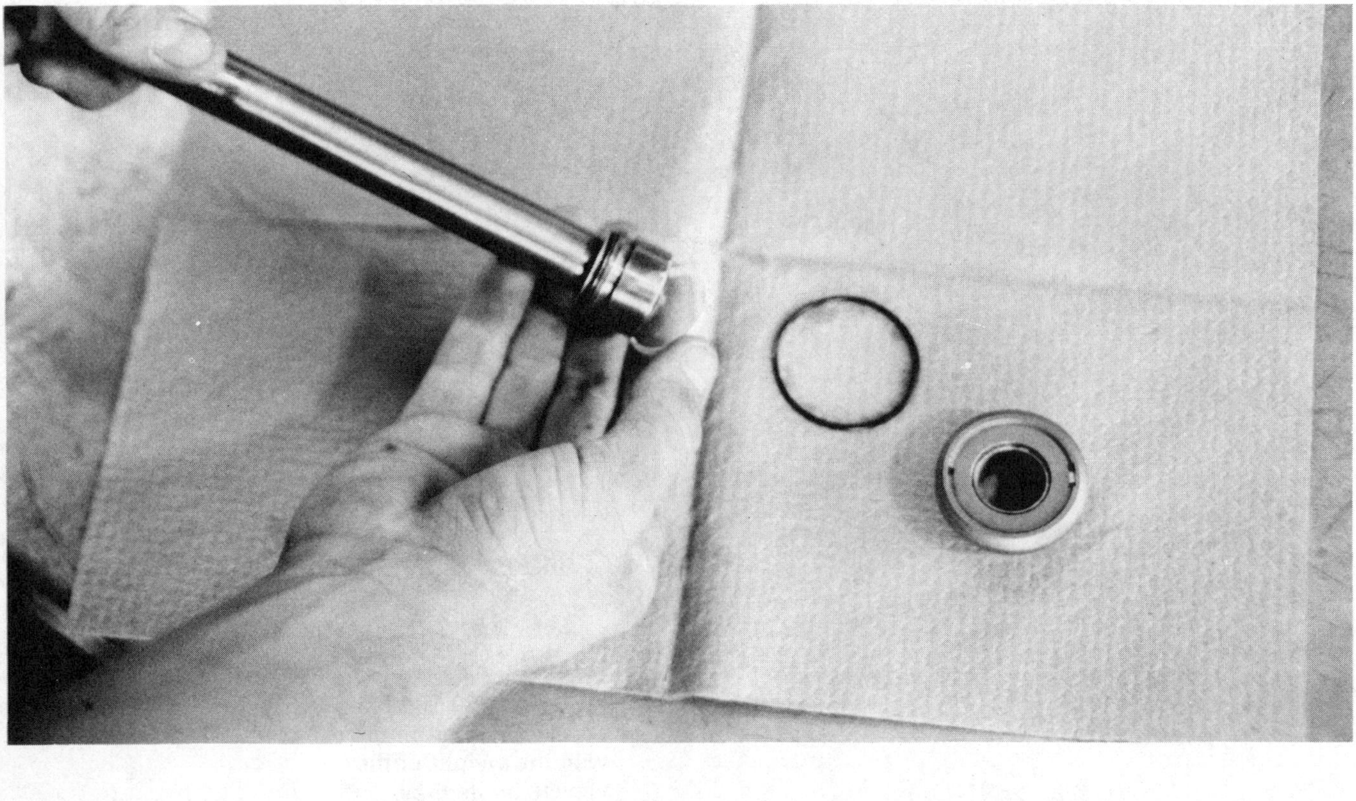

JOB REPORT SHEET II-H1

SHOCK ABSORBER SERVICE

Student name____________________ Date________________

Make of car__________ Model__________ Year__________

Time started__________ Time finished__________ Total time__________

Flat-rate time ____________________

Checkup Questions:

1. How much vehicle weight do standard shocks support?____________________

2. Shock absorbers are purged to remove: ____________________

3. Explain why it is important to tighten shock mounting bolts after the weight of the vehicle is on the tires: ____________________

4. Can oil be used to lubricate rubber parts if soapy water or rubber lube cannot be found? _____

DATE COMPLETED____________________ INSTRUCTOR'S SIGNATURE ____________________

TEST II-H

Name________________________________Date ________________________

Directions: Select the best answer to each question and circle the correct letter at the left.

1. The basic principle by which shock absorbers operate is:
   I. Conversion of the energy of movement into heat
   II. Dampening hydraulic movement
      A. I only
      B. II only
      C. Both I and II.
      D. Neither I nor II
2. Which of the following statements about standard shock absorbers is true?
   I. Shock absorbers dampen spring oscillations.
   II. Shock absorbers absorb most of the road shock
      A. I only
      B. II only
      C. Both I and II.
      D. Neither I nor II
3. When the wheel moves away from the body of a vehicle, this action is called:
   I. Extension
   II. Rebound
      A. I only
      B. II only
      C. Both I and II.
      D. Neither I nor II
4. When the body of a vehicle moves toward the wheels, this action is called:
   I. Jounce
   II. Compression
      A. I only
      B. II only
      C. Both I and II.
      D. Neither I nor II
5. Shock absorbers have rubber mounts at both ends to:
   I. Make replacement easier
   II. Absorb some of the road noise
      A. I only
      B. II only
      C. Both I and II.
      D. Neither I nor II
6. On coil front suspensions the shock absorbers are nearly always mounted:
   I. Alongside the front spring
   II. To the lower control arm
      A. I only
      B. II only
      C. Both I and II.
      D. Neither I nor II

7. For a shock to be replaced because of fluid loss it must:
   I. Show heavy (wet) oil loss
   II. Show *any* evidence of oil loss whatsoever
      A. I only
      B. II only
      C. Both I and II.
      D. Neither I nor II
8. When a vehicle continues to rock and pitch after going through a dip, it probably:
   I. Needs the springs replaced
   II. Has too much tire pressure
      A. I only
      B. II only
      C. Both I and II.
      D. Neither I nor II
9. When physically testing each shock absorber, which of the following is a reason(s) for replacement?
   I. Little or no resistance during compression or rebound movements
   II. A skip in resistance in the mid-position
      A. I only
      B. II only
      C. Both I and II.
      D. Neither I nor II
10. When installing new shock absorbers, it is very important to:
   I. Tighten mounting bolts only after the weight of the vehicle is resting on the ground
   II. Use plenty of penetrating oil to lubricate rubber bushings
      A. I only
      B. II only
      C. Both I and II.
      D. Neither I nor II

DATE COMPLETED ______________________ INSTRUCTOR'S SIGNATURE ______________________

unit III

Steering

Steering linkage and steering gears are discussed in this unit. Individual components and their relationships to front suspension are explained as are service procedures for manual and power steering systems.

OBJECTIVE III-A (written test)

After studying the preparation assignment you will be expected to rectly answer a minimum of seven out of ten multiple choice quest in a written test without reference materials on the following stee linkage topics:

1. Steering arms
2. Tie rod ends
3. Tie rod adjusting sleeves
4. Center link (drag link, relay rod)
5. Idler arm assembly
6. Pitman arm

Preparation:

Study pages 223 through 241.

STEERING LINKAGE DESIGNS

Steering linkage designs for modern passenger cars and light trucks fall into two categories:

1. Parallelogram
2. Rack and pinion

PARALLELOGRAM

Most mid- and full-size American automobiles use the so-called *parallelogram* steering linkage. It connects both front wheels to the steering gear through a series of parts as follows (Fig. III-1):

1. Two steering arms are attached to the wheels by means of the backing plates which are part of the steering knuckle.
2. Two sets of tie rod ends, each consisting of an outer and an inner tie rod end, are connected to an adjusting sleeve which is positioned between the two. The outer tie rod end is connected to the steering arm, while the inner tie rod end is connected to the center link.

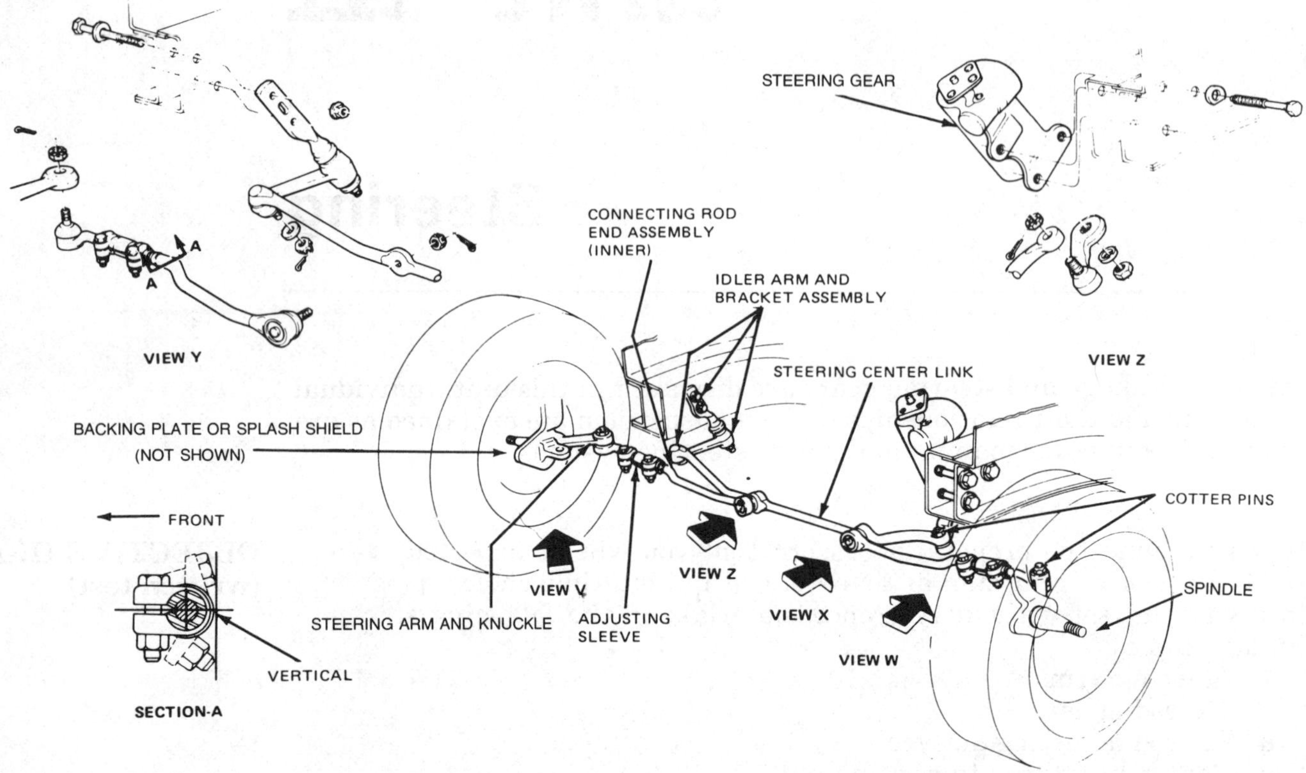

FIGURE III-1 This is a typical parallelogram steering linkage (steering damper optional). (Courtesy of Cadillac Motor Car Division)

3. The center link (see Fig. III-1) connects with the two inner tie rod ends. It has an idler arm and a pitman arm attached to it.
4. The idler arm is attached to the right side of the center link and to the frame.
5. The pitman arm is attached to the left side of the center link and to the steering gear box. The pitman arm and idler arm are always parallel to each other and move through equal arcs.

RACK AND PINION

The steering linkage on many imported and small domestic vehicles is the rack and pinion design. This system combines the steering linkage directly with the steering gears to provide a simple, lightweight steering mechanism (Fig. III-2). The major components are:

1. Steering arms (there are 2) are joined with the steering knuckle and turn the wheels to the desired angle.
2. Two tie rods with tie rod ends connect the steering arms to the rack.
3. The rack is a steel bar with teeth that mesh with the pinion gear to give lateral movement to the steering system.
4. The pinion gear is part of the steering gears that connect the rack with the steering column.

STEERING ARM

The steering arm is usually an integral part of the steering knuckle forging. It turns the wheels under pressure that is applied through the tie rods.

The steering arm has a predetermined bend away from the tire which sets one of the steering-geometry angles, known as *turning radius* or *toe-out on turns* (Fig. III-3). This angle and the condition of the steering arms are checked during the alignment procedure of the front end. The only way to correct a defective steering arm is to replace it. *(Do not try to bend the steering arms in order to adjust the turning radius.)*

TIE ROD ENDS

There are two to four tie rod ends on a modern vehicle, depending on its steering linkage design (Fig. III-4). The rod ends are small ball joints that allow the suspension system to move up and down while controlling the lateral movement of the linkage. Clamps or lock nuts are used to hold the threaded end, but the stud end is held in a tapered hole by a nut and cotter pin.

The types of tie rod ends in use today can be divided into two basic categories: (1) those that use a coil spring preload and show vertical clearance when compressed and (2) those that do not use a coil spring for preload and show no vertical clearance when in operative condition.

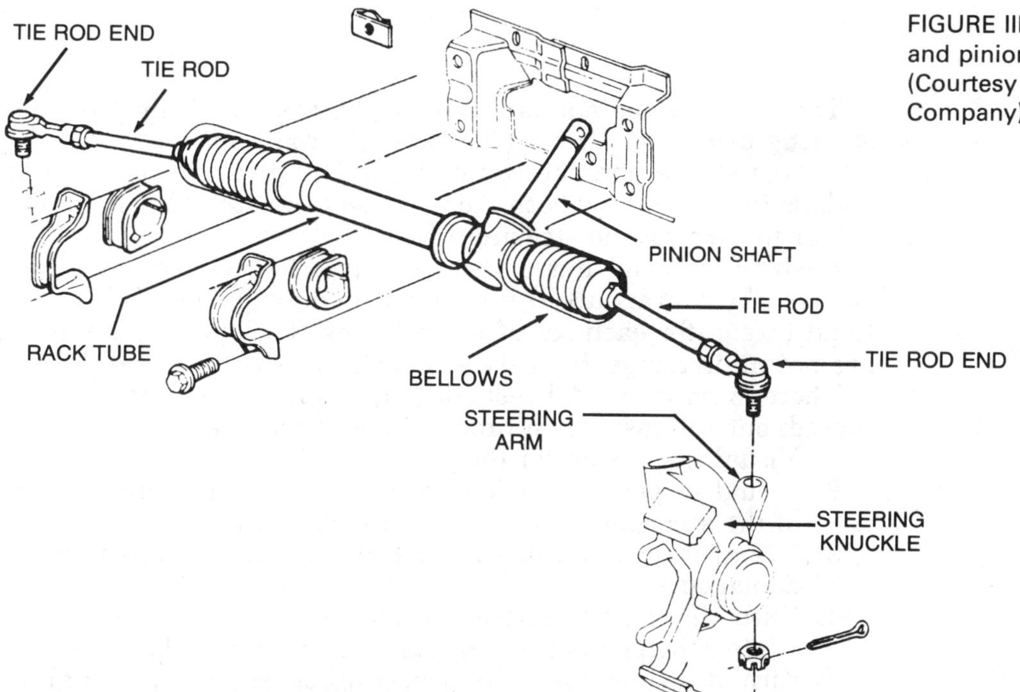

FIGURE III-2 Manual rack and pinion steering gear. (Courtesy of Ford Motor Company)

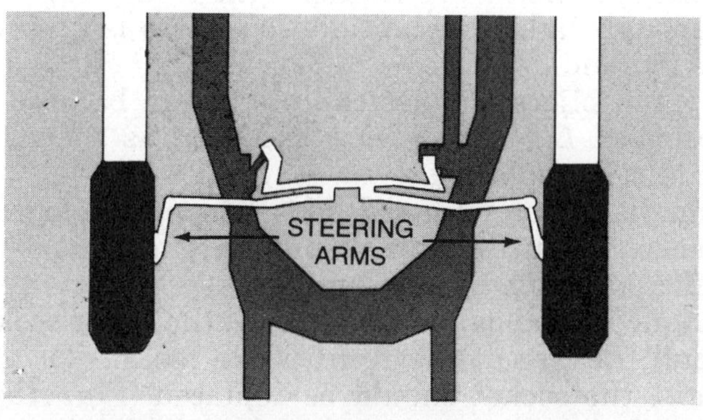

FIGURE III-3 Drawing of steering arms (note the predetermined bend in each one). (Courtesy of Chrysler Motors Corp.)

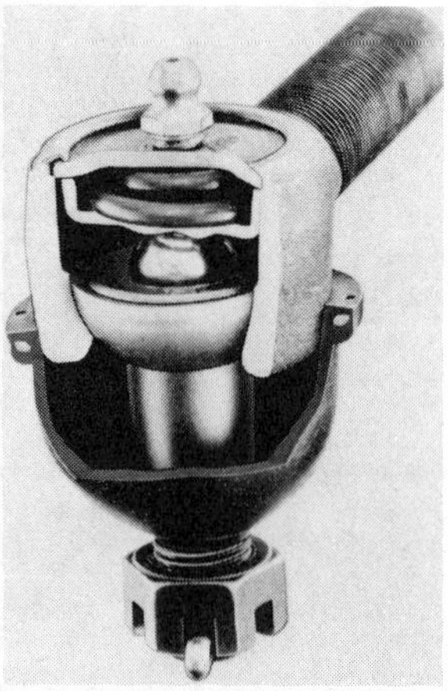

FIGURE III-4 Cutaway of typical tie rod end, showing ball stud and socket. (Courtesy of Moog Automotive, Inc.)

FIGURE III-5 Threaded adjusting sleeve (connects the two tie rod ends) allows for toe-in adjustment. (Courtesy of Chevrolet Motor Division)

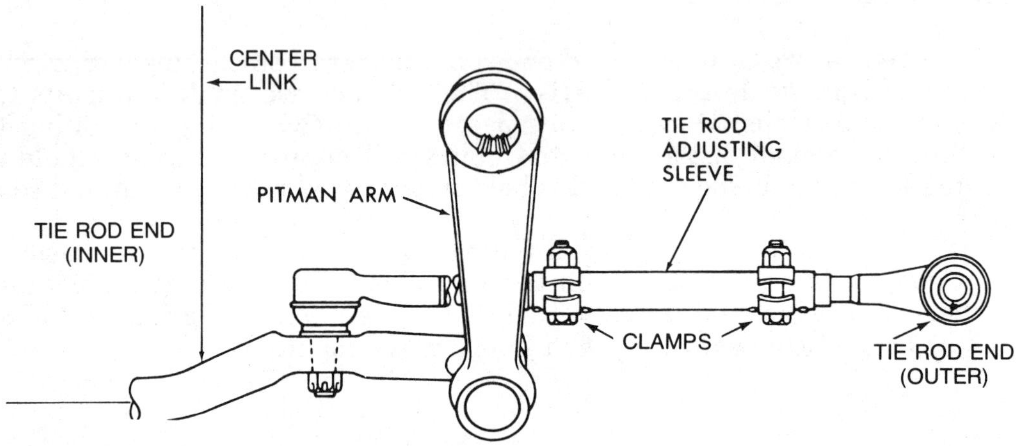

The outer tie rod end is connected between the adjusting sleeve and the steering arm (Fig. III-5). The inner tie rod end is connected between the adjusting sleeve and the center link. The primary function of the tie rod ends is to transmit to both front wheels the steering motions made by the driver in turning the steering wheel.

Each tie rod end has a threaded arm for connection into an adjusting sleeve. These arms are threaded with one right-hand thread and one left-hand thread for each set of tie rod ends. The inner and outer tie rod ends are not interchangeable between each other or from side to side.

There is no universal test for determining when a tie rod should be replaced, but the inspection process should include the following:

1. Visual inspection for damaged threads.
2. Visual inspection for bent or distorted tie rod ends (compare one side of the steering system to the other).
3. Visual inspection for rubber boots or closure caps that are missing or damaged.
4. With the vehicle setting on the floor or on an alignment rack, and while a helper is turning the steering wheel back and forth about ⅛ turn at a time, visually inspect all tie rod ends (and all other suspension joints) for perceptible looseness. Use a drop light or flashlight to provide yourself with enough light. Perceptible looseness is grounds for replacement of a tie rod end.
5. A large pair of water pump pliers or dust cap pliers can be used to check for excessive looseness. *Lightly* squeeze the joint as shown in Fig. III-6. If loose, it should be replaced.
6. Rack and pinion systems have outer tie rod ends like parallelogram types and should be checked as outlined in Steps 1 through 5.

Inner tie rod ends on a rack and pinion system are checked by lifting the front wheels off the ground. Squeeze the bellows boot until the inner socket can be felt, then push and pull the tire at the front while feeling for any looseness. If there is looseness, the socket should be replaced (Fig. III-7).

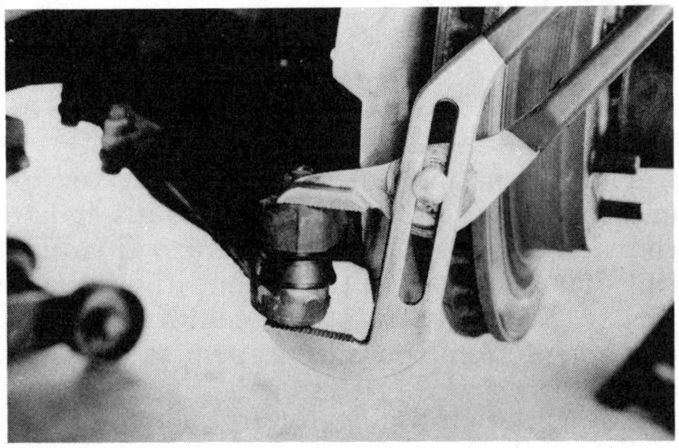

FIGURE III-6 A large pair of water pump pliers can be used to check for looseness in the tie rod end. Squeeze lightly while looking for looseness.

ADJUSTING SLEEVES

The adjusting sleeve is threaded from both ends with a left- and a right-hand thread. When the sleeve is turned, the tie rod ends are both moving either in or out of the sleeve. This allows for a change of distance between the two tie rod end centers (Fig. III-8). The only thing that needs to be checked in the adjusting sleeve is the condition of the clamps and the threads inside the sleeve. If too much pressure is applied to the sleeve while adjusting the distance between the two tie rod end centers, the sleeve may collapse in the center. Replacement of the adjusting sleeve is the only repair to be made.

FIGURE III-7 Check the inner socket by squeezing the bellows boot until the inner socket can be felt. Then push and pull the tire at the front. If looseness can be felt, replace the socket. Also, if any openings are found in the bellows boot, replace it.

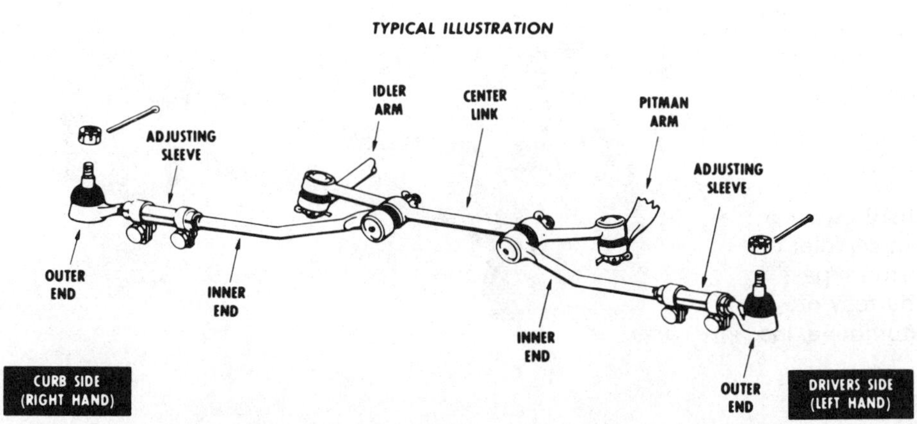

FIGURE III-8 Tie rod ends are connected between the center link and the steering arms by adjusting sleeves on each side. (Courtesy of Moog Automotive, Inc.)

CENTER LINK

The center link may also be called the *drag link* or *relay rod.* The term "center link" will be used here.

The center link has one end connected to the pitman arm and the other end connected to the idler arm. Both ends usually have a joint (similar to the tie rod ends) that allows for a turning motion (Fig. III-9). The center link has two tapered holes that allow the two inner tie rod ends to be attached. The purpose of the center link is to evenly transmit the turning movement from the pitman arm to both front wheels, through the tie rod ends.

The joints on the center link should be checked in the same way as the joints on the tie rod ends.

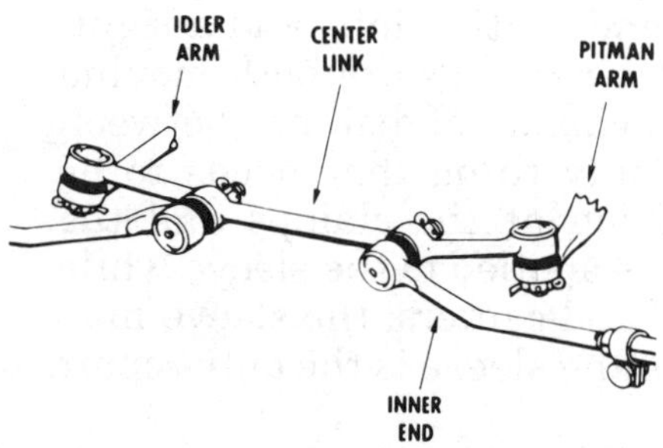

FIGURE III-9 The center link is connected between the idler arm and pitman arm, with both tie rod ends attached. (Courtesy of Moog Automotive, Inc.)

IDLER ARM

The idler arm is connected between the frame and the passenger side of the center link (Fig. III-10). Its purpose is to support and hold the passenger side of the center link in its proper horizontal plane. If the center link is allowed to move into a different plane as it turns back and forth, the toe-in (distance between the center of the tread at the front of the tire) would be in constant change. The amount of change would depend on the position and amount of pressure being applied to the front wheel.

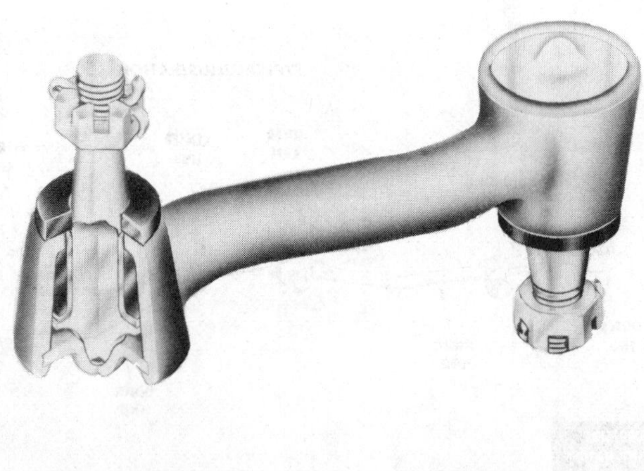

FIGURE III-10 Cutaway view of typical idler arm with ball stud-type joints. (Courtesy of Moog Automotive, Inc.)

The idler arm also adds some friction to the linkage to help the car track, and, ideally, it will have some resiliency, to help dampen road shock.

Idler arms must be checked for worn sockets or deteriorated bushings. This can be done by grasping the idler arm and applying an upward and downward pressure of approximately 25 lbs. (10 kg). If the idler arm moves up or down more than ¼ inch (6.3 mm) with little pressure, it should be replaced. Keep in mind, however, that even a new idler arm can be forced to show movement when excessive pressure is applied. Sample of different idler arms are shown in Fig. III-11.

Service work on the idler arm may require a complete new idler arm or only the replacement of the bushings. This will depend on the design of the arm itself.

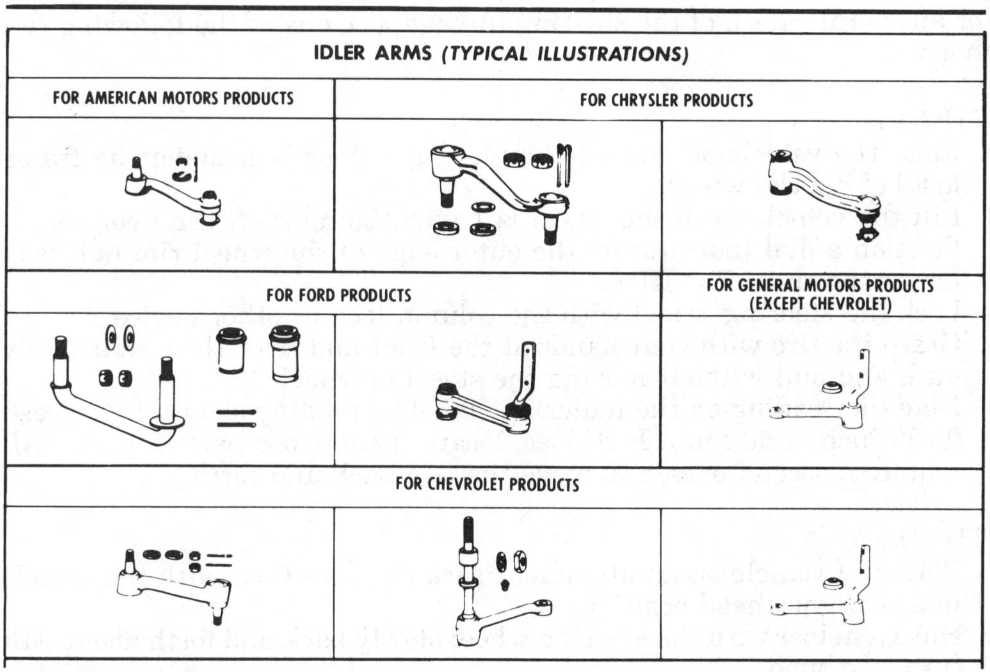

FIGURE III-11 Samples of idler arm designs used on American built cars. (Courtesy of Moog Automotive, Inc.)

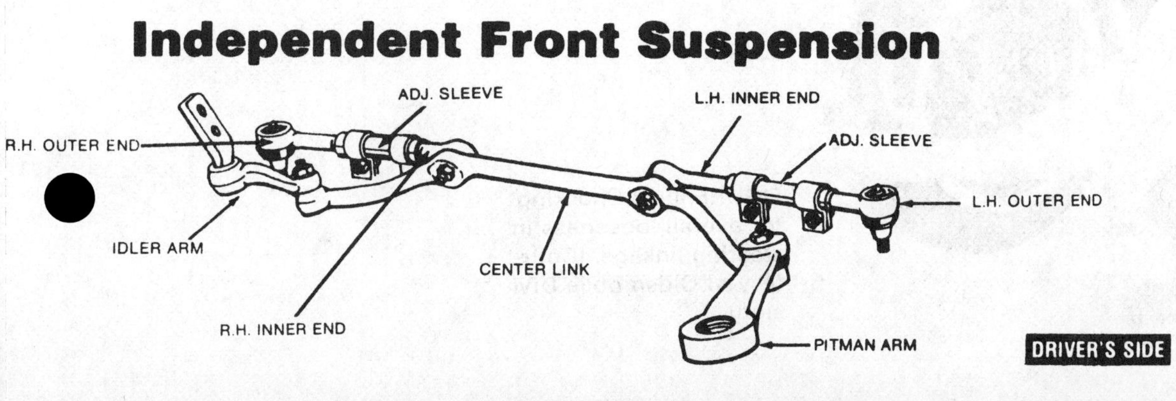

FIGURE III-12 The pitman arm is connected between the steering gear pitman shaft and the center link. (Courtesy of Oldsmobile Division)

PITMAN ARM

The pitman arm connects the steering gear to the center link. When the steering wheel is turned, the pitman arm moves the steering linkage (Fig. III-12). The pitman arm is the most heavily stressed part in the steering-linkage system.

On many late model vehicles, the pitman arm will have ball-joint sockets or rubber bushings. The sockets should be checked by the method described for tie rod ends. The rubber bushings should be checked for deterioration or looseness.

COMPLETE STEERING LINKAGE INSPECTION

For an overall check of the steering linkage, use one of the following two methods.

METHOD 1

1. Raise the vehicle on one side by placing a floor jack under the frame just behind the wheel.
2. Lift the vehicle until the wheel is 1 inch (25 mm) off the floor.
3. Position a dial indicator on the outer edge of the wheel rim half way up, as shown in Fig. III-13.
4. Lock the steering wheel with the column lock or other device.
5. Grasp the tire with your hands at the front and rear, then alternately push and pull without moving the steering wheel.
6. Note the reading on the indicator dial. The reading should not exceed 0.180 inch (4.58 mm). If it does, locating the loose part or parts will require a second person to move the tire back and forth.

METHOD 2

1. Place the vehicle on an alignment rack or clean floor with the wheels in a straight ahead position.
2. Have a helper turn the steering wheel slowly back and forth about ⅛th turn at a time.
3. With a drop light or flashlight, look for noticeable looseness at the linkage joints.
4. If the looseness cannot be eliminated by the tightening of the attaching nuts, new parts will have to be installed.

FIGURE III-13 Checking for overall looseness in steering linkage. (Courtesy of Oldsmobile Division)

OBJECTIVE III-A1 (manipulative)

Working as a team with one other student, you will remove, replace, and adjust, where necessary, at least two of the steering components listed below. Each job will be done in accordance with manufacturer's recommended procedure and specifications. Time recommended: Twice the flat-rate time listed in the Motors or Chilton Flat-Rate manual.

1. Replace steering arms
2. Replace tie rod ends, inner or outer
3. Replace tie rod adjusting sleeve
4. Replace center link (drag link or relay rod)
5. Replace idler arm assembly
6. Replace pitman arm

Preparation:

1. Review pages 223 through 230.
2. Study pages 231 through 243.
3. Study manufacturer's or general repair manual procedure for the specific automobiles to which you are assigned.
4. Carefully observe instructor's demonstration on steering component replacement procedure.

Materials and Tools Needed:

1. Diagonal pliers
2. Ball peen hammer (possibly two)
3. Tie rod-end puller
4. Box/open end combination wrench set
5. Ratchet and socket set
6. Torque wrench
7. New cotter pins (select proper diameter and length)
8. Jack or hoist
9. Jack (safety) stands
10. Chassis grease gun
11. Wheel bearing packer
12. Hub-cap and dust-cap remover
13. Lug wrench
14. Channel lock (water pump) pliers
15. New grease seals

STEERING ARM REPLACEMENT

PREPARATION

Steering arm replacement is seldom necessary except when visual inspection reveals obvious damage or when front end alignment checks indicate improper turning radius angles. The steering arm is usually an integral part of the steering knuckle forging, however some vehicles have a removable steering arm.

To replace the integral type steering arm-knuckle follow the procedures in Unit II under Spring Removal. For replacement of the removable type, the following general steps can be used.

REMOVAL OF STEERING ARMS

1. Raise the vehicle with a jack or hoist and place jack (safety) stands under the frame or front crossmember.
2. Remove the wheel and rotor or drum.
3. Remove any cotter pins or safety wire that secures the steering arm to the knuckle nuts and bolts.
4. Disconnect the tie rod end from the steering arm by removing the cotter pin and loosening the nut one or two turns.
5. If a puller is available, attach it to the steering arm over the stud as shown in Figure III-14. Turn the screw shaft of the puller until it is tight. Then strike the taper area of the steering arm with a hammer to separate the two pieces.

Note: *Do not strike the tie rod end stud.*

6. A hammer may be used in the manner shown in Fig. III-15 when the right puller is not available. Always use two hammers, holding one against the steering arm and striking with the other on the opposite side. Be sure to leave the loosened stud nut in place to protect the stud threads from missed hammer blows.
7. A tie rod end fork may be used, but with this method, damage nearly always occurs to the rubber boot on the tie rod end (Fig. III-16). Replacement of the steering arm is a reversal of the removal process.

A few tips to remember are:

1. Care should be taken to ensure that the hole, ball stud, and nut are free of dirt and grease before tightening the nut.
2. Always use a torque wrench when assembling front suspension parts.
3. Always use new cotter pins on castellated nuts.
4. Double check all parts and their condition.

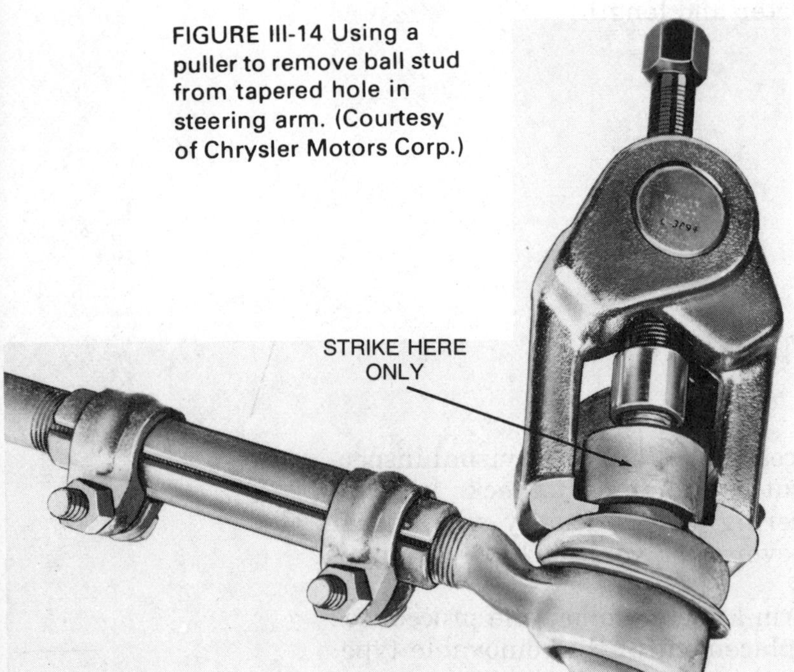

FIGURE III-14 Using a puller to remove ball stud from tapered hole in steering arm. (Courtesy of Chrysler Motors Corp.)

FIGURE III-15 Remove tie rod end ball stud by using one hammer to back up steering arm and the other to strike the area of the tapered hole. (Courtesy of Chevrolet Motor Division)

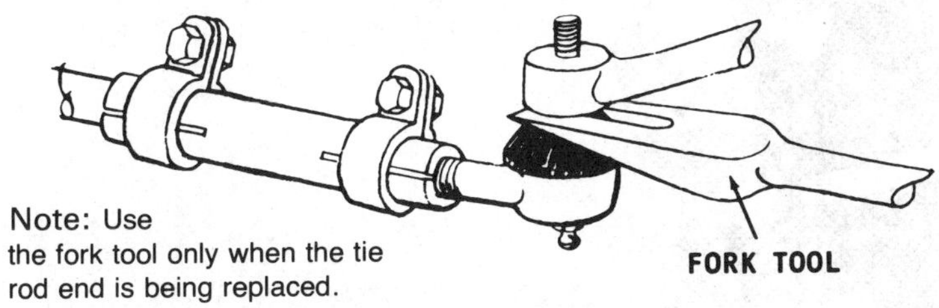

FIGURE III-16 Tie rod end fork and a hammer will remove the ball stud from the tapered hole. (Courtesy of Moog Automotive, Inc.)

TIE ROD END AND ADJUSTING-SLEEVE REPLACEMENT

PREPARATION:

Tie rod ends on most automobiles are self-adjusting for wear and require no attention in service other than periodic lubrication and occasional inspection to see that ball studs are tight (Fig. III-17). *Replacement of tie rod ends should be made when excessive up-and-down movement is evident.* Loss of motion or stiffness in turning would also indicate tie-rod replacement is necessary.

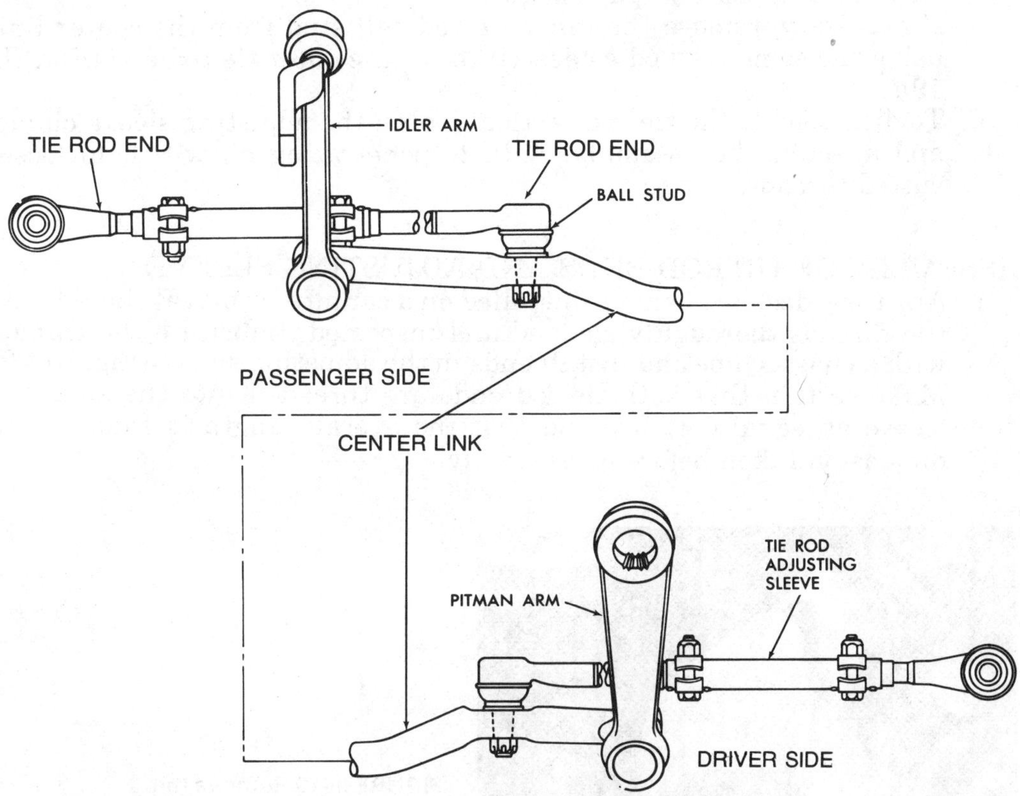

FIGURE III-17 Chevrolet tie rod assembly for both sides of the automobile. (Courtesy of Chevrolet Motor Division)

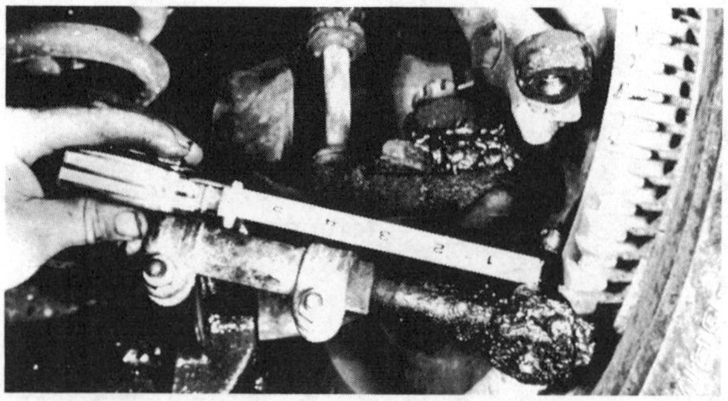

FIGURE III-18 Measure overall length of tie rod asembly before removal from car. (Courtesy of Moog Automotive, Inc.)

REMOVAL OF TIE ROD-END ASSEMBLY

The procedure for removal of the tie rod assembly should follow the basic steps listed below:

1. Raise the vehicle with a jack or hoist and place jack (safety) stands under the frame.
2. Measure and record the dimension of each tie rod from the adjusting sleeve to the end (Fig. III-18). Also measure and record the dimensions of tie rod and adjusting sleeve assembly.
3. Remove the cotter pins from the ball studs, then remove the castellated nuts.
4. Separate the tie rod end from the steering arm by using one of the methods discussed on page 232.
5. If necessary, remove the inner tie rod ball stud from the center link, using the same method as described for the outer tie rod end (Fig. III-19).
6. To disassemble the tie rod section, loosen the adjusting sleeve clamps and unscrew the assembly. A little penetrating oil will help loosen rusted threads.

ASSEMBLY OF TIE ROD ENDS AND ADJUSTING SLEEVE

1. Any threaded part being reinstalled on a car after removal should have the threads thoroughly cleaned and inspected. Lubricate the threads with a chassis lube and install ends in the adjusting sleeve (Fig. III-20). Make certain that both tie rod ends are threaded into the adjusting sleeve an equal distance and that the overall length is equal to the dimension taken before disassembly.

FIGURE III-19 Remove the inner tie rod end ball stud from center link, using the same methods as for the outer tie rod end. (Courtesy of Oldsmobile Division)

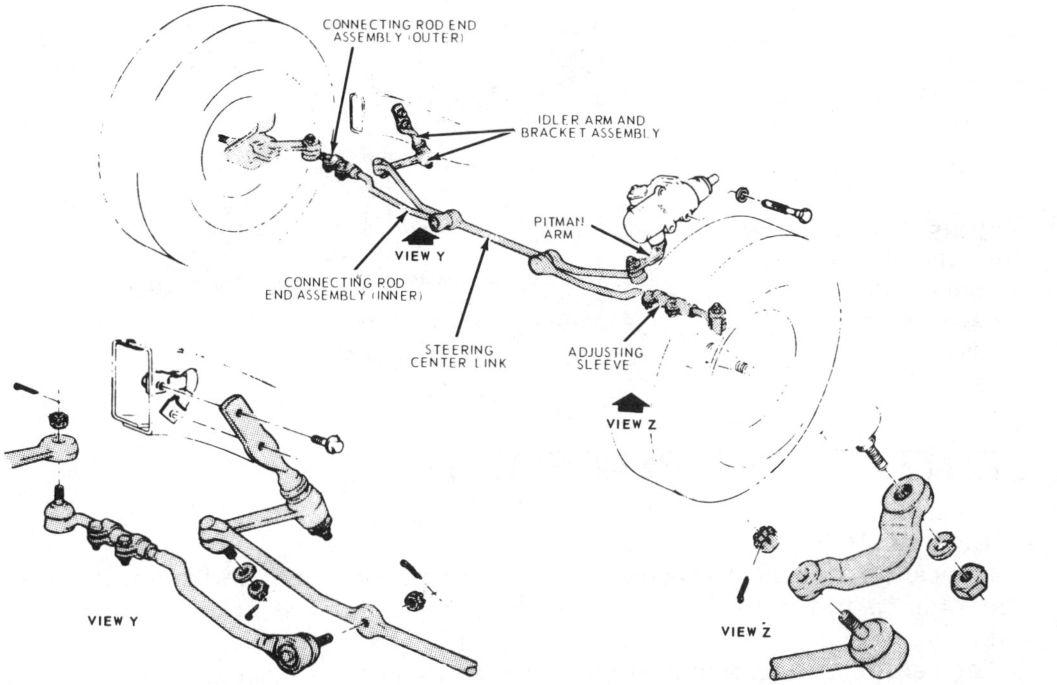

FIGURE III-20 Steering linkage for a Lincoln Continental. (Courtesy of Ford Customer Service Division)

2. Inspect threads on ball studs and ball stud nuts. If damaged or dirty, clean or replace as necessary. Install new rubber boots or seals on ball studs if needed.
3. Install ball studs in the steering arms and the center link. Torque the ball stud nuts to manufacturer's torque specifications. Always advance nuts just enough beyond torque specifications to insert cotter pins.
4. Lubricate the tie rod ends before removing the automobile from the hoist and adjust toe-in to manufacturer's specifications.

Caution: *Before tightening the tie rod adjusting sleeve clamp bolts:*

A. *Check for manufacturer's recommended position of the clamps. Specific positions are required to prevent the clamps from making contact with other suspension or frame components possibly causing erratic steering control.*
B. *Place the clamps between the locating dimples at each end of the sleeve.*
C. *Maintain the relationship of the clamp slot with the slit in the sleeve to prevent damage to the sleeve. The slot in the sleeve may be in any position except at the edges of the clamp jaws.*
D. *Center all the tie rod end studs in their sockets (Fig. III-23).*

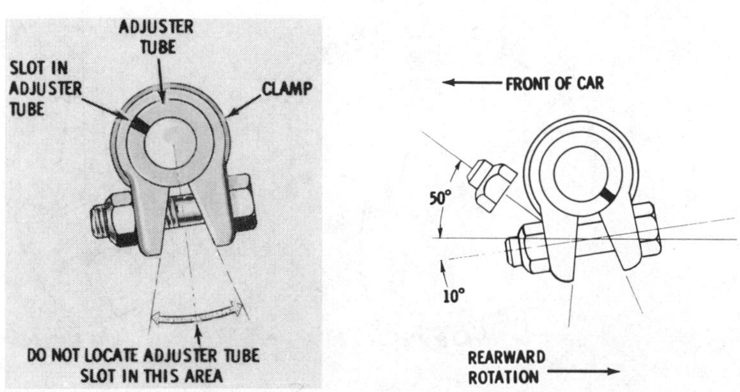

FIGURE III-21 Positioning adjusting sleeve clamp should follow manufacturer's recommendations. (Courtesy of Oldsmobile Division)

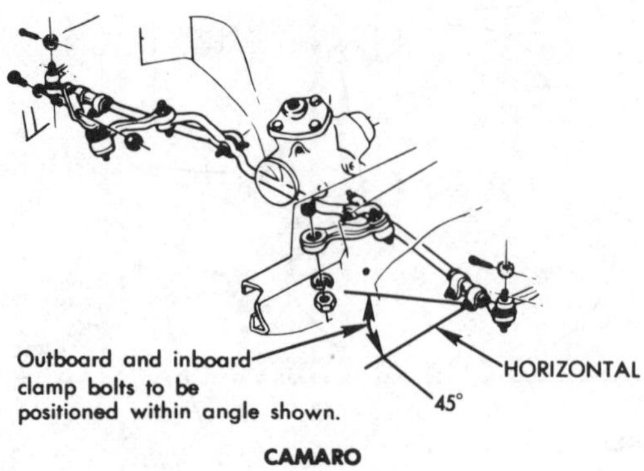

FIGURE III-22 Sample of adjusting sleeve clamp positions. (Courtesy of Chevrolet Motor Division)

CENTER LINK REPLACEMENT

PREPARATION:

The center link is connected between the pitman arm and the idler arm. The inner tie rod end for each side of the car is usually attached to the center link.

The center link is replaced when there is obvious damage such as bent or broken ends or unusual differences in shape from side to side.

REMOVAL OF CENTER LINK

1. Raise front wheels off floor with hoist or jack.
2. Place jack (safety) stands under frame or lower control arms.
3. Remove and discard cotter pins from ball-stud and/or bushing-bolt nuts to include inner tie rod ends, pitman arm, and idler-arm connections.
4. Remove the inner tie rod ends using one of the methods described in the steering arm section.
5. Remove the center link ball stud at both the pitman arm (Fig. III-24) and the idler arm by using a tie rod end type puller.

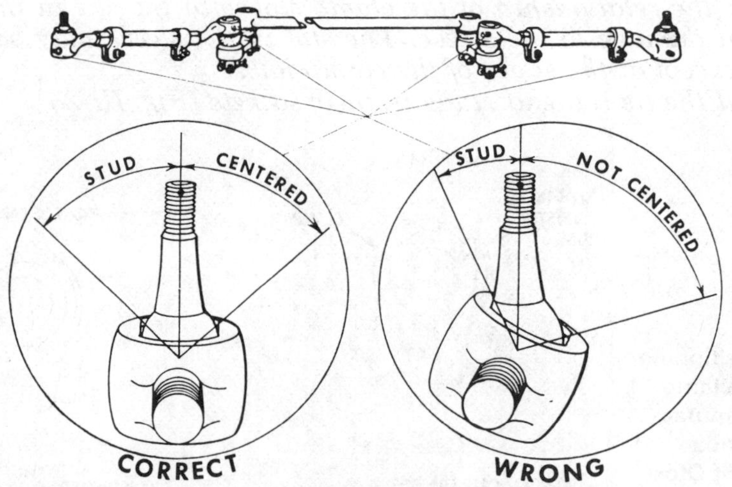

FIGURE III-23 When replacing tie rod ends or performing wheel alignment on Mustang, Comet, or Falcon cars, special attention should be given to positioning of components. Failure to center tie rod end studs to permit full stud swing can cause binding of the steering linkage, resulting in unnecessary wear and in extreme cases failure of the tie rod end.

6. Detach both ends by shifting the steering linkage as required to remove the ball studs from the tapered hole. In some cases, the connection between the idler arm and center link will be a rubber bushing and not a ball stud. In this case, remove the attaching bolts and nuts as necessary.

INSTALLING CENTER LINK

1. Install the center link to the idler arm, making certain that the idler stud seal is in place (Fig. III-25). In the case of rubber bushings, be sure they are in good condition. Tighten all nuts to manufacturer's specified torque and install a cotter pin.
2. On the pitman arm side of the center link, connect it to the pitman arm and tighten to recommended torque. Install the cotter pin by turning the nut just enough to line the hole in the stud with one of the slots in the castellated nut.
3. Install the inner tie rod ends to the center link, as described in the tie rod end section.
4. Lubricate all joints as prescribed by the manufacturer.

Caution: *Do not use any lubricant except soap and water on rubber bushings. Be certain all tapered studs and holes are free of dirt and grease before assembling.*

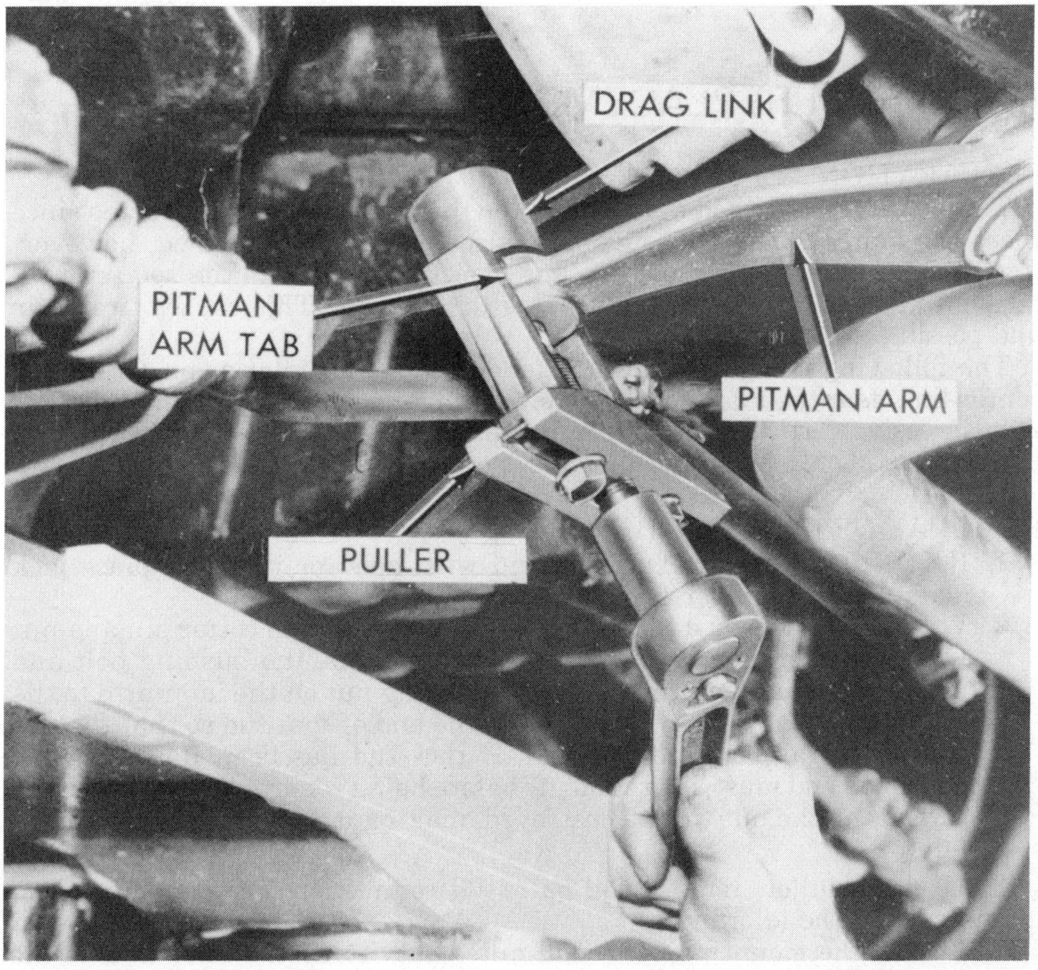

FIGURE III-24 Puller being used to remove the center link from the pitman arm. (Courtesy of Cadillac Motor Car Division)

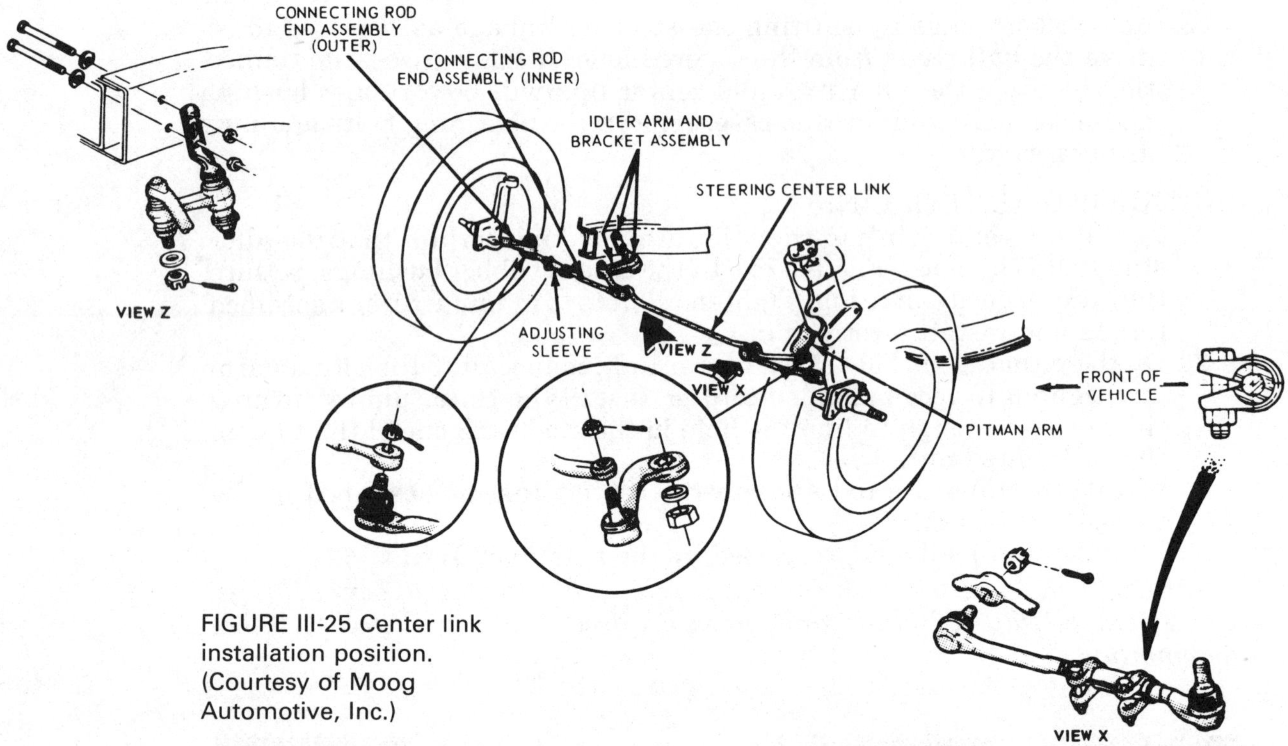

FIGURE III-25 Center link installation position. (Courtesy of Moog Automotive, Inc.)

IDLER-ARM REPLACEMENT

PREPARATION

There is no established set of rules for the inspection of idler arms since there are many different designs and sizes. It should be known, however, that when the idler arm moves in any direction other than its required rotational movement, the toe setting (toe-in) changes. This causes tire wear and possible steering instability.

The following is a general recommended check for idler arms. Grasp the center link near the idler arm and move it up and down vigorously (Fig. III-26). Excessive vertical looseness indicates that the idler arm or its bushing is worn and should be replaced.

REMOVAL OF IDLER ARM

1. Raise front wheels off the ground with hoist or jack and place jack (safety) stands under frame or lower control arms.
2. With the vehicle on a hoist or on stands, remove the cotter pin and nut from the idler arm ball stud or, in some cases, the bushing bolt and nut. Some manufacturers use a self-locking nut on the idler arm to the center link stud bolt. Others flatten the end of the stud so that the nut is locked into place (Fig. III-26). If the stud has been flattened, the deformed end must be ground off before being removed. Otherwise, the threads on the nut will be destroyed, making it necessary to buy a new nut.
3. Using a puller, remove the ball stud connection between the center link and the idler arm.
4. Loosen the mounting bolts and nuts that attach the idler arm to the frame. The idler arm can then be removed from the vehicle.

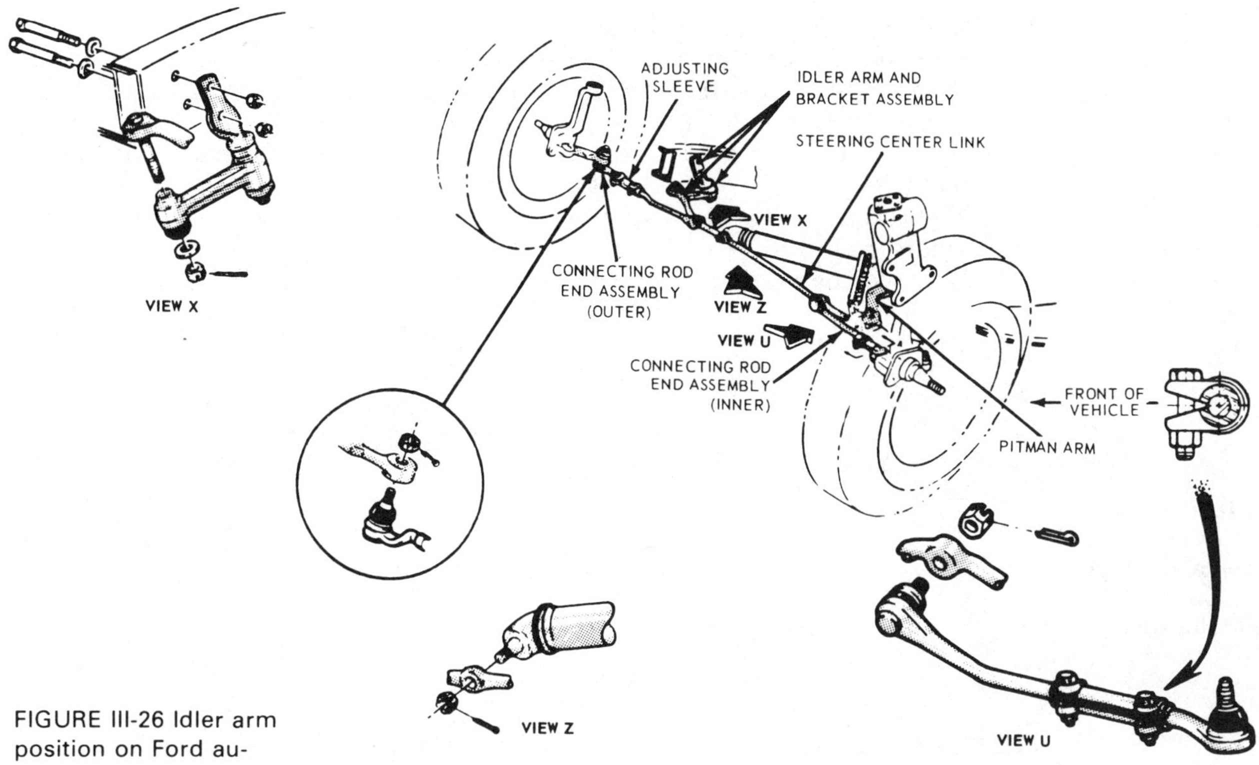

FIGURE III-26 Idler arm position on Ford automobiles. (Courtesy of Ford Customer Service Division)

Idler arms vary in design (Fig. III-27). On some it is necessary to replace the complete idler arm; on others, rubber bushings may be replaced. Depending on idler arm design, bushings may be positioned in the idler arm and, in some cases, in the center link.

When possible, press old bushing out with an arbor press. When a press is not available, an air hammer with a chisel attachment may be used.

New bushings must be pressed into position to avoid damage. Do not try to install them using a hammer. Make certain that the bushing is positioned in the idler arm or center link in the exact position as the old one. *Do not lubricate rubber bushings with anything but soapy water.*

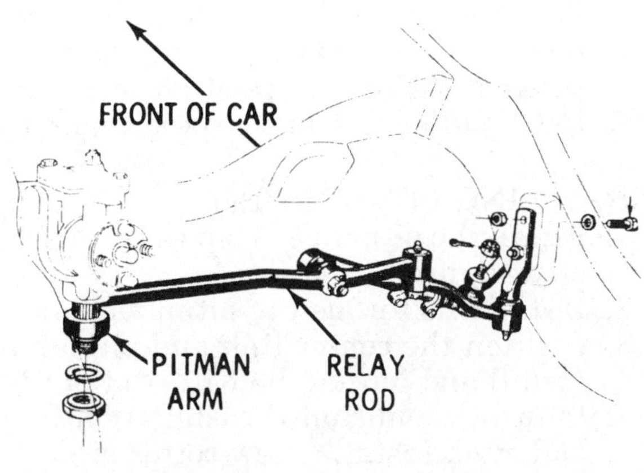

FIGURE III-27 Idler arm assembly for Oldsmobile. Idler arms vary in design. (Courtesy of Oldsmobile Division)

INSTALLATION OF IDLER ARM

1. Position the idler arm on the center link and attach with ball stud nut. When rubber bushings are involved, attach the idler arm on the center link with the bolt and nut (Fig. III-28).
2. Secure the idler arm to the frame with idler arm to frame attaching bolts.
3. Tighten all bolts and nuts to specified torque, and lock with cotter pins where required.
4. Remove the vehicle from the hoist and align the front end.

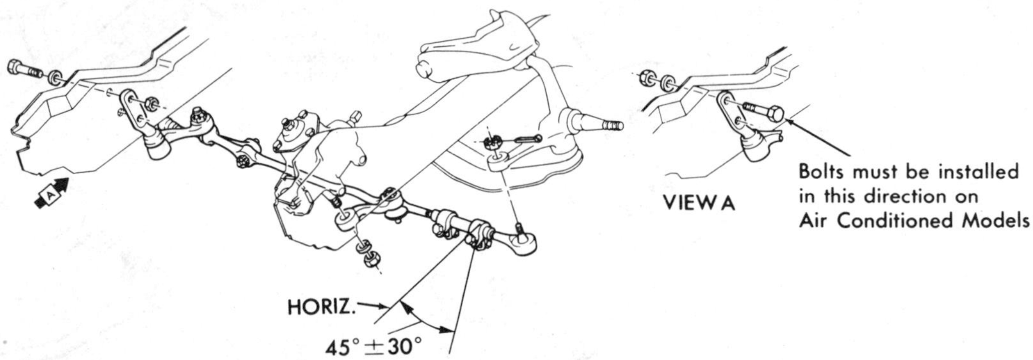

FIGURE III-28 Idler arm assembly for Chevrolet, showing bolt and nut attachment. (Courtesy of Chevrolet Motor Division)

REPLACEMENT OF PITMAN ARM

PREPARATION:

A pitman arm is replaced when it is bent or when the attachment openings are worn.

REMOVAL OF PITMAN ARM

1. Raise front wheels off floor with hoist or jack and place jack (safety) stands under frame or lower control arms.
2. Center the steering wheel to bring the wheels to a straight ahead position.
3. Mark the pitman arm and pitman shaft, using a center punch or chisel. This will provide reference for proper assembly later on.
4. Remove the cotter pin from the pitman arm ball stud, then loosen the nut from the ball stud one or two turns.
5. Using a puller or two hammers, break ball stud loose from tapered hole and remove the nut.
6. Pull down on the center link to allow for removal.
7. Remove the pitman arm nut from the pitman shaft and mark the position of the arm in relation to the pitman shaft.
8. Use a puller to remove the pitman arm from the pitman shaft.

INSTALLING PITMAN ARM

1. Install the pitman arm on the pitman shaft with reference marks properly aligned.
2. Install and torque the pitman shaft nut to recommended torque.
3. Position the center link and pitman arm together with the ball stud. Install and torque the ball stud nut to specified torque, then continue to turn the nut until castellation lines up with the cotter hole in the ball stud. Install a new cotter pin.

Note: *Toe-in should be adjusted before driving.*

JOB REPORT SHEET III-A1

STEERING COMPONENT REPLACEMENT

Student name______________________________Date____________________

Replace or repair a minimum of *two* of the following five steering-component assignments.

1. Replace steering arms.
   Make of car_______________Model_______________Year _______________

   Time started_______________Time finished_______________Total time __________

   Flat-rate time_______________

   a. Installation of new steering arms to steering knuckle. Torque specifications of attaching bolts

   and nuts: ____

Use new cotter pins or mechanic's wire to secure nuts when called for by manufacturer.

   Notes: ____

   ____

   b. Attach tie rod ends to steering arms.
   Torque specifications of tie rod end stud nuts:____________________________

   Use new cotter pin to secure tie rod end stud nut and make certain that there is no looseness in

   tie rod end mounting. Notes:________________________________

   ____

   ____

   c. Repack and install wheel hub drum and bearings.
   See instructions on Objective I-A1.

   Spindle nut torque: _______________

   Lug nut torque: _______________

   d. Lubricate all steering joints.

2. Replace tie-rod ends (inner or outer) and/or, Replace tie-rod adjusting sleeve.

   Make of car_______________Model_______________Year _______________

   Time started_______________Time finished_______________Total time __________

   Flat-rate time _______________

a. Tie rod measurements (before disassembly).
Inner tie rod measurement:

Left side:________________ Right side:________________

Outer tie rod measurement:

Left side:________________ Right side:________________

Total tie rod assembly measurement:

Left side:________________ Right side:________________

b. Install tie rod end assembly.

Torque specifications of ball-stud nuts________________________________

Install new cotter pins; be sure to lubricate all steering joints. Notes: ________________

____

____

Total tie rod assembly measurement after assembly.

Left side:________________ Right side:________________

c. Toe in adjustment.
Toe-in setting before adjustment: ________________

Factory specifications for toe-in: ________________

Toe-in setting after adjustment: ________________

3. Replace center link (drag link or relay rod).

Make of car________________ Model________________ Year ________________

Time started________________ Time finished________________ Total time ____________

Flat-rate time________________

a. Replace center link.

Ball stud and/or bushing bolt-nut torque specifications: ________________

Install new cotter pins; lubricate all steering joints. Notes: ________________

____

____

b. Lubricate all steering joints.

4. Replace idler arm.

Make of car__________Model__________Year __________

Time started__________Time finished__________Total time __________

Flat-rate time__________

a. Idler arm to center link mounting bolt and nut or ball stud-nut torque: __________

____

b. Idler arm to frame mounting bolts and nuts torque specifications: __________

c. Toe-in adjustment.

Toe-in setting before adjustment: __________

Factory specifications for toe-in: __________

Toe-in setting after adjustment: __________

5. Replace pitman arm.

Make of car__________Model__________Year __________

Time started__________Time finished__________Total time __________

Flat-rate time__________

a. Pitman arm to pitman shaft nut torque specifications: __________

____

b. Pitman arm to center link ball stud nut torque specifications: __________

____

DATE COMPLETED__________INSTRUCTOR'S SIGNATURE __________

TEST III-A

Student name________________________________Date ____________________

Directions: Select the best answer to each question and circle the correct letter at the left.

1. American automobiles have steering linkage that can be classified as:
   I. Rack and pinion design with tie rod ends
   II. Parallelogram design with an idler arm and a pitman arm
   A. I only
   B. II only
   C. Either I or II
   D. Neither I nor II
2. The steering arm:
   I. controls the turning radius angles.
   II. is mounted between the steering knuckle and the tie rod ends.
   A. I only
   B. II only
   C. Both I and II
   D. Neither I nor II
3. Tie rod ends are designed with:
   I. Coil spring-loaded ball studs with no end play
   II. Preloaded ball studs with no end play
   A. I only
   B. II only
   C. Both I and II
   D. Neither I nor II
4. The tie rod assembly consists of the following parts:
   I. Outer tie rod ends and inner tie rod ends
   II. Outer tie rod ends, inner tie rod ends, and adjusting sleeves
   A. I only
   B. II only
   C. Both I and II
   D. Neither I nor II
5. Tie rod ends must allow for:
   I. Up-and-down movement but no turning
   II. Turning movement but little or no up-and-down movement
   A. I only
   B. II only
   C. Both I and II
   D. Neither I nor II
6. The center link is held level by the:
   I. Idler arm and pitman arm
   II. Pitman arm and tie rod ends
   A. I only
   B. II only
   C. Both I and II
   D. Neither I nor II

7. The center link transmits turning pressure to:
   I. The left side only
   II. The right side only
      A. I only
      B. II only
      C. Both I and II
      D. Neither I nor II
8. The idler arm is connected between the frame and the:
   I. Passenger side of the center link
   II. Rack gear of the rack and pinion
      A. I only
      B. II only
      C. Both I and II
      D. Neither I nor II
9. A bad idler arm will allow for constant:
   I. Change in toe-in
   II. Change in the horizontal plane of the center link
      A. I only
      B. II only
      C. Both I and II
      D. Neither I nor II
10. Overall looseness in the steering linkage can be checked with—
   I. A dial indicator
   II. A micrometer
      A. I only
      B. II only
      C. Both I and II
      D. Neither I nor II

DATE COMPLETED____________________ INSTRUCTOR'S SIGNATURE ____________________

After studying the following assignment you will be expected to correctly answer a minimum of seven out of ten multiple choice questions in a written test without reference materials on the design and theory of the manual steering gear.

OBJECTIVE III-B (written test)

Preparation:

Study pages 247 through 263.

STEERING GEARS

Various steering gear designs have been developed and used in the past, but vehicle manufacturers have opted for general types for the modern car. The worm and recirculating ball nut steering gear is used on most mid-size and larger vehicles (Fig. III-29). The rack and pinion design is standard on most small cars especially those with the MacPherson strut suspension (Fig. III-30). Both types are available in manual or power assist designs and both are found on domestic as well as imported vehicles.

The purpose of the steering gear is to give the driver control of the automobile and, at the same time, reduce the steering effort through a gear ratio referred to as *steering ratio* (Fig. III-31).

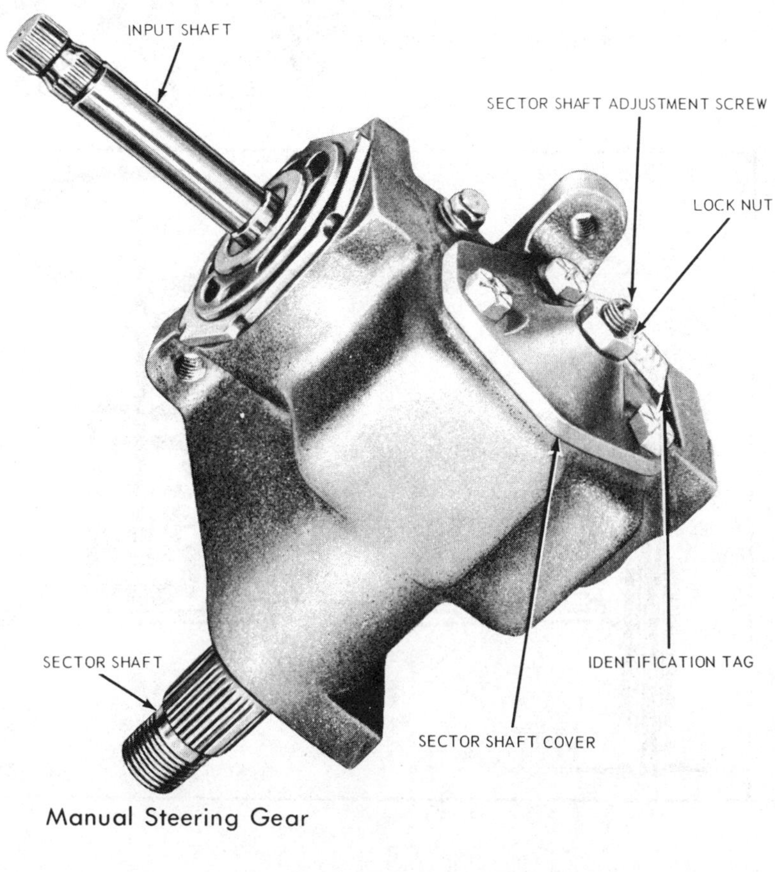

FIGURE III-29 Manual worm and recirculating ball steering gear has been most accepted by American auto makers. (Courtesy of Ford Customer Service Division)

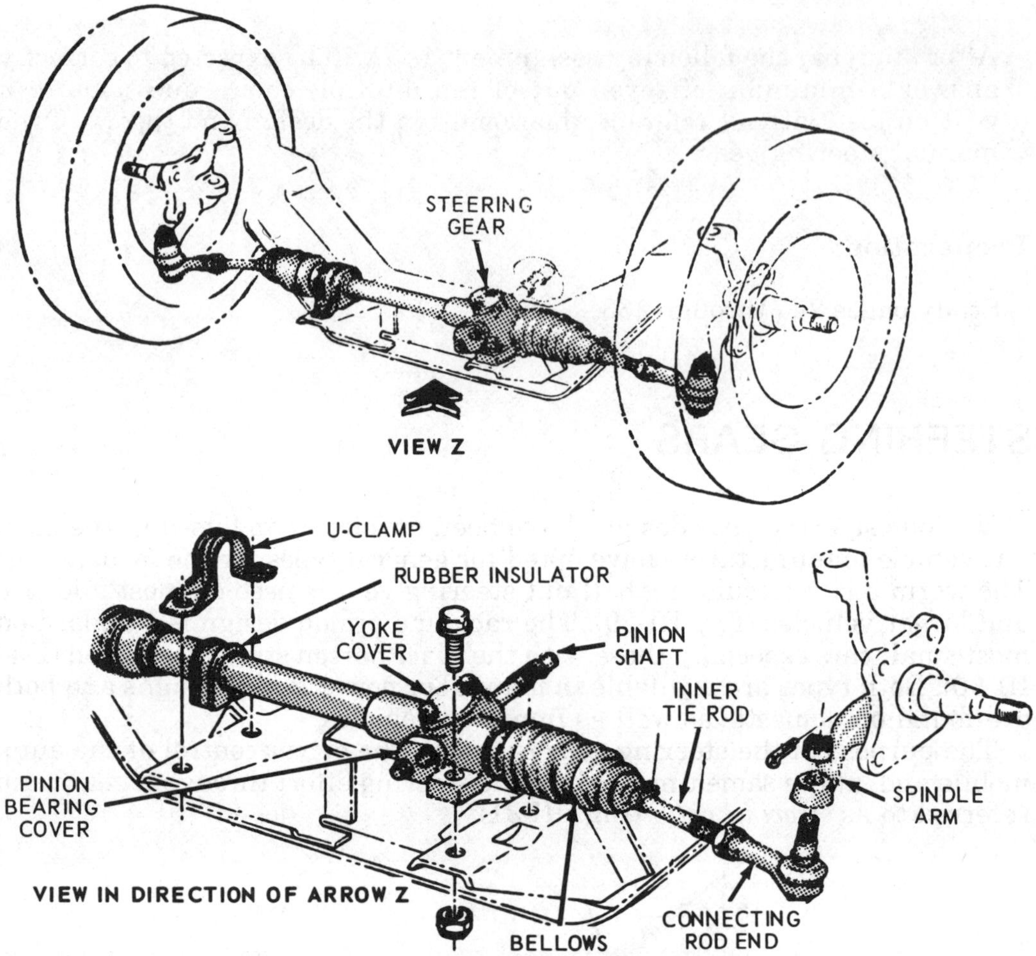

FIGURE III-30 Rack and pinion steering gear and linkage, as used on the Ford Pinto. (Courtesy of Ford Customer Service Division)

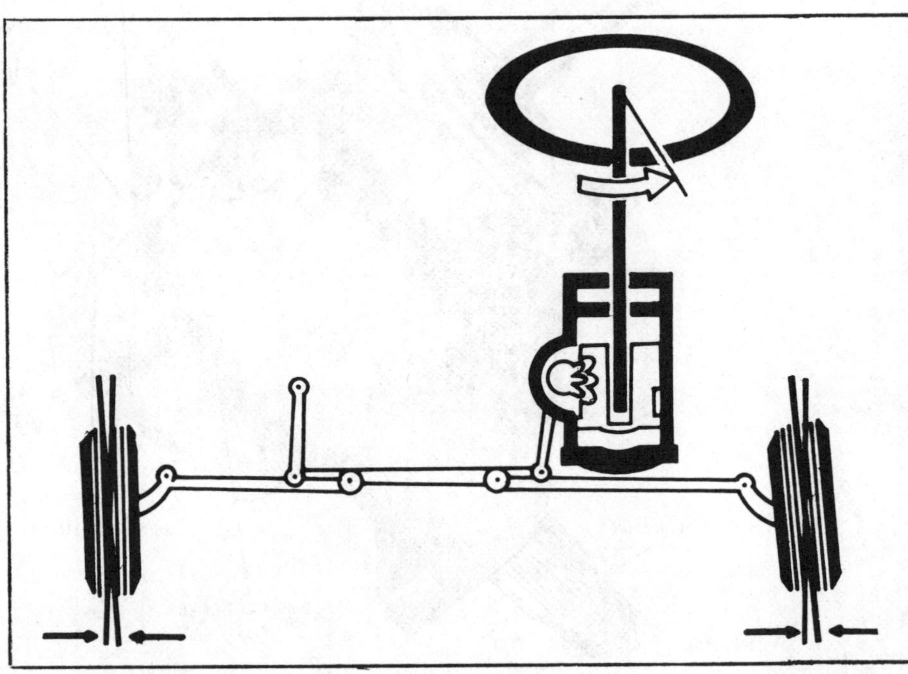

FIGURE III-31 A good steering gear ratio gives the driver good control and reduces effort. (Courtesy of Buick Motor Division)

Steering ratio is a term used to describe the rate of movement of the steering wheel as compared to the front wheel movement. This ratio is normally described in degrees (Fig. III-32). Some sports cars have a *quick* (10:1) *ratio,* which means the steering wheel must be turned 10 degrees for every 1 degree of front wheel movement. Other passenger cars have ratios as high as 28:1. This would be a *slow steering ratio.*

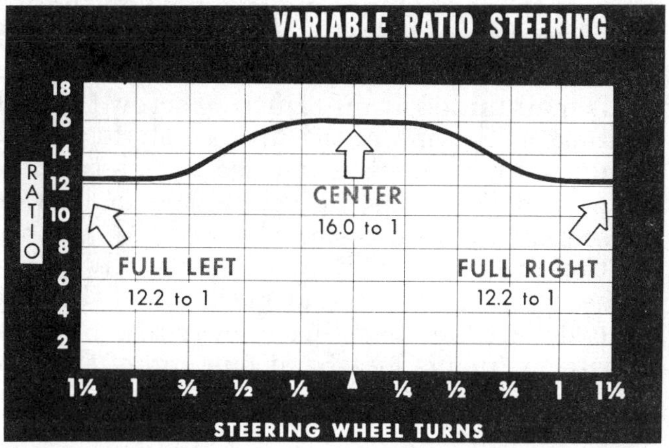

FIGURE III-32 This graph illustrates variable steering gear ratio. (Courtesy of Buick Motor Division)

In most power-steering systems we have a *variable steering ratio* (Fig. III-33). This means a high (16:1) ratio in the straight-ahead position while on the turns, the ratio is reduced to 13:1. The ratio varies mostly as a result of the design of the steering box itself.

The steering gears in Figs. III-34 and III-35 are two different views of the same steering unit and are typical of all ball bearing worm and nut gearboxes.

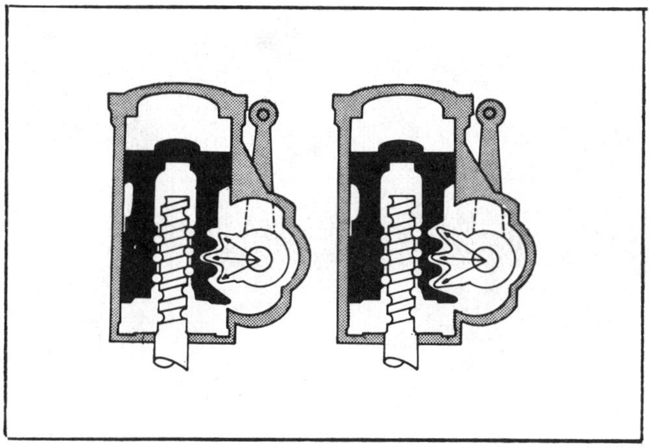

FIGURE III-33 Steering gear tooth design sets the steering ratio in most cars. (Courtesy of Buick Motor Division)

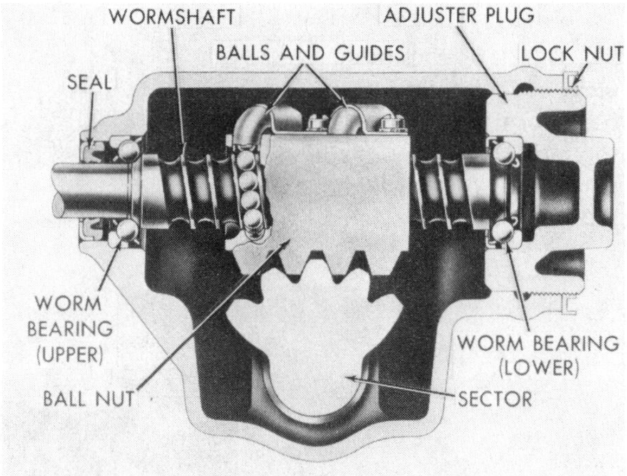

FIGURE III-34 Cutaway view of recirculating ball worm and nut steering gearbox. (Courtesy of Chevrolet Motor Division)

The ball nut is mounted on the worm and is driven through steel ball bearings that circulate in helical grooves in both the worm and nut (Fig. III-34). The balls are allowed to circulate because of the two ball return guides attached to the nut. The worm is connected to the steering shaft through a coupling that permits removal of the gear assembly or steering shaft independent of each other. As the worm is turned, the ball nut moves up or down: the ball nut moves up when the steering wheel is turned to the right and down when the steering wheel is turned to the left.

The ball nut is meshed to the pitman shaft teeth and the lash between the two is maintained by an adjusting screw through the side or top cover that is retained in the end of the pitman shaft. The teeth of the ball nut and the pitman shaft are both machined with a taper so that as the shaft is pushed down with the adjusting screw, the fit between the two becomes tighter. These teeth are also designed so that the teeth fit tighter in a straight-ahead position than they do in a turn (Fig. III-36).

The worm is protected at each end of the case with a thrust bearing. The preload on these two thrust bearings is controlled by the worm bearing adjuster, which is located at one end of the steering gearbox.

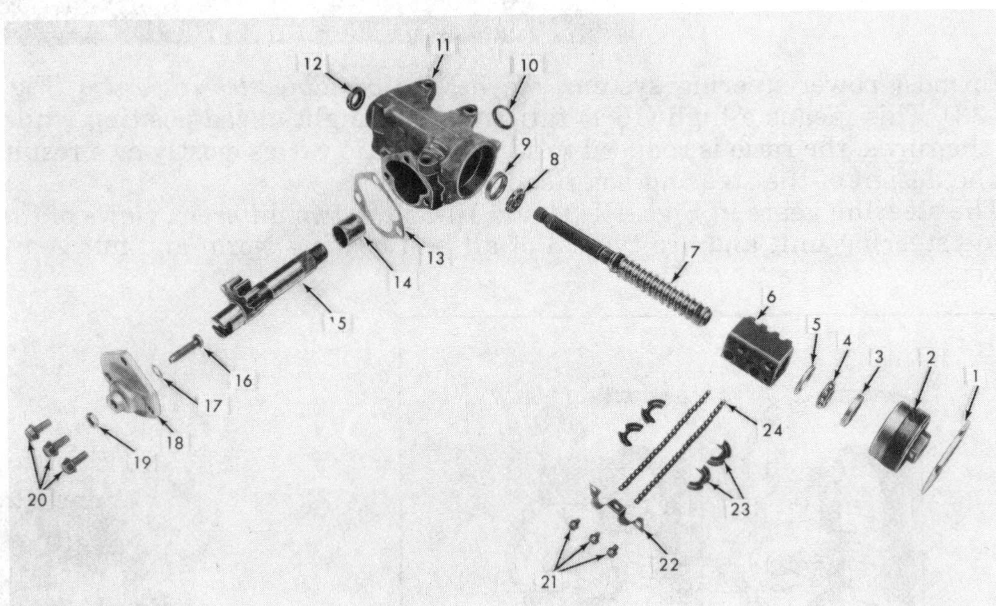

FIGURE III-35 Exploded view of recirculating ball worm and nut steering gearbox. (Courtesy of Chevrolet Motor Division)

FIGURE III-36 Steering gear tooth contact. (Courtesy of Buick Motor Division)

Working as a team with one other student, you will adjust the manual steering gearboxes on a minimum of three instructor assigned late model automobiles. Use manufacturer's recommended procedures and specifications. Time limit: Twice the time allowed in the Motors or Chilton Flat Rate manual.

OBJECTIVE III-B1 (manipulative)

Preparation:

1. Study pages 251 through 256.
2. Study the manufacturer's recommended procedure for the specific automobile to which you are assigned.
3. Carefully observe instructor's demonstration on adjusting manual steering gears.

Materials and Tools Needed:

1. Jack or hoist
2. Jack (safety) stands
3. Box/open-end combination wrench set
4. Socket and ratchet set
5. Pitman arm puller
6. Screwdrivers—both regular and phillips
7. Inch-lb. torque wrench
8. Foot-lb. torque wrench

DIAGNOSING STEERING GEAR PROBLEMS

Before adjustments on the gearbox are attempted, the following diagnosis chart should be studied as an example. This particular chart is for the Chevrolet manual steering gear. For more specific instructions on each make and model, see the manufacturer's repair manual.

DIAGNOSIS

Cause	Solution
a. Excessive play or looseness in steering system	1. Excessive pitman shaft to ball nut lash 2. Worm bearings loosely adjusted 3 Steering wheel loose on shaft, loose pitman arm, tie rods, steering arms, or steering linkage ball studs 4. Worn upper ball joints 5. Front wheel bearings loosely adjusted
b. Hard steering—excessive effort required at steering wheel	1. Low or uneven tire pressure 2. Insufficient or improper lubricant in steering gear or front suspension. 3. Front wheel alignment incorrect 4. Steering gear adjusted too tight

DIAGNOSIS	
Cause	Solution
c. Rattle or chuck in steering gear	1. Pitman arm loose on shaft or steering-gear mounting bolts loose 2. Loose or worn steering shaft bearing 3. Insufficient or improper lubricant in steering gear 4. Excessive lash between ball nut and pitman shaft in straight-ahead position or worm thrust bearings adjusted too loose. **Note:** *On turns, a slight rattle may occur, due to the increased lash between ball nut and pitman shaft as gear moves off the center of "high point" position. This is normal, and lash must not be reduced to eliminate this slight rattle.*
d. Poor returnability	1. Steering gear adjusted too tight 2. Steering column misaligned 3. Front wheel alignment incorrect 4. Insufficient or improper lubricant in steering gear or front suspension

Courtesy of Chevrolet Motor Division

STEERING GEAR ADJUSTMENTS

There are only two adjustments that can be made on the ball-bearing worm and nut steering gearbox (Fig. III-37). They are worm-bearing-thrust preload and the mesh load of the ball nut and pitman (or sector) shaft.

WORM BEARING PRELOAD

The worm bearing preload adjustment must be checked and corrected, if necessary, before attempting to adjust the mesh load between the ball nut and pitman shaft.

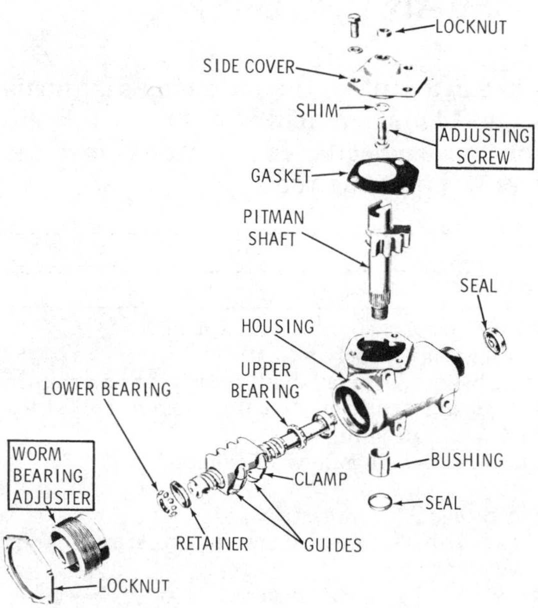

FIGURE III-37 Steering gear adjustments are made at the adjusting screw for pitman shaft depth. Also at the worm bearing adjuster for bearing preload. (Courtesy of Oldsmobile Division)

The worm bearing thrust preload is adjusted by tightening the worm bearing adjuster until the proper inch-pound reading is obtained at the steering shaft nut (Fig. III-38). The following is a brief outline of the procedure (for more details read the instructions for the specific automobile on which you are working.)

1. Remove horn ring button.
2. Remove steering linkage from pitman arm.
3. Turn steering wheel off center by two turns.
4. Loosen pitman (sector) shaft adjusting screw.
5. Place inch-lb. torque wrench on steering shaft nut and turn one turn towards the straight-ahead position (Fig. III-38). Compare torque reading with specifications. If torque requirement is off, loosen worm bearing adjuster locknut with a brass drift punch and turn the worm bearing adjuster until torque is within the required setting.

FIGURE III-38 Adjust the steering shaft nut with a torque wrench. (Courtesy of Chevrolet Motor Division)

6. Adjust by turning worm bearing adjuster until proper torque reading is obtained (Fig. III-39).
7. When adjustment is correct, retighten locknut to 100 ft-lb. and recheck bearing preload.

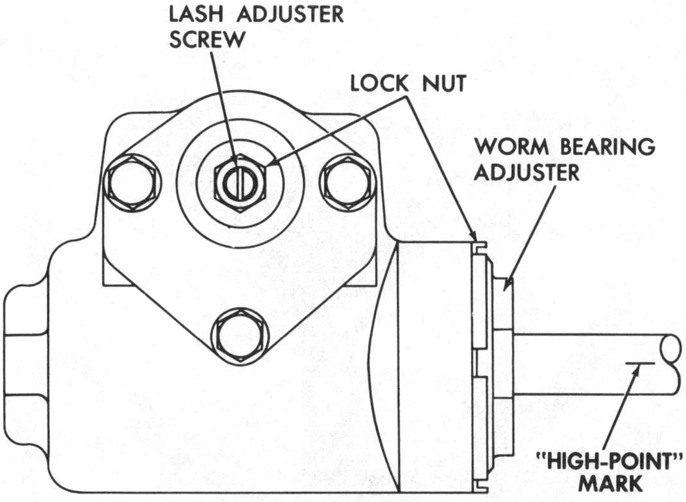

FIGURE III-39 Manual steering gear adjustments may be made to the lash-adjuster screw and worm bearing adjuster nut. (Courtesy of Chevrolet Motor Division)

BALL NUT TO PITMAN MESH LOAD

The ball nut to sector-shaft adjustment is made by following this brief outline:

1. Steering linkage and horn button must be removed for worm-bearing preload adjustment.
2. Turn steering wheel to straight-ahead position and loosen adjusting-screw locknut.
3. Tighten pitman-shaft adjusting screw (lash adjuster screw) until specified torque or pounds of pull reading is found during a turnoff from center (Fig. III-40 and 41.)
4. Tighten locknut and replace all parts removed.

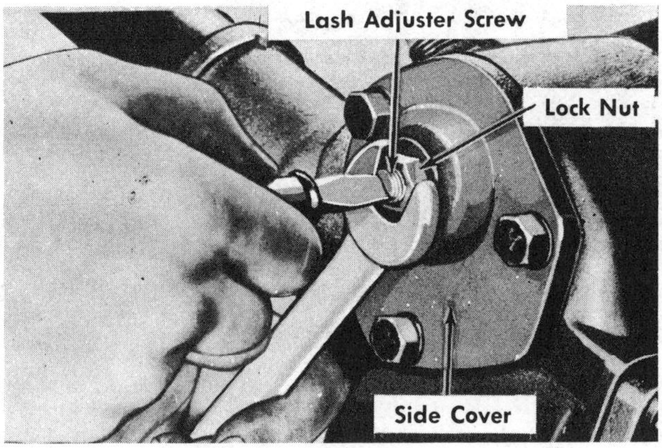

FIGURE III-40 Adjusting lash by changing the depth of the pitman shaft with the lash-adjuster screw. (Courtesy of Chevrolet Motor Division)

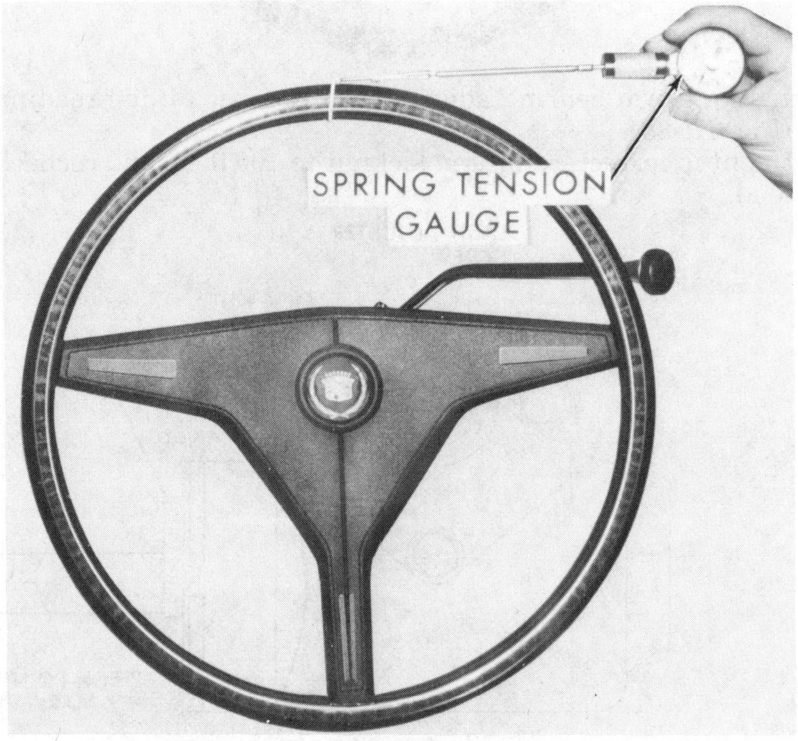

FIGURE III-41 Use a spring tension gauge to check steering gear adjustments. (Courtesy of Cadillac Motor Car Division)

JOB REPORT SHEET III-B1

STEERING GEAR ADJUSTMENT

Student name______________________________Date____________________

ASSIGNMENT #1

Make of car________________Model________________Year________________

Time started________________Time finished________________Total time________________

Flat-rate time ________________

A. Worm bearing preload adjustment
   1. Torque specification for steering wheel turning (off center by two turns): ________________
   2. Worm bearing adjustor locknuts torque specifications: ________________

B. Ball nut to Pitman mesh load
   1. Torque specifications for steering wheel turning (over center): ________________
   2. Pitman (sector) shaft adjusting screw locknut specifications: ________________

DATE COMPLETED________________INSTRUCTOR'S SIGNATURE ________________

ASSIGNMENT #2

Make of car__________ Model__________ Year__________

Time started__________ Time finished__________ Total time__________

Flat-rate time __________

A. Worm bearing preload adjustment
   1. Torque specification for steering wheel turning (off center by two turns): __________
   2. Worm bearing adjustor lock nuts torque specifications: __________

B. Ball nut to Pitman mesh load
   1. Torque specifications for steering wheel turning (over center): __________
   2. Pitman (sector) shaft adjusting screw lock nut specifications: __________

DATE COMPLETED__________ INSTRUCTOR'S SIGNATURE __________

ASSIGNMENT #3

Make of car__________ Model__________ Year__________

Time started__________ Time finished__________ Total time__________

Flat-rate time __________

A. Worm bearing preload adjustment
   1. Torque specification for steering wheel turning (off center by two turns): __________
   2. Worm bearing adjustor lock nuts torque specifications: __________

B. Ball nut to Pitman mesh load
   1. Torque specifications for steering wheel turning (over center): __________
   2. Pitman (sector) shaft adjusting screw locknut specifications: __________

DATE COMPLETED__________ INSTRUCTOR'S SIGNATURE __________

Working as a team with one other student, you will disassemble, inspect, reassemble, and adjust one instructor assigned ball worm and nut type of manual steering gear. Follow the manufacturer's recommended procedure and specifications. Time recommended: Twice the flat-rate time allowed in the Motors or Chilton Flat Rate manual.

OBJECTIVE III-B2 (manipulative)

Preparation:

1. Review pages 257 through 263.
2. Study manufacturer's procedure and specifications on the specific automobile to which you are assigned.
3. Carefully observe the instructor's demonstration on manual steering gear overhaul and adjustment.

Materials and Tools Needed:

1. Jack or hoist
2. Jack (safety) stands
3. Box/open-end combination wrench set
4. Socket and ratchet set
5. Pitman arm puller
6. Screwdrivers—both regular and Phillips
7. Inch-lb. torque wrench
8. Foot-lb. torque wrench
9. Hammer

MANUAL STEERING-GEAR REMOVAL

Removal of the manual steering gear varies with different car makes and models. However, there are a few common instructions for removing all gearboxes (Fig. III-42).

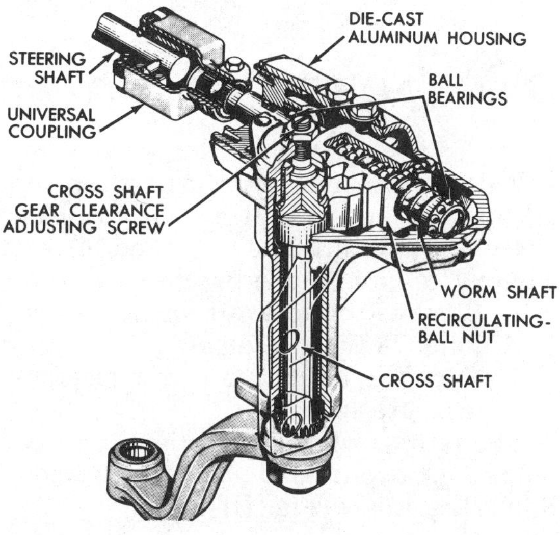

FIGURE III-42 Cutaway of a recirculating ball, worm and nut steering-gearbox. (Courtesy of Chrysler Motors Corp.)

Review the procedure outlined on page 224 for the removal of the pitman arm from the sector shaft (Fig. III-43). Disconnect the steering gear from the steering column at the flexible coupling (Fig. III-44). Remove the gear to frame bolts, position steering linkage out of the way, and withdraw the gearbox from the car.

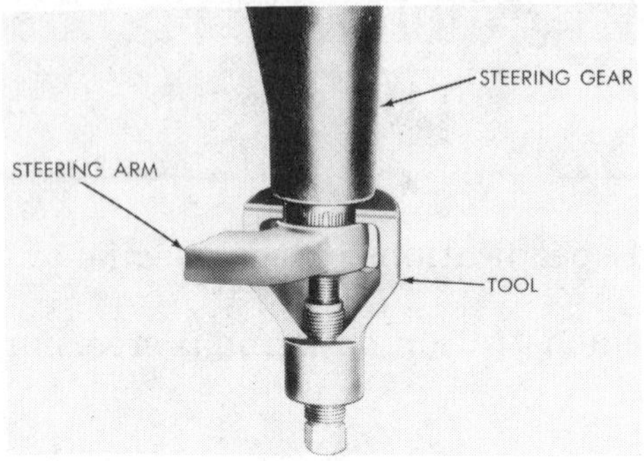

FIGURE III-43 Removal of pitman arm in preparation for removal of steering gear from automobile. (Courtesy of Chrysler Motors Corp.)

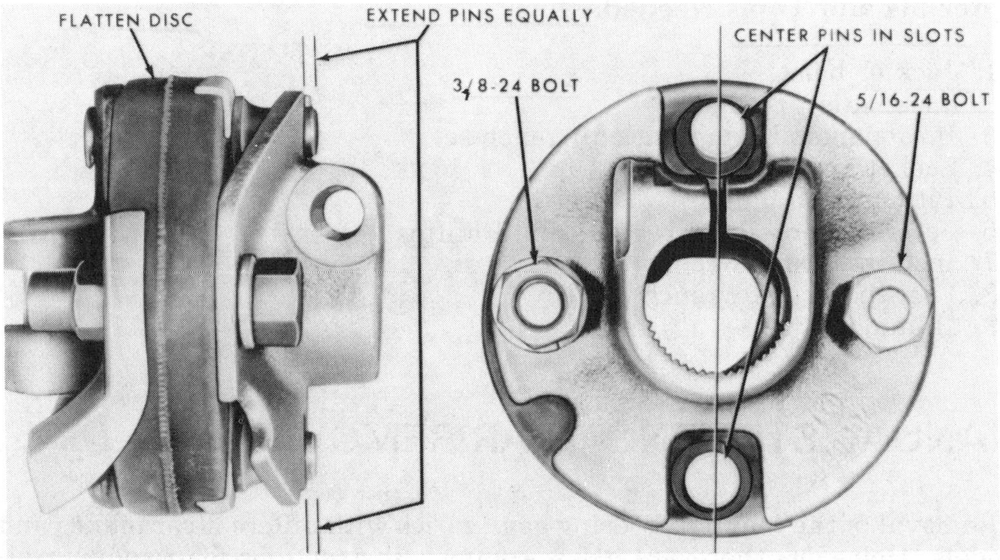

FIGURE III-44 Disconnect flexible coupling between steering column and steering gearbox. (Courtesy of Cadillac Motor Car Division)

MANUAL STEERING-GEAR DISASSEMBLY

With the steering gearbox cleaned and mounted in a holding fixture (Fig. III-45), center the ball nut by rotating the worm shaft between three and three-and-a-half turns from end of travel.

Next, remove the side cover by doing the following:

1. Remove the locknut on the adjusting screw.
2. Remove the three side cover bolts.
3. Turn the adjusting screw clockwise until the side or top cover is free (Fig. III-46).

The pitman shaft and adjusting screw can now be pulled up through the side-cover opening. Be certain to keep the shim or washer in place on the adjusting screw (Fig. III-47).

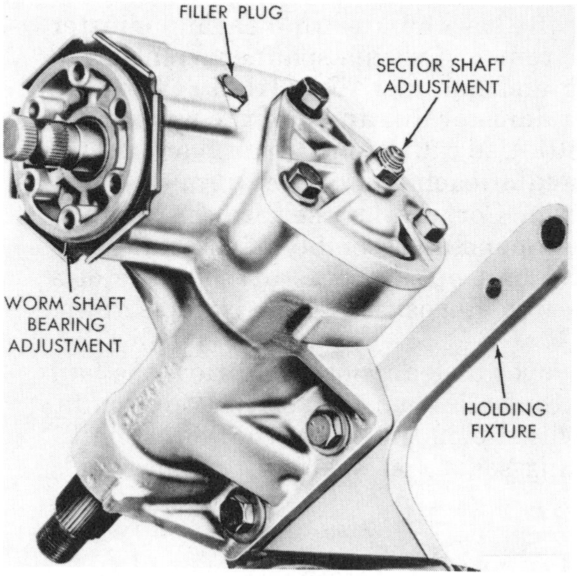

FIGURE III-45 A holding fixture is helpful when overhauling a steering gearbox. (Courtesy of Chrysler Motors Corp.)

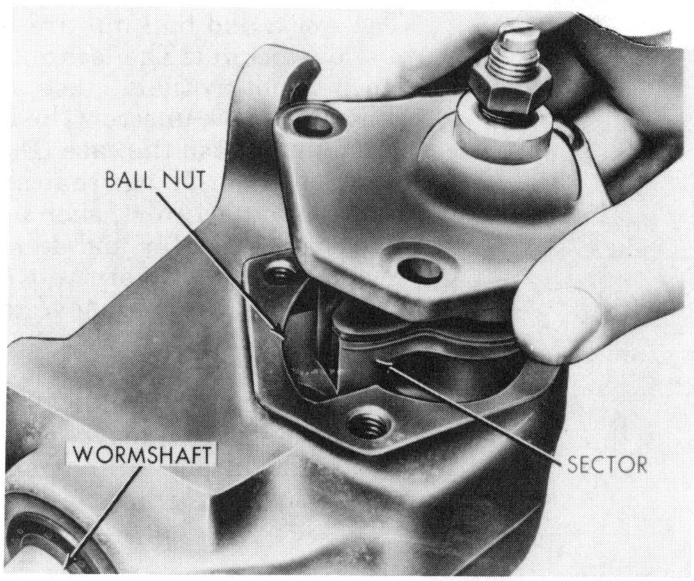

FIGURE III-46 Turn the adjusting screw clockwise to remove the side or top cover. (Courtesy of Oldsmobile Division)

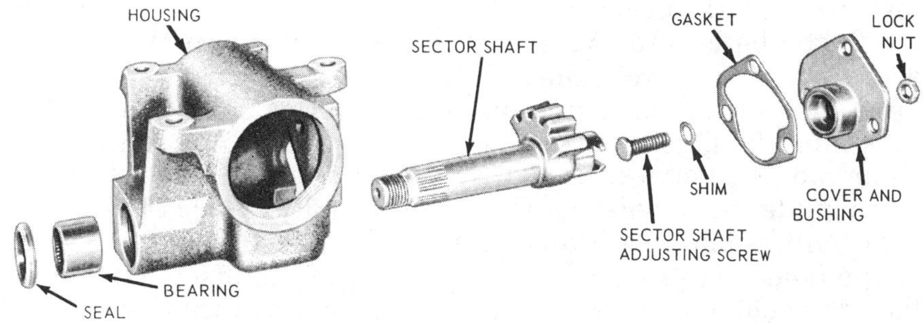

FIGURE III-47 Keep the shim in place when the sector shaft is removed from the housing. (Courtesy of Ford Customer Service Division)

Before removing the flexible coupling from the worm shaft, scribe a locating mark on both parts. Then by removing the locking bolt from the flexible coupling, the coupling can be removed from the worm shaft (Fig. III-48).

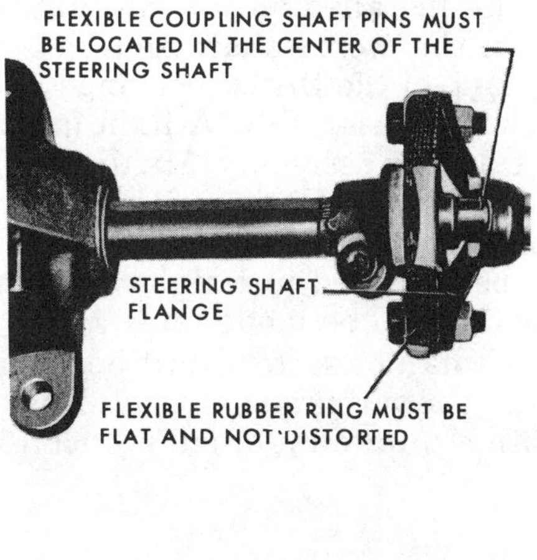

FIGURE III-48 Mark position of flexible coupling before removing it. (Courtesy of Buick Motor Division)

The worm and ball nut are held in the case by a worm-bearing adjuster nut and a locknut. The locknut can be removed with a spanner wrench. If a wrench is not available, use a punch and hammer (Fig. III-49). With the locknut removed, unscrew the bearing adjuster nut and pull the worm shaft and ball nut out of the case (Fig. III-50). Use caution not to drop or damage the two thrust bearings that are located at each end of the worm.

Some manufacturers, such as Chrysler Corporation, service the ball and worm as an assembly and do not recommend disassembly of the ball nut.

When disassembly of the worm and ball nut is necessary because of a binding or tightness, remove the ball nut. Reposition guide-clamp screws, clamp, and guides.

Invert ball nut and worm assembly over a clean cloth, and turn the ball nut slowly from side to side, to remove the balls from each of the two circuits. After the balls are removed, the nut will slip off the end of the worm shaft.

Clean and inspect all parts, including seals and bushings; replace seals and also bushings when necessary.

REASSEMBLY OF STEERING GEAR

Reassemble the ball nut and worm using the following steps:

Make certain that the worm nut is positioned on the worm shaft in the proper direction. The teeth on the ball nut must have the narrow but deep teeth at the top of the case (Fig. III-51). Before installing the ball nut, check the mesh of the ball nut with the teeth on the sector (or pitman) shaft.

Separate the balls into two equal piles so that an equal number can be placed in each of the two guides (Fig. III-52). Center the worm nut guide holes over the worm groove, and place at least two balls in each guide hole. The balls must be pushed to the bottom of the ball nut. This will prevent them from lodging between the two return guides on top of the ball nut. If the return guides have holes for inserting the balls, place the guides into the ball nut openings and insert an equal number of balls in each return guide.

When no holes are provided in the return guides, insert an equal number of balls into each side of the ball nut. Make certain that none of the balls move into the worm grooves between the two guides on the top of the ball nut during this step. Coat the inside of the return guide with a light grease. The balls that cannot be placed in the ball nut may be inserted into the return guide. The grease will hold the balls in position while the return guides are installed on the ball nut.

Return the worm shaft and ball nut to the case as an assembly (Fig. III-53). One of the thrust bearings will be placed in the case with the worm shaft and ball assembly. A light grease will help to hold the thrust bearing in place during assembly. Make certain that the worm nut is in the position necessary for installation of the sector or pitman shaft. Add the other thrust bearing to the assembly, and install the worm thrust adjuster. The adjuster should be installed just tight enough to hold the bearings in place. Final adjustment will be made later.

The pitman or sector shaft adjusting screw and selective shim should be placed in the pitman shaft slot. The screw must be able to turn free in the slot with a minimum of 0.002-inch (0.0508 mm) end play (Fig. III-54).

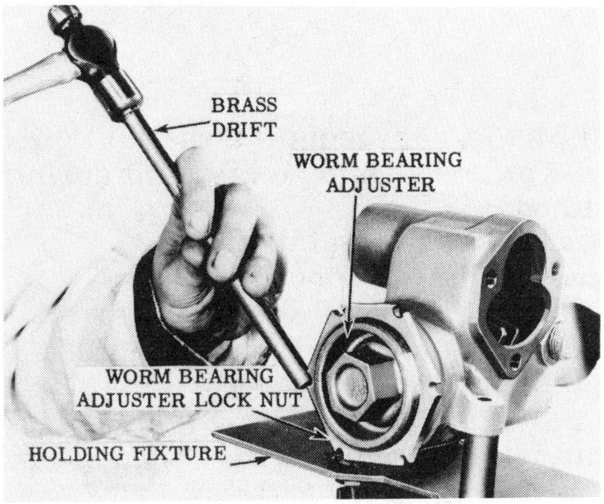

FIGURE III-49 Using hammer and punch, remove worm bearing adjustor lock nut as shown. (Courtesy of Oldsmobile Division)

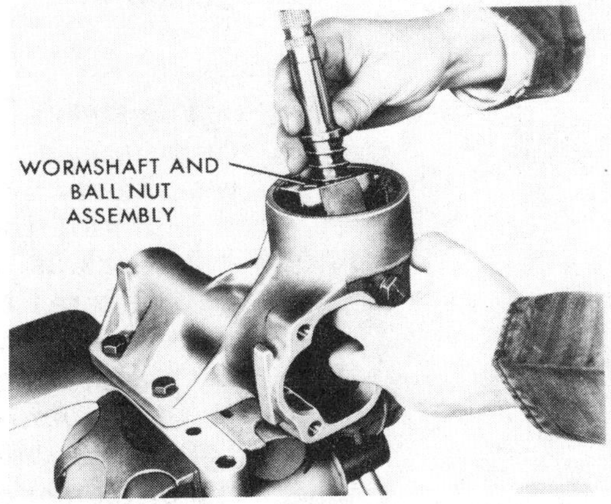

FIGURE III-50 After removal of worm shaft and ball nut assembly, remove guide clamp screws, clamp, and guides. (Courtesy of Chrysler Motors Corp.)

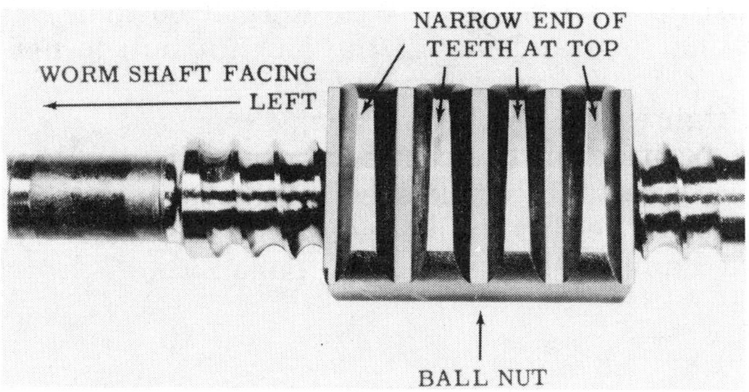

FIGURE III-51 Position ball nut on worm shaft so that narrow but deep teeth are at the top. (Courtesy of Oldsmobile Division)

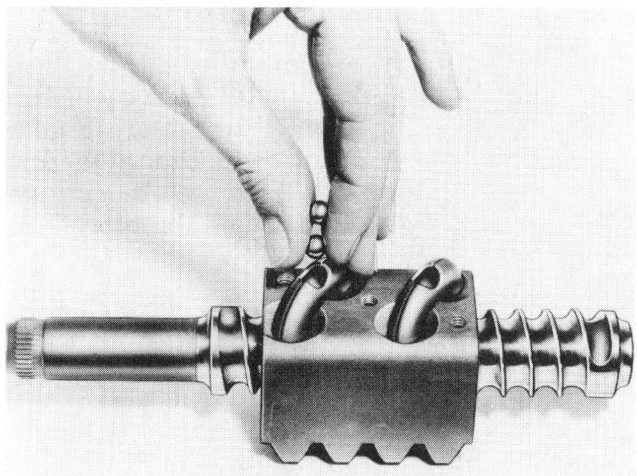

FIGURE III-52 An equal number of balls should be placed in each side of the ball nut. (Courtesy of Oldsmobile Division)

FIGURE III-53 Installing worm shaft and ball nut assembly in the case. (Courtesy of Oldsmobile Division)

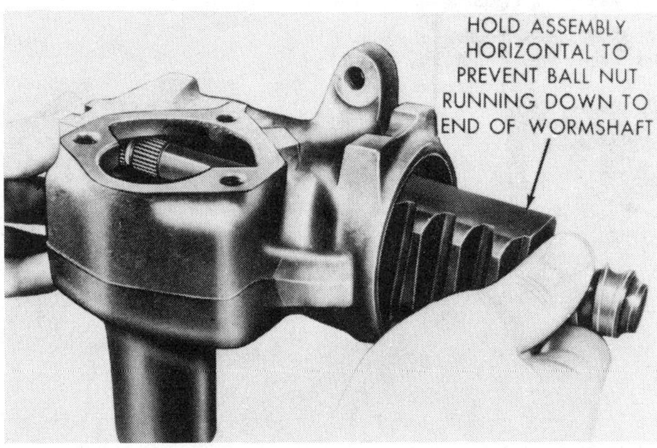

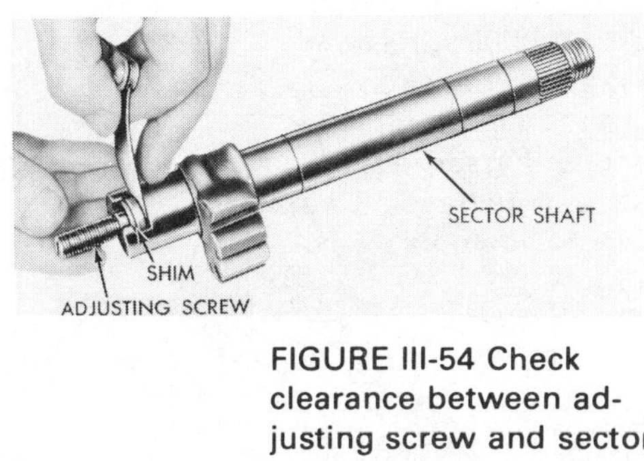

FIGURE III-54 Check clearance between adjusting screw and sector (pitman) shaft. (Courtesy of Chrysler Motors Corp.)

The steering gears used on General Motors cars require a special lubricant (grease) packed into the case. This grease should be installed during reassembly by using the following method:

1. Rotate the shaft until the ball nut is at one end of its travel.
2. Pack as much of the grease as possible into the housing.
3. Rotate the shaft so that the ball nut is at the opposite end of its travel.
4. Repeat the same process until the ball nut and worm shaft are packed with the special lubricant.

Note: *See manufacturer's specifications for amount and type of lubricant.*

Turn the worm shaft until the ball nut is centered in the case; then insert the pitman or sector shaft. Use care that the splined end of the shaft does not damage the oil seal or bushing as it slides through the case opening. The sector must be centered with the ball nut upon installation. The remainder of the lubricant for General Motors products, or all of the lubricant on other makes, may be added at this time. Using a new side cover gasket, install the side cover and start the pitman-shaft adjusting screw into the cover (Fig. III-55). This can be done by inserting a thin screwdriver through the adjusting screw hole in the case and turning the adjusting screw counterclockwise. Thread the adjusting screw into the case until it bottoms on the side cover, then back off one turn and install the locknut but do not tighten it.

Install side cover bolts and tighten to specified torque.

Return steering gear to its position on the car, being careful to follow individual manufacturer's procedures and torque specifications.

With the steering gear in place, the final adjustments may be made by following the procedure outlined under Objective III-B1, page 251.

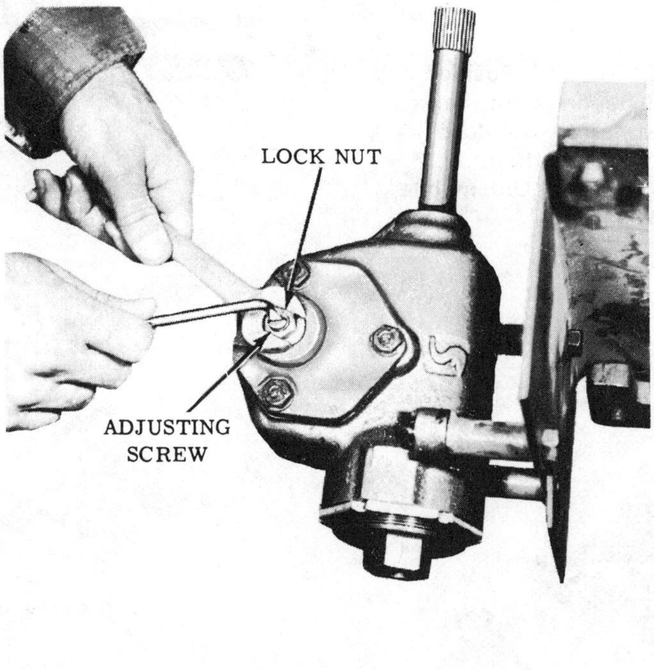

FIGURE III-55 Adjusting of the lash between sector and ball nut must be done with gear in the center of its travel. (Courtesy of Oldsmobile Division)

JOB REPORT SHEET III-B2

STEERING GEAR OVERHAUL

Student name____________________ Date________________

Make of car____________ Model____________ Year____________

Time started____________ Time finished____________ Total time____________

Flat-rate time ____________

1. Record the number of balls in the ball nut: ____________
2. Clearance between adjusting screw and sector shaft should be: ____________
3. Actual measurement between adjustment screw and sector shaft is: ____________
4. Type of lubricant used in gearbox: ____________
5. Worn shaft bearing preload specifications: ____________
6. Ball nut to Pitman arm mesh load: ____________

DATE COMPLETED____________ INSTRUCTOR'S SIGNATURE ____________

TEST III-B

Student name______________________________Date__________________

Directions: Select the best answer to each question and circle the correct letter at the left.

1. Steering ratio is designed to:
   I. Reduce steering effort
   II. Increase steering feel
      A. I only
      B. II only
      C. Both I and II
      D. Neither I nor II
2. A steering ratio of 10 degrees to 1 degree means that the:
   I. Steering wheel turns 1 degree to 10 degrees at the wheels
   II. Steering wheel turns 10 degrees to 1 degree at the wheels
      A. I only
      B. II only
      C. Both I and II
      D. Neither I nor II
3. A steering ratio of 16 to 1 at center and 13 to 1 at a turn is:
   I. Not possible
   II. A variable ratio
      A. I only
      B. II only
      C. Both I and II
      D. Neither I nor II
4. The ball bearing, worm and nut type steering has:
   I. Two worm bearings
   II. One sector
      A. I only
      B. II only
      C. Both I and II
      D. Neither I nor II
5. The pitman shaft and the sector shaft are:
   I. Different names for the same thing
   II. Two different parts in the steering gearbox
      A. I only
      B. II only
      C. Both I and II
      D. Neither I nor II
6. There are two adjustments on the steering gearbox. Which of the two listed below is adjusted first?
   I. Lash adjustment
   II. Worm bearing preload adjustment
      A. I only
      B. II only
      C. Both I and II
      D. Neither I nor II

7. The ball nut has teeth that are cut with one end narrow and the other end wide. Which end is at the top when properly installed?
   I. The narrow end
   II. The wide end
      A. I only
      B. II only
      C. Both I and II
      D. Neither I nor II
8. The lash adjusting screw must have ________________ clearance between it and the sector shaft:
   I. Zero
   II. A maximum of 0.002 inch
      A. I only
      B. II only
      C. Both I and II
      D. Neither I nor II
9. Worm bearing preload adjustment is made with the steering wheel:
   I. In the straight-ahead position
   II. All the way to the left or the right
      A. I only
      B. II only
      C. Both I and II
      D. Neither I nor II
10. The lash adjustment is made with the steering wheel:
   I. In the straight-ahead position
   II. At least one turn off center
      A. I only
      B. II only
      C. Both I and II
      D. Neither I nor II

DATE COMPLETED________________ INSTRUCTOR'S SIGNATURE ________________________

OBJECTIVE III-C (written test)

After studying the preparation assignment, you will be expected to correctly answer a minimum of seven out of ten multiple-choice questions in a written test without reference materials on the following power-steering topics:
1. Power steering pump
2. Hose design
3. Theory of operation

Preparation:

1. Study pages 267 through 294.
2. Study manufacturer's manuals on power steering pumps for Ford, Chrysler, and General Motors.

THEORY OF HYDRAULICS

Hydraulics is a science of liquids such as water, oil, and brake fluid. In the study of hydraulics, we find that pressure and motion can be transmitted through liquid as though the liquid were solid. Thus, a hydraulic brake line can carry as much pressure as a steel tie rod.

TRANSMITTING MOTION AND PRESSURE

The transmitting of motion and pressure through a hydraulic system is accomplished by the fact that liquid cannot be compressed. With this fact in mind, visualize a basic hydraulic system as illustrated in Fig. III-56. This system must be a completely sealed unit for both motion and pressure to be transmitted to their full capacity. Piston *A* is the applying piston and piston *B* is the output piston. If both of these pistons are of the same diameter, any motion applied to piston *A* will be transmitted through the hydraulic line to piston *B*. If piston *A* moves 1 inch, piston *B* will move 1 inch. The distance piston *B* moves will always be in direct relation to the size of the two pistons. For example, if piston *A* is 1 inch in diameter and piston *B* is 2 inches in diameter, the movement of the output piston (piston *B*) will always be one-half of the movement of the applied piston (piston *A*) (Fig. III-57).

The opposite is the case when the applied piston (piston *A*) is larger than the output piston (piston *B*) as in Fig. III-58.

The same examples can be used to explain the transmitting of pressure. Pressure, like motion, is transmitted pound for pound for each square inch of piston surface area. For example, if both the applied piston (piston *A*) and the output piston (piston *B*) have 1 square inch of surface area, the pressure transmitted will be equal to the input pressure (Fig. III-59). If we apply 100 lb. pressure to piston *A*, piston *B* will have an output of 100 lb. pressure.

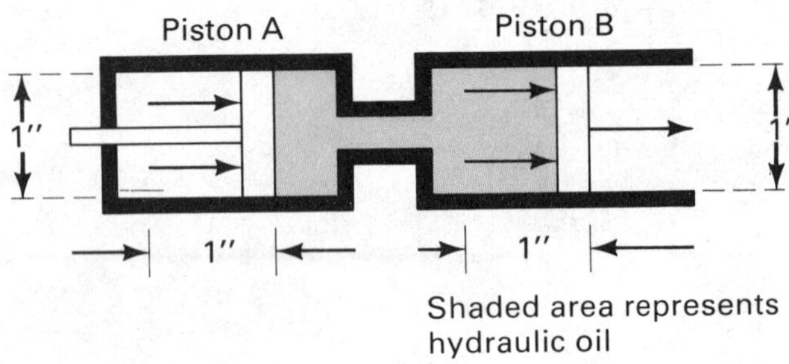

FIGURE III-56 Hydraulic motion when the pistons are the same size.

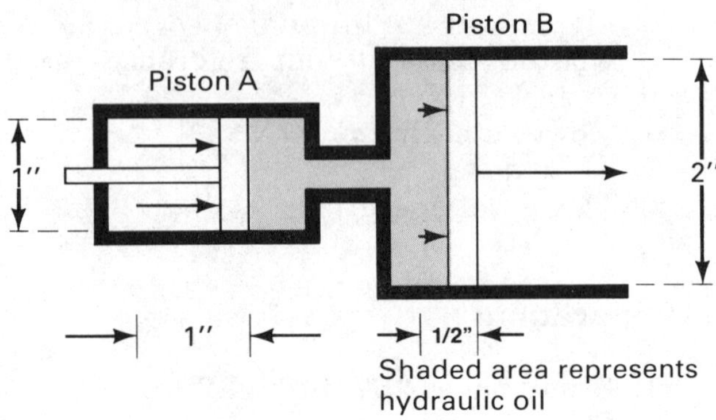

FIGURE III-57 Hydraulic motion when the applying piston is smaller than the output piston.

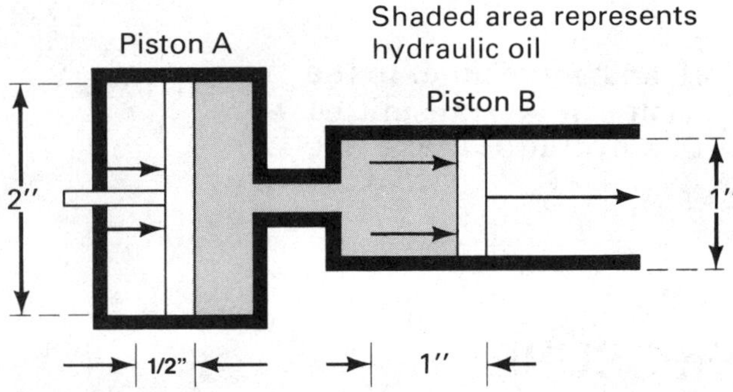

FIGURE III-58 Hydraulic motion when the applying piston is larger than the output piston.

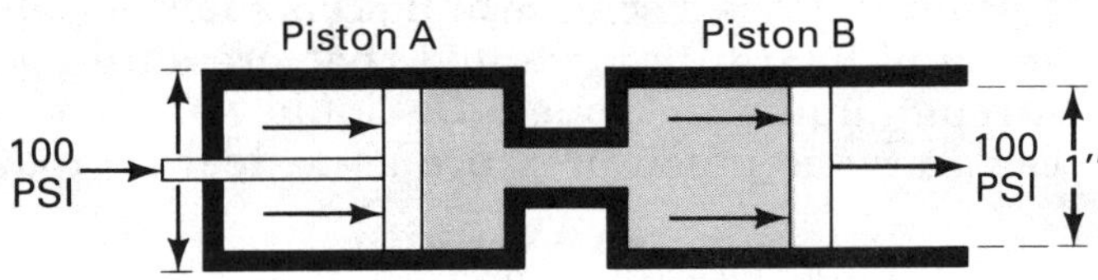

FIGURE III-59 Pressure is transmitted pound for pound for each square inch of piston surface.

The mechanical devices (cylinders and pistons) used thus far to explain transmitting motion and pressure in hydraulics resembles the system used in hydraulic brakes. For power steering, piston *A* or the applying piston is replaced with a pressure pump capable of producing pressures in excess of 1300 lb. per square inch. The power steering pump will supply the necessary pressure for transmitting pressure and motion through the hydraulic oil.

American manufacturers have used many different power steering pumps in the past. However, the three illustrated in this text are the only ones that are in use on current production model automobiles. They are: The Saginaw Constant Displacement Vane Pump, Fig. III-60; the Thompson Slipper Pump, used on some Ford automobiles, Fig. III-61; and the Chrysler Corporation Roller Pump, Fig. III-62.

All three pumps accomplish the same purpose: they take hydraulic oil and place it under pressure for use as a power assist in the steering system.

SAGINAW VANE

The Saginaw Constant Displacement Vane Pump is identified by the long oval shape of the filler neck. The drive pulley is secured to the drive shaft by a large nut. Rectangular pumping vanes, carried by a shaft driven rotor, move hydraulic oil from the intake to the pressure cavities of the cam ring.

As the rotor turns, centrifugal force throws the vanes against the inside surface of the cam ring (Fig. III-63). Here the vanes pick up the oil and force it into high pressure areas. As more oil is picked up by the vanes, the oil is forced into the thrust plate cavities through two crossover holes in the cam ring and pressure plate. These holes empty into the high pressure area between the pressure plate and the housing end-plate. When the high pressure areas are filled, it causes oil to flow under the vanes in the slots of the rotor. The oil forces the vanes out against the oval surface of the cam ring. This helps to pick up larger quantities of oil. As the oil is moved around to the smaller area of the cam ring by the vanes, it is forced into the high pressure line. The system contains a flow control valve that is pressure regulated. This valve allows excessive pressure to be dumped back into the pump reservoir.

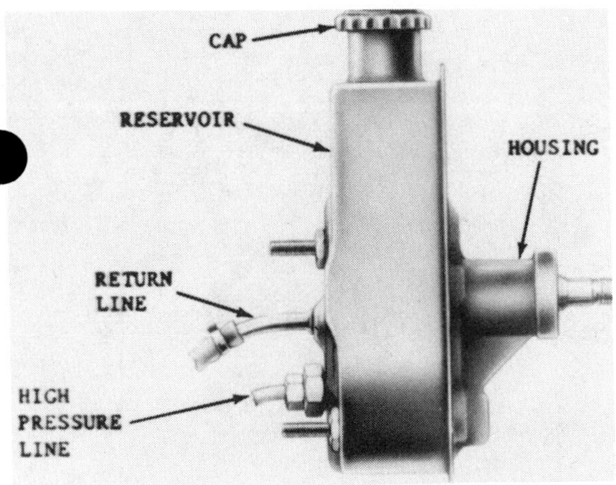

FIGURE III-60 Saginaw constant displacement power steering pump. (Courtesy of Oldsmobile Division)

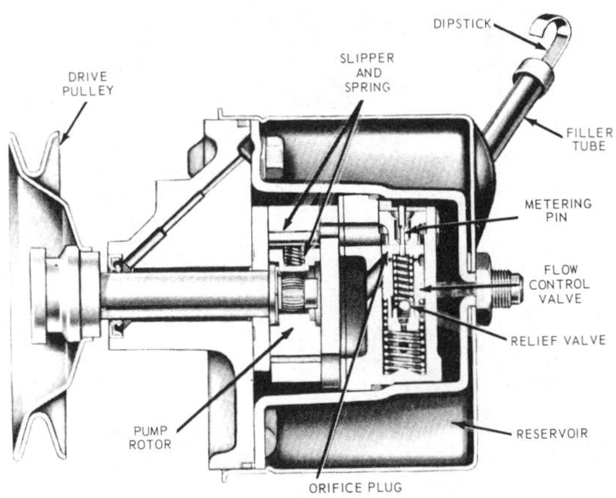

FIGURE III-61 Sectional view of the Thompson Slipper power steering pump. (Courtesy of Ford Customer Service Division)

FIGURE III-62 Chrysler Corporation Roller Power Steering Pump. (Courtesy of Chrysler Motors Corp.)

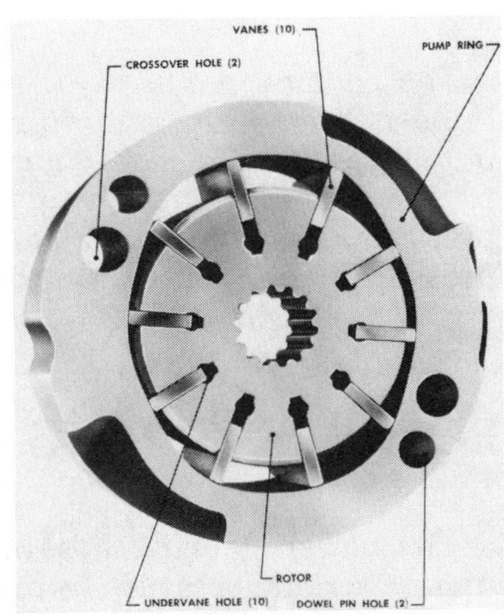

FIGURE III-63 Photo showing pump ring and rotor of the Saginaw constant displacement power steering pump. (Courtesy of Oldsmobile Division)

THE CHRYSLER CORPORATION ROLLER PUMP

This pump can be identified by a ⅜-inch threaded hole in the pulley end of the drive shaft. The rotor in the pump is star shaped and, upon rotation, propels 12 steel rollers against the inside surface of the cam ring. (Fig. III-64). The rollers serve the same function as the vanes used in the Saginaw Vane Pump. As the rollers move around the eccentric surface of the cam ring, oil is drawn through the inlet ports. As the rollers are forced into the V-shaped cavities of the rotor, the oil is exhausted through the discharge ports. The flow control valve permits a regulated amount of oil to return to the intake side of the pump.

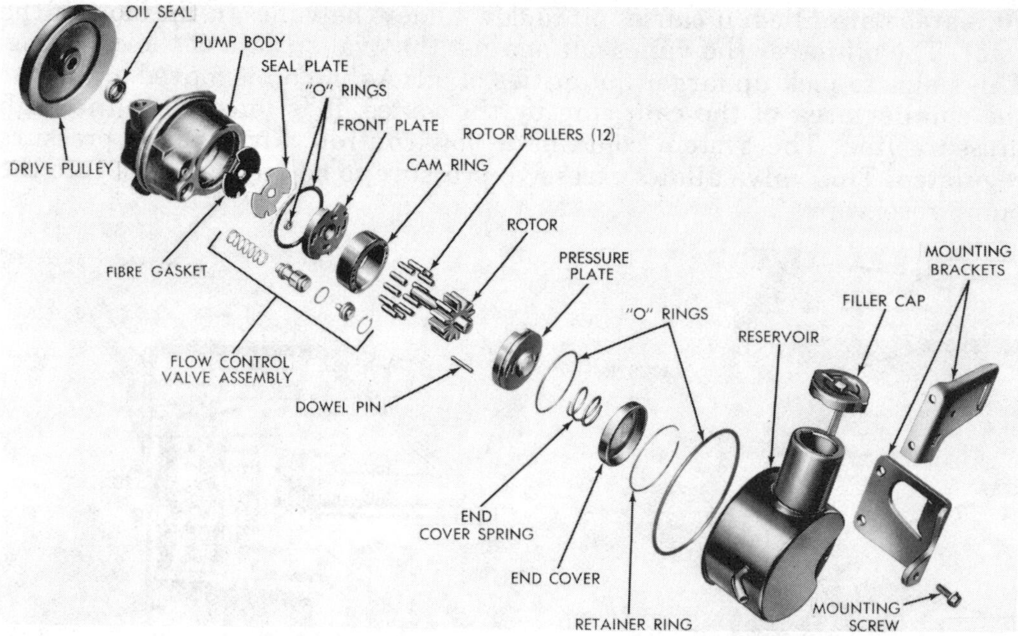

FIGURE III-64 Exploded view of Chrysler Corp. roller power steering pump. (Courtesy of Chrysler Motors Corp.)

THE THOMPSON SLIPPER PUMP

The Thompson pump can be identified by its identification tag mounted on the back side of the reservoir. It has an integral reservoir mounted with the pump (Fig. III-65). The Thompson Slipper Pump is a constant displacement rotary pump having spring loaded slippers in place of vanes or rollers. A pressure relief valve is also mounted in the pump body to allow excessive pressure to be returned to the intake side of the pump.

POWER STEERING HOSE

The power steering hose must be capable of carrying the hydraulic oil under extreme pressure from the power steering pump to the power steering gearbox or ram. This hose has steel ends that are heavily plated to resist rust (Fig. III-66). These steel ends are connected to couplings that are pressure bonded to the rubber hose to prevent leakage. The pressure hose is reinforced to prevent bursting, while still allowing for expansion. This assures safe and proper resistance to fluid surge during operation. The return hose is the low pressure return from the gearbox to the pump reservoir.

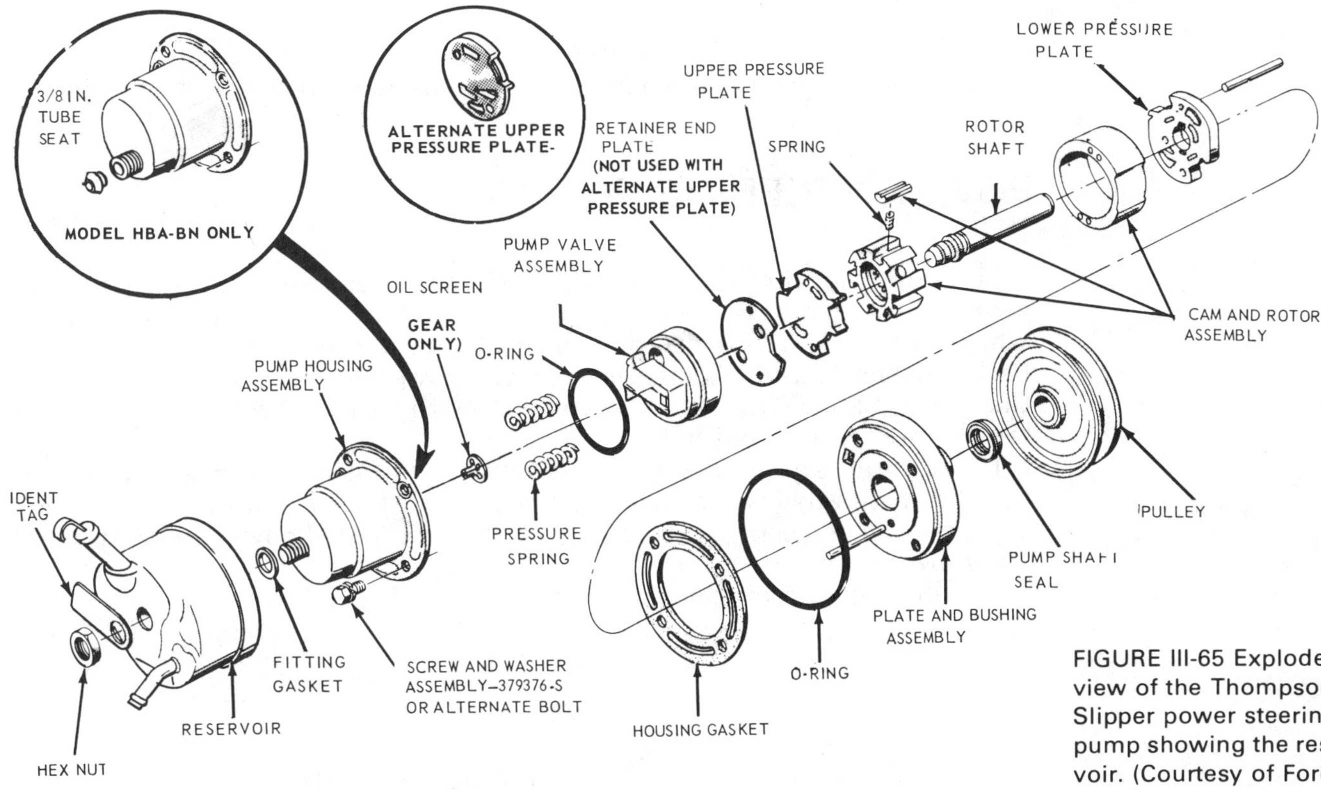

FIGURE III-65 Exploded view of the Thompson Slipper power steering pump showing the reservoir. (Courtesy of Ford Customer Service Division)

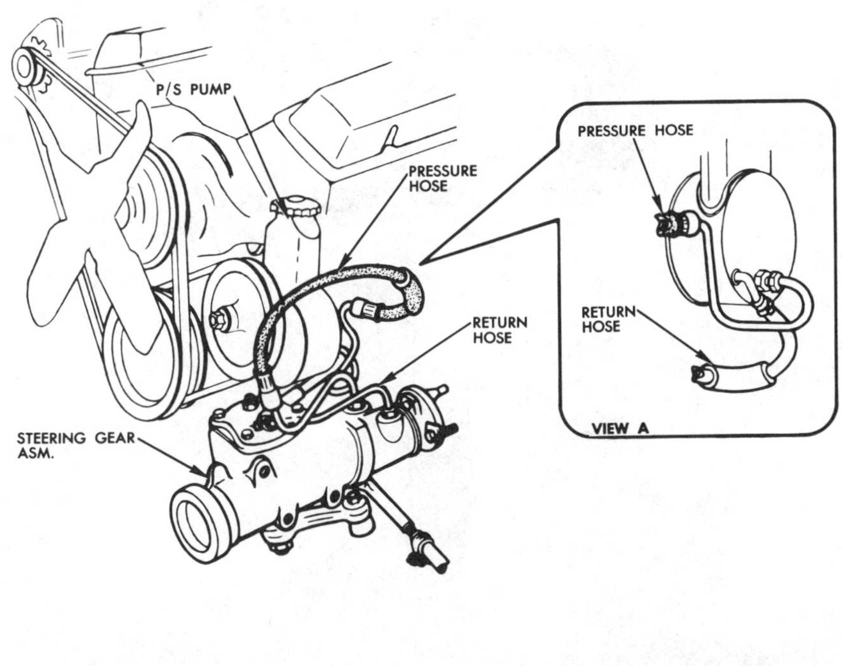

FIGURE III-66 Typical power steering arrangement showing pressure and return hoses. (Courtesy of Chevrolet Motor Division)

OBJECTIVE III-C1 (written test)

In a written test and without reference material, you will correctly answer a minimum of seven out of ten multiple-choice questions on the design and theory of the power steering gear, after studying the preparation assignment:

Preparation:

1. Study pages 272 through 289.
2. Study manufacturer's reference manuals on each power-steering gear.

POWER STEERING GEARS

Mechanically, there is very little difference between the power steering gear and the manual steering gear. Both use a recirculating ball system in which steel balls act as a rolling thread between the steering worm and the ball nut. The ball nut is one piece on the power steering gear and includes the power piston. The sector teeth mesh with teeth broached into the combined power piston and ball nut.

For all practical purposes, the power steering gear is a mechanical gear that has had a hydraulic assist and control system added to it.

On today's domestic car we find three basic types of power steering gears. They are:

1. Chrysler Corporation Constant Control (Fig. III-67)

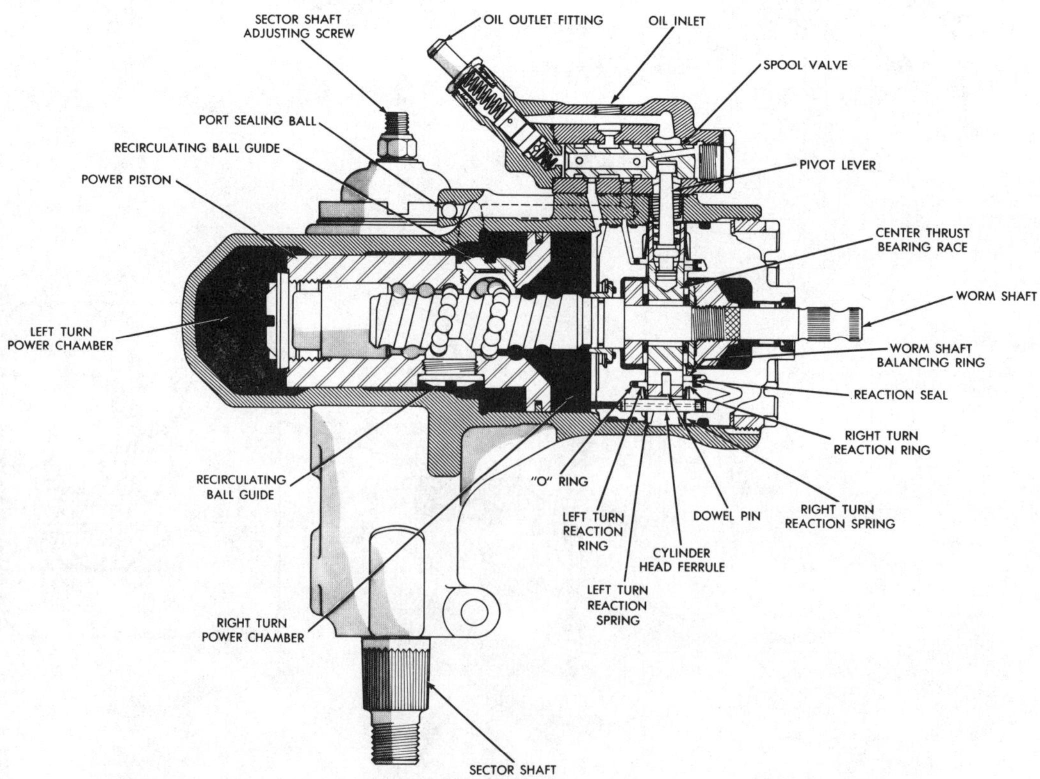

FIGURE III-67 Cutaway view of the constant control power steering gearbox. (Courtesy of Chrysler Motors Corp.)

2. Ford Motor Company Torsion Bar (Fig. III-68)
3. Saginaw's Rotary Valve (Fig. III-69)

All three of these power steering gears use an integral power steering unit with a hydraulic assisted system. These systems furnish power to reduce the amount of turning effort required to steer the automobile. They also help to reduce road shock and vibrations felt at the steering wheel.

If the power assist system fails because of a hydraulic problem, the mechanical system will still function and the driver will still have manual control of the car.

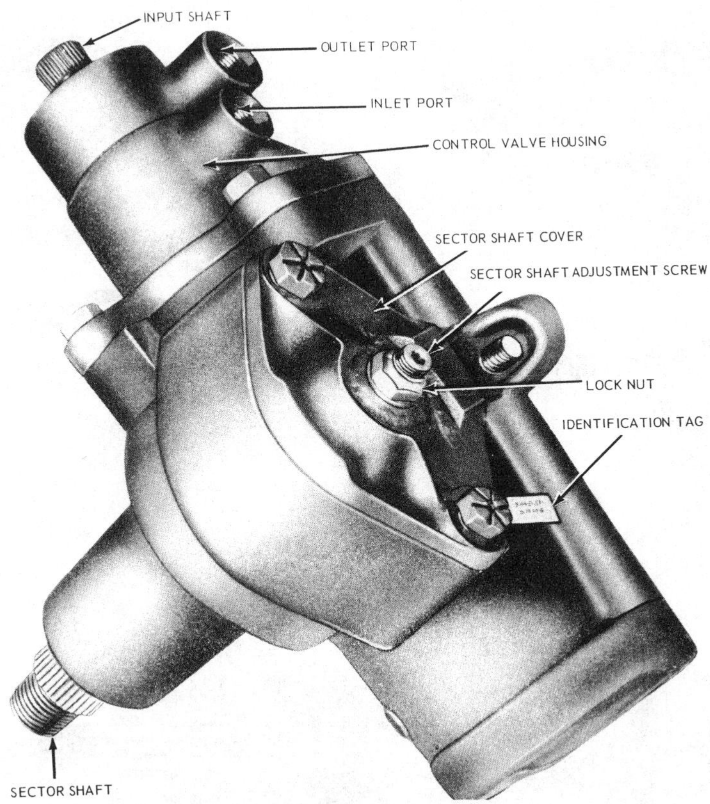

FIGURE III-68 Ford's integral power steering gear torsion bar. (Courtesy of Ford Customer Service Division)

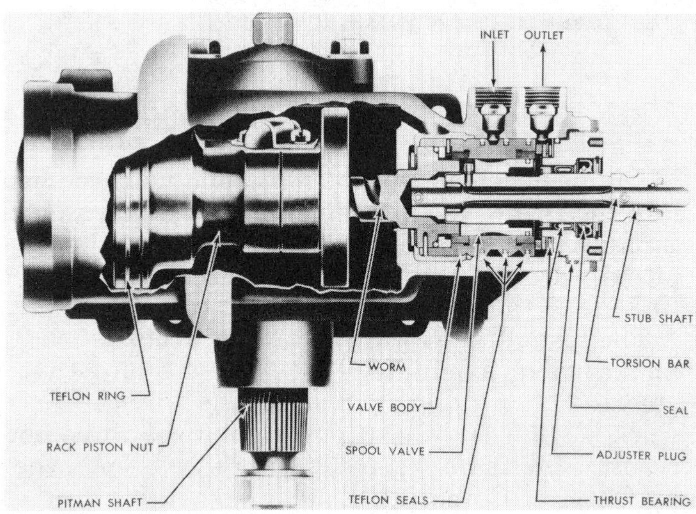

FIGURE III-69 Cutaway view of Saginaw's rotary valve power steering gear. (Courtesy of Chevrolet Motor Division)

POWER STEERING PISTON OPERATION

When the steering wheel is turned, the worm shaft of the power steering gear moves the power piston up or down in exactly the same way that the worm shaft of a manual gear moves the ball nut up or down (Fig. III-70). The direction of the turn depends on the cut of the worm gear and the placement of the pitman arm.

For example, in this Chrysler steering gear, the piston moves up in a left turn and down in a right turn. Just the opposite happens in the Buick: the piston moves up in a right turn (Fig. III-71) and down in a left turn (Fig. III-72).

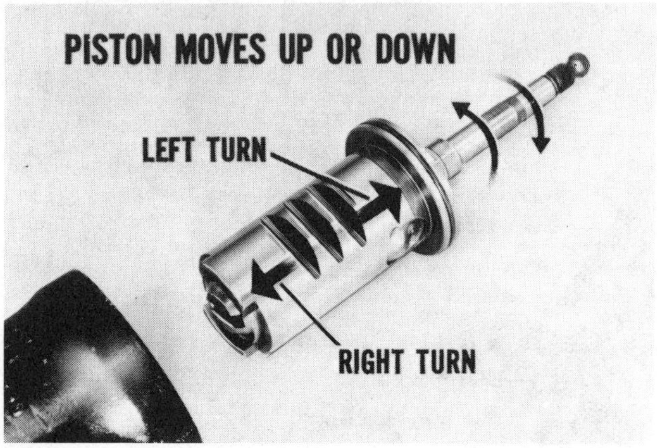

FIGURE III-70 Movement of the power piston in a right or left turn. (Courtesy of Chrysler Motors Corp.)

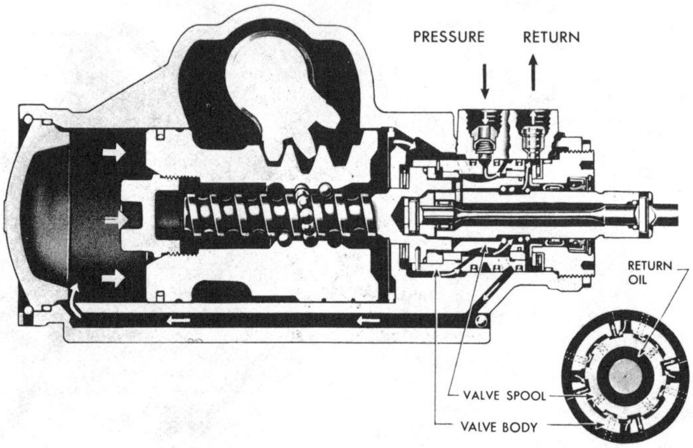

FIGURE III-71 The piston moves up in a right turn and down in a left turn. (Courtesy of Buick Motor Division)

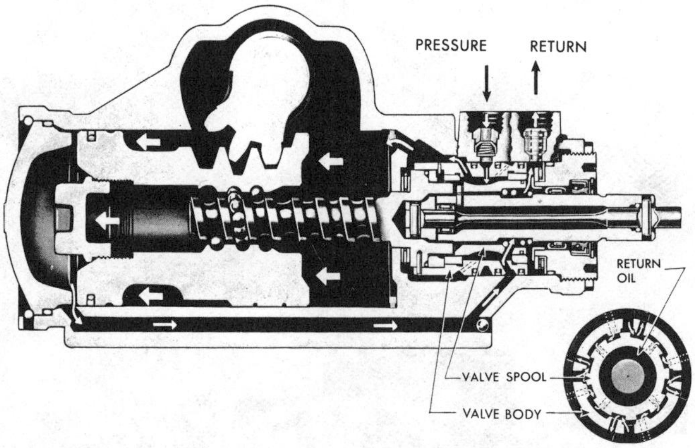

FIGURE III-72 This piston moves down in a left turn and up in a right turn. (Courtesy of Buick Motor Division)

The sector shaft teeth are meshed with the gear teeth that are broached into the side of the piston. The up-and-down movement of the power piston causes the sector shaft to rotate. The rotation of the sector shaft moves the pitman arm in an arc. This motion is transmitted through the steering linkage to the front wheels of the car.

The rest of the power steering gearbox is made up of a reaction assembly and a valve assembly. It is at this point that the three types of power steering units differ in design. The way the reaction and valve assemblies work is slightly different in each of the three power steering units, but the result is the same. This text will describe only the constant control unit used by Chrysler Corporation.

THE REACTION ASSEMBLY

The reaction assembly is made up of the following basic parts (Fig. III-73):

1. Cylinder head
2. Spring washers or reaction springs
3. Right and left turn reaction rings
4. Center race and spacer
5. Housing head
6. Pivot lever

First the cylinder head is placed on the worm shaft with accompanying seals. Next comes the left turn reaction ring and spring washer. After the left or lower reaction ring comes the center race and spacer. The right turn reaction ring and its spring washer are installed next. This much of the assembly is held in place by an adjusting nut. When the housing head is installed, it helps the cylinder head hold the reaction springs, washers, and reaction rings in place against the center race and spacer. The pivot lever is placed into a groove on the outer edge of the center race.

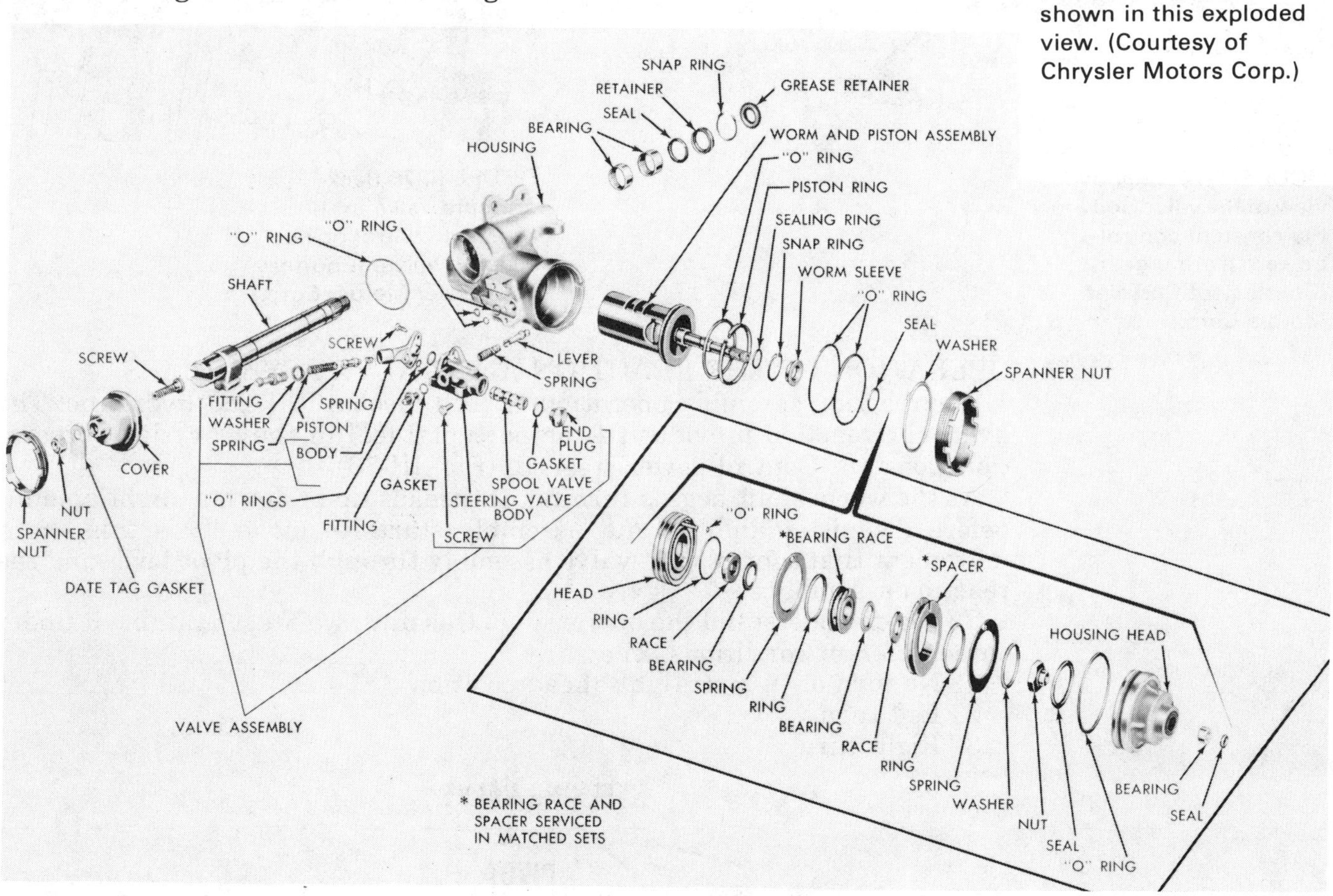

FIGURE III-73 The reaction assembly and its relation to the power piston. The Chrysler Corporation constant control power steering gear is shown in this exploded view. (Courtesy of Chrysler Motors Corp.)

THE VALVE ASSEMBLY

The valve assembly consists of two very basic valves and oil passages. The main valve is the steering or spool valve. It is this valve, with its connection to the pivot lever, that controls the direction of the fluid flow (Fig. III-74).

The back pressure valve is placed in the return line to give what its name implies, back pressure. It is because of back pressure that the driver is able to obtain the proper steering feel (Fig. III-75).

The two fluid passages extend from the steering valve mounting surface to the interior of the cylinder. One fluid passage leads from the upper part of the housing to the lower half of the power piston. The lower fluid passage goes to the upper half of the power piston.

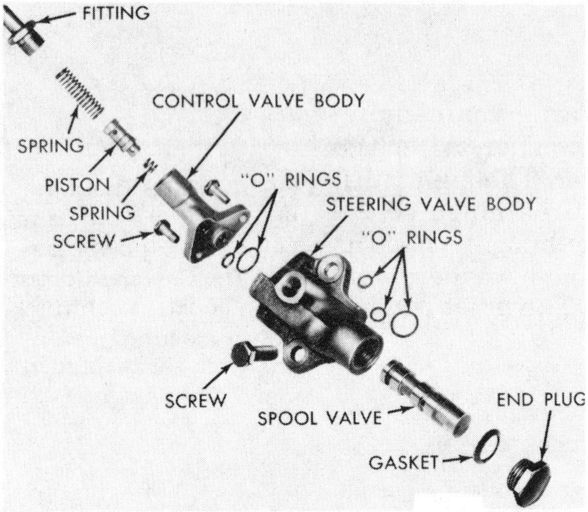

FIGURE III-74 Exploded view of the valve body for the constant control power steering gear. (Courtesy of Chrysler Motors Corp.)

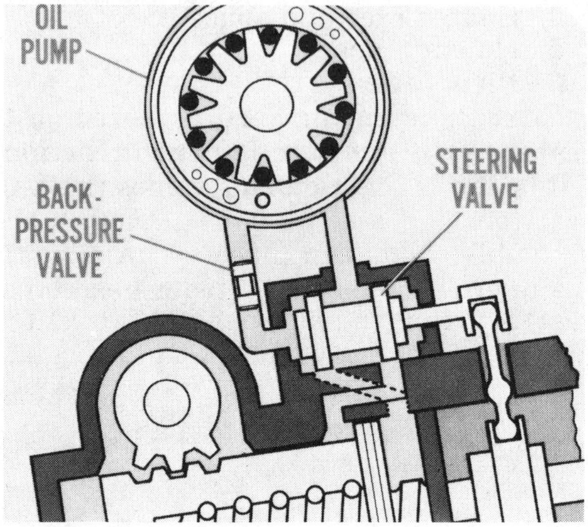

FIGURE III-75 Back-pressure valve in the constant control power steering valve. (Courtesy of Chrysler Motors Corp.)

OPERATION OF THE REACTION AND VALVE ASSEMBLIES

The reaction assembly operation is both mechanical and hydraulic. The hydraulic reaction provides the proportional feel in the Chrysler Corporation Constant Control power steering (Fig. III-76).

As the worm shaft begins turning, it threads up or down a slight amount before the piston and ball nut assembly starts to move. It is this slight movement that controls the valve assembly through the pivot lever and the reaction assembly.

To better understand the operation of this unit, we must examine it under three different conditions. They are:

1. No turn or in a straight-ahead position
2. Left turn
3. Right turn

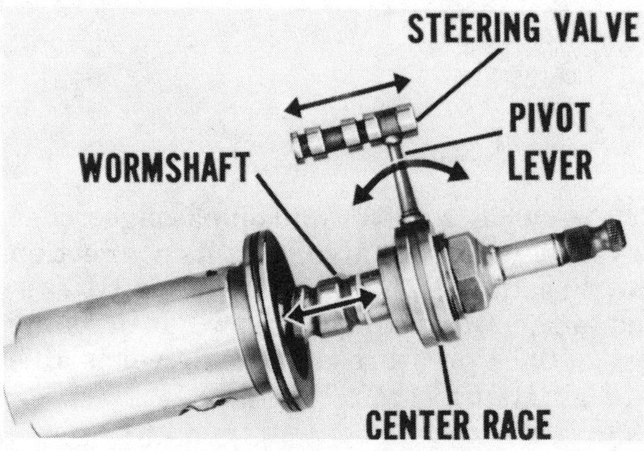

FIGURE III-76 The threading up or down of the worm shaft moves the reaction ring up or down and thus moves the valve up or down. (Courtesy of Chrysler Motors Corp.)

No Turn When there is no movement of the steering wheel, the steering valve is in the center position and pressure on the upper and lower side of the piston is equal (Fig. III-77). The return passage is open and the hydraulic oil returns to the power-steering pump. Both sides of the power piston have equal pressure, so there is no power assist.

Power Assist In A Left Turn When the driver turns the steering wheel for a left turn, the worm shaft threads itself down into the power piston and ball nut assembly (Fig. III-78). The movement of the worm shaft moves the reaction ring down and causes the pivot lever to move the steering valve up (Fig. III-79). With the steering valve up, the passage to the lower side of the power piston is open to the pressure side of the power steering pump. At the same time, the passage to the upper side of the power piston is open to the return line. This creates a difference in pressure, and the power piston is pushed up because of the mechanical gear on the worm shaft and the hydraulic pressure on the lower side of the power piston.

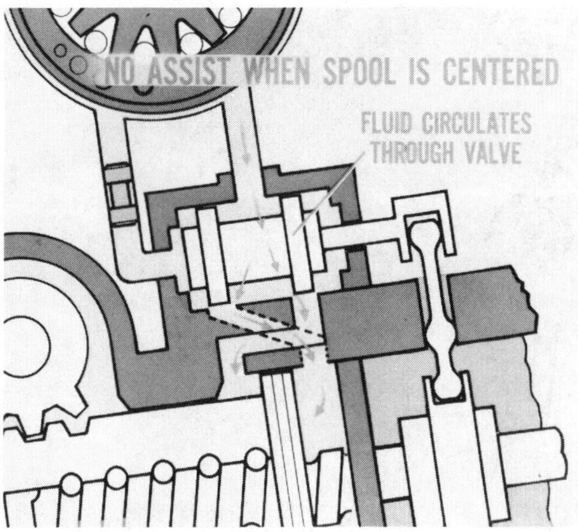

FIGURE III-77 When there is no driver's effort on the steering wheel, the reaction ring is centered, and the valve allows equal pressure to each side of the power piston. (Courtesy of Chrysler Motors Corp.)

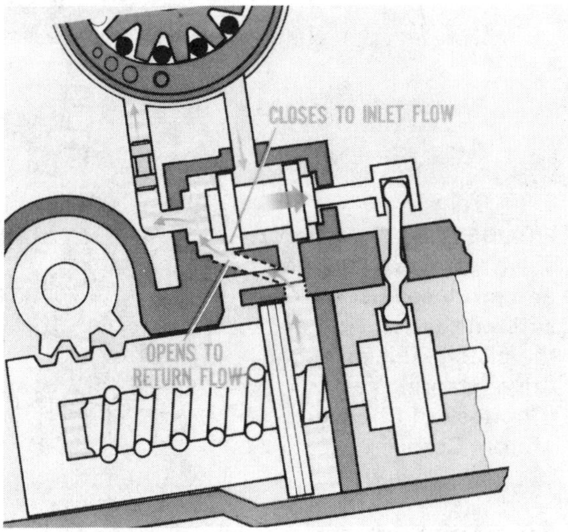

FIGURE III-78 Return port opens to the top of the power piston during a left turn. (Courtesy of Chrysler Motors Corp.)

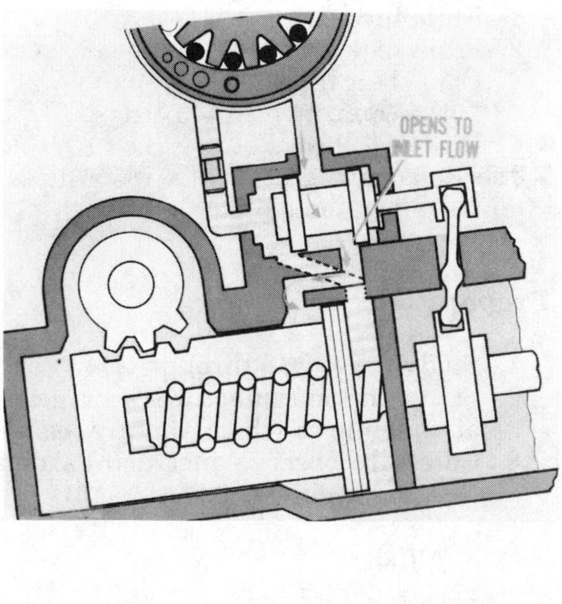

FIGURE III-79 Inlet flow is open to the bottom side of the power piston during a left turn. (Courtesy of Chrysler Motors Corp.)

If the driver wants to hold the steering wheel in any position, the downward pressure on the worm shaft is removed, and the steering valve recenters itself. Thus, the driver is able to hold the steering wheel in any position without fighting hydraulic pressure.

During the turn, the spring-loaded control valve restricts the return line flow so that a back pressure develops in the upper side of the power piston (Fig. III-80). This back pressure keeps the turning motion smooth and helps to provide the desirable steering feel, especially in finding the straight-ahead position again.

Power Assist In A Right Turn The same hydraulic assist action takes place for a right turn as in a left turn, but the steering valve and the power piston movements are reversed. The power assist continues as long as turning force is applied to the steering wheel.

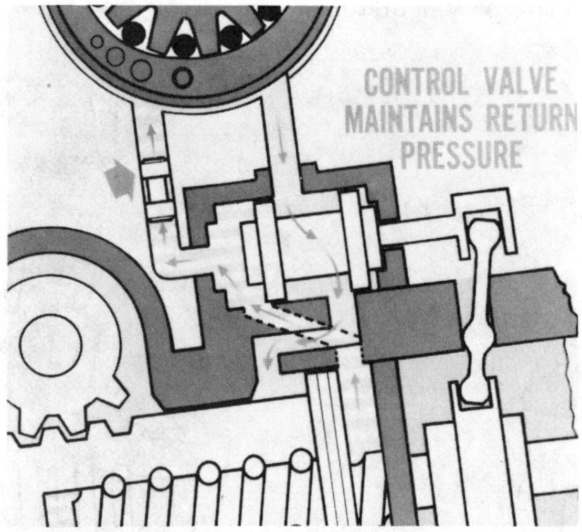

FIGURE III-80 The control valve restricts return flow and produces back pressure on the top side of the piston. This gives the driver steering feel. (Courtesy of Chrysler Motors Corp.)

OBJECTIVE III-C2 (manipulative)

Working alone, you will diagnose, repair or adjust, and report in writing the condition of the power steering system on a minimum of three instructor-assigned automobiles. The diagnosis will include the following:

1. Fluid level
2. Fluid leakage from the power steering system
3. Drive belt tension
4. Loose component mountings
5. Pressure checks

Job Report Sheet III-C2 will be used for your report. Time limit: 20 minutes per automobile.

Preparation:

1. Study pages 278 through 294.
2. Study the manufacturer's or general repair manual for the specific automobile to which you are assigned.
3. Carefully observe instructor's demonstration on diagnosing power-steering problems.

Materials and Tools Needed:

1. Set of box/open-end combination wrenches
2. Belt-tension gauge
3. Pressure gauges
4. Power steering fluids

Complaints of power-steering problems are usually the result of something other than the power steering gear or pump. You should therefore always check those items that relate to the steering system before condemning the power steering pump or gear.

Conditions such as hard or loose steering, road shock, or vibrations are often the result of such factors as low tire pressure, front end misalignment, or worn or damaged suspension—or steering linkage parts. These parts should always be checked and corrected before any adjustment or disassembly of the power steering gear or pump is attempted.

FLUID LEVEL CHECK

The following are the recommended procedures and levels of fluid for each of the three major manufacturers of power steering systems:

CHRYSLER CORPORATION POWER STEERING SYSTEMS Fluid should be checked every three months or at every oil change. To check the power steering fluid level, start the engine and turn the steering wheel from stop to stop several times, to expel air from the system. On a cold engine, oil level should be only to the bottom of the filler neck (Figs. III-81 and III-82). However, if the engine is hot, oil should be halfway up in the filler neck.

Do not use transmission oil. Use *Chrysler Corporation's power steering fluid #2084329 or its equivalent.*

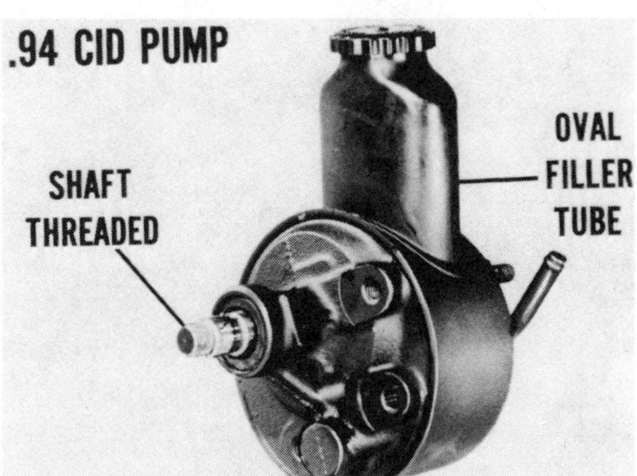

FIGURE III-81 Fill the power steering unit only with approved power steering fluid. If no dipstick is provided, fill only to the bottom of the filler tube. (Courtesy of Chrysler Motors Corp.)

FIGURE III-82 With this type of unit, fill to the bottom of the filler tube. (Courtesy of Chrysler Motors Corp.)

FORD MOTOR COMPANY POWER STEERING SYSTEMS Fluid level should be checked every 6,000 miles, using the following procedure: Run the engine until fluid is at normal operating temperature. Then turn the steering wheel all the way from one stop to the other, to expel any air (Fig. III-83). With the engine off, check fluid level. The fluid level must show on the cross-hatching between the bottom of the dipstick and the "full" mark. **Caution:** *Do not overfill. Use only Ford power steering fluid C1A2-19582-A or its equivalent.*

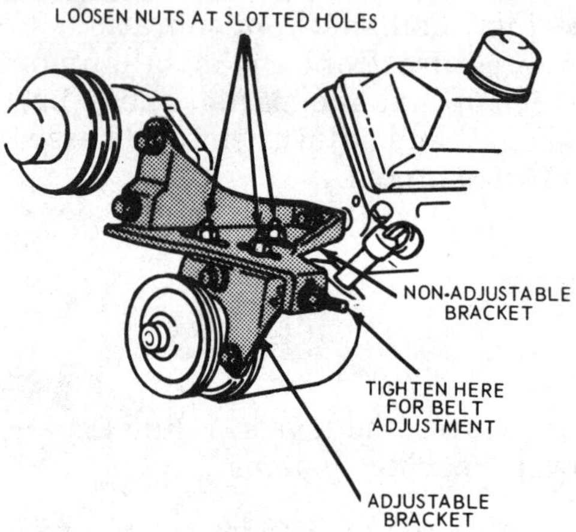

FIGURE III-83 Ford Motor Company power steering systems should not be overfilled. Use approved power steering fluid only. (Courtesy of Ford Customer Service Division)

SAGINAW COMPANY POWER-STEERING SYSTEMS Fluid level should be checked every 6,000 miles, using the following procedure: Run the engine until fluid is at normal operating temperature (Fig. III-84). With the engine off, check fluid level. The fluid level should be to the "full" mark on the dipstick. *Use only Saginaw power steering fluid #1050017 or its equivalent.*

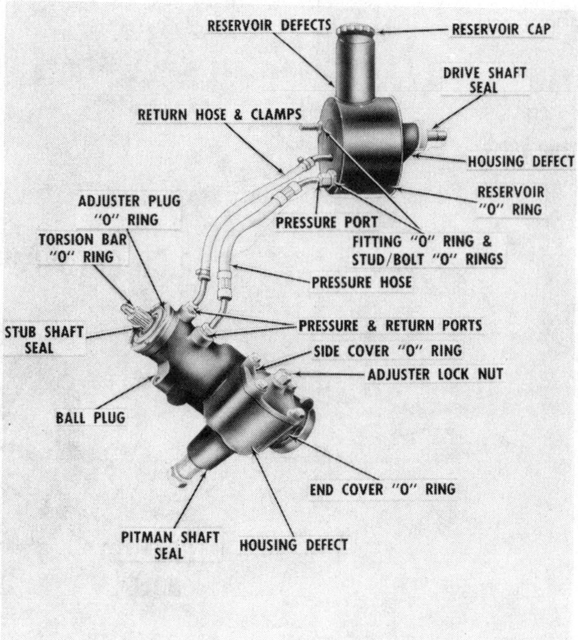

FIGURE III-84 The illustration shows a Saginaw Company power steering system. Run engine until operating temperature is reached before checking fluid level. (Courtesy of Chevrolet Motor Division)

STEERING SYSTEM LEAKAGE CHECKS

Each type of steering system will be different in its leakage problems. However, as a guide, the text will show those likely spots for leaks in the Saginaw system.

Figs. III-85 and III-86 show those spots where leaks may occur in both the power steering pump and gear. The following procedure should be used in checking for fluid leaks:

1. With the vehicle's engine off, wipe or steam clean the entire power steering system: gear pump, hoses, and connections.
2. Check oil level in pump's reservoir and fill to specified level with approved power steering fluid.
3. Start engine and have someone turn the steering wheel from stop to stop, several times. Do not hold the steering wheel against the stop for any length of time as this may result in power steering-pump damage. As the steering wheel is being turned, watch for leakage at those spots shown in Fig. III-85 and III-86.

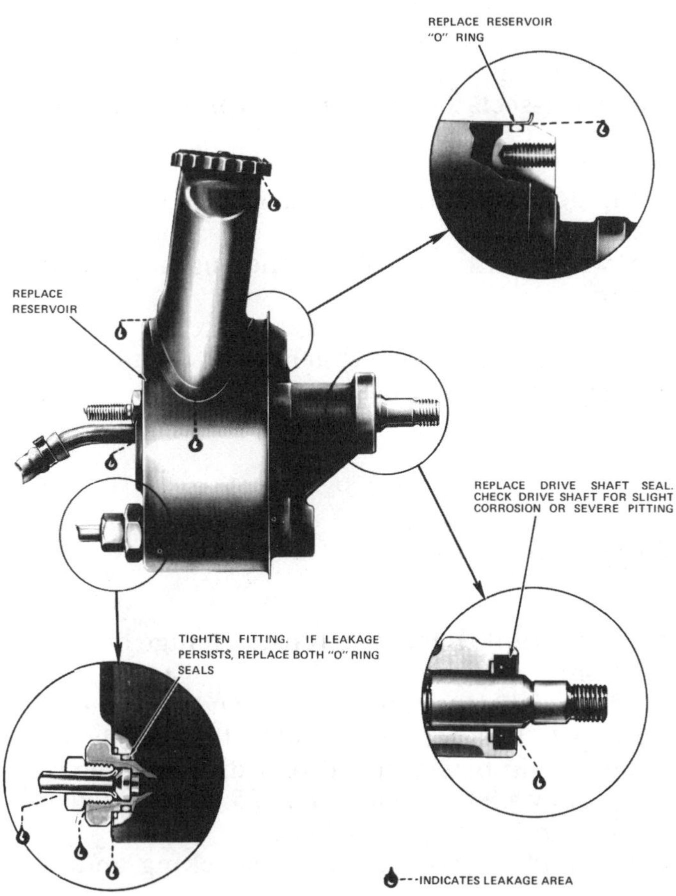

FIGURE III-85 Power steering pump leak diagnosis may be accomplished by visual inspection. (Courtesy of Pontiac Motor Division)

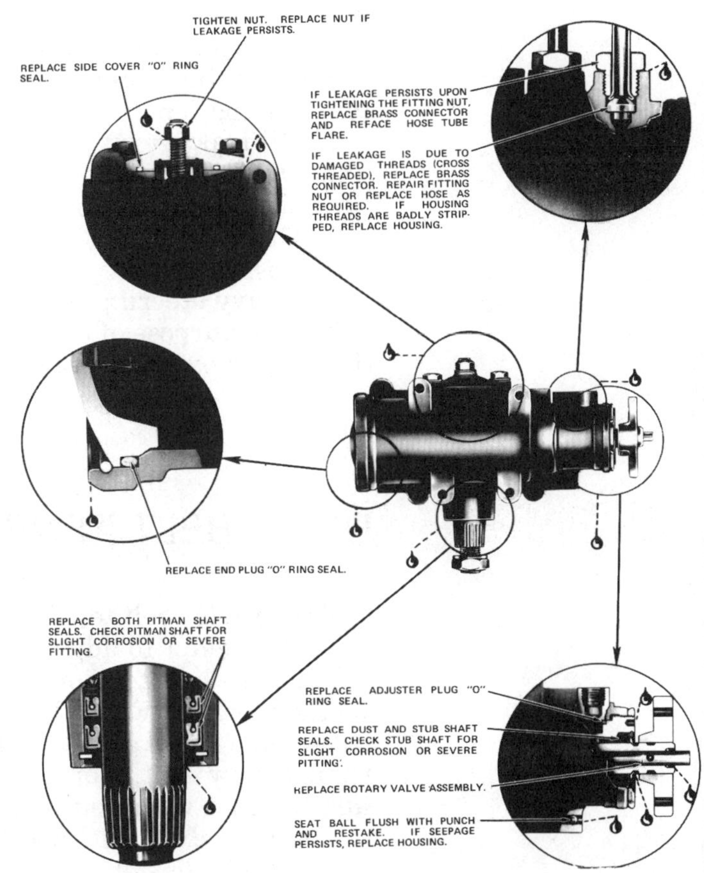

FIGURE III-86 Power steering gear-link diagnosis requires careful observation. (Courtesy of Pontiac Motor Division)

POWER STEERING LEAKAGE DIAGNOSIS CHEVROLET DIVISION OF GENERAL MOTORS CORPORATION

EXTERNAL LEAKAGE

General Procedure

1. Wipe suspected area dry.
2. Check for overfilled reservoir.
3. Check for oil aeration and overflow.
4. Check hose connections—tighten if necessary.
5. Verify exact point of leakage.
   Example: Torsion bar, stub shaft, and adjuster seals are close together; the exact leakage point could be confused.
   Example: The point from which oil drips is not necessarily the leakage point: oil overflowing from the reservoir, for instance.
6. When service is required:
   a. Clean leakage area before disassembly.
   b. Replace leaking seal.
   c. Check component sealing surfaces for damage.
   d. Reset bolt torque to specifications where required.

Some of the customer complaints associated with the power steering system may be reported as:

1. Oil leakage on garage floor
2. Oil leaks visible on steering gear, pump, or anywhere else on the left side of engine compartment
3. Growling noise, especially when parking or when the engine is cold
5. Loss of power when parking
5. Heavy steering effort

For the purpose of troubleshooting complaints of this nature, assume that there is an external leak in the power steering system.

DRIVE BELT TENSION

To check for belt tension, it is important that a belt tension gauge be used. Do not attempt to adjust by feel only.

Place the belt tension gauge midway between the pulleys on the drive belt being checked (Fig. III-87). If a belt has been previously tensioned, it is considered to be a used belt and should be tightened to a different setting than a new belt. For example, Chevrolet recommends 75 lb. tension be applied on a used belt and 125 lb. tension on a new belt.

The procedure outlined here should be followed while checking and adjusting belt tension:

1. Loosen the pump attaching bolts and adjust the belt to correct tension by moving the pump outward away from the engine. (**Caution:** *Do not pry or put pressure on the pump reservoir or pull against the filler neck. Always pry against the pump housing or other spots specified by the manufacturer.)*

2. Snug all pump mounting bolts and remove the pry bar.
3. Tighten all pump mounting bolts to specified torque.
4. Recheck belt tension and remove belt tension gauge.

FIGURE III-87 Always use a belt tension gauge, as shown, to check power steering pump belts. (Courtesy of Chevrolet Motor Division)

LOOSE COMPONENT MOUNTINGS

Check all bracket and mounting bolts for both the power steering pump and gear. The pump will have a variety of attachments, and each manufacturer's recommended tightening specifications should be followed (Fig. III-88). The power steering gear is always mounted to the side of the frame with a minimum of three bolts. If the bolts are loose, check for spacer shims that may have fallen out. It is important that the correct alignment of the gearbox is maintained.

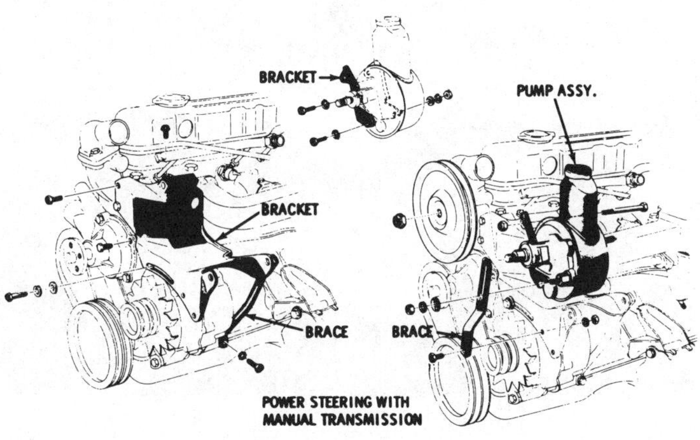

FIGURE III-88 Typical power steering pump mounting brackets. (Courtesy of Oldsmobile Motor Division)

PUMP PRESSURE CHECKS

If all other checks fail to produce the cause of a problem, a pressure check of the system should be performed. To make a pump pressure check, the vehicle's front wheels must be on the ground with the vehicle weight on them. **Caution:** *Do not hold the steering wheel at a full turn for over five seconds (Fig. III-89). This will prevent damage to the pump.*

To make the pressure check, you will need a pressure gauge sufficient to check pressures of at least 900 lb. psi. Connect this pressure gauge in the high pressure line between pump and steering gearbox.

Start the engine and allow the system to reach operating temperature. Then shut the engine off and check fluid level. Fill if necessary.

To check the pressure, turn the steering wheel slowly to the left or right until wheel is at full turn position. Read the pressure gauge while you hold the steering wheel in this position. Compare the pressure gauge reading with the specifications.

To determine the cause of a low pressure reading, slowly close the hand valve on the valve gauge. If the pressure comes up to a normal reading, the problem is in the power steering gear. If the pressure remains low, the problem is in the power steering pump.

Caution: *Do not hold the hand valve on the gauge closed for over five seconds during this test or pump damage may result.*

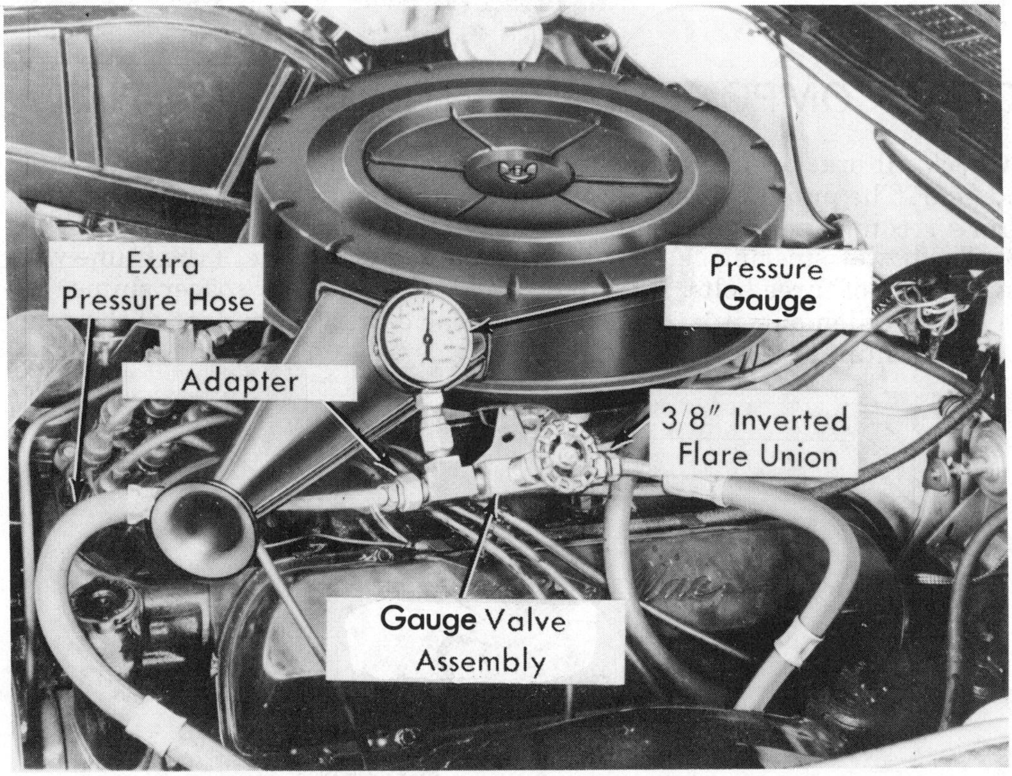

FIGURE III-89 A pressure gauge setup for checking power steering pump pressure. (Courtesy of Cadillac Motor Car Division)

POWER STEERING DIAGNOSTIC CHART
CHEVROLET DIVISION OF GENERAL MOTORS CORPORATION

CONDITION	*POSSIBLE CAUSE*	*CORRECTION*
System Noise: There is some noise in all power steering systems. Common complaints are listed as follows:		
Pump noise-chirp	Loose belt	Adjust belt tension to specification
Belt squeal	Loose belt	Adjust belt tension to specification
Gear noise ("hissing" sound)	There is some noise in all power-steering systems. One of the most common is a hissing sound most evident at "standstill" parking. There is no relationship between this noise and performance of the steering. *Hiss* may be expected when steering wheel is at end of travel or when slowly turning at standstill.	Do not replace valve unless *hiss* is extremely objectionable. Slight hiss is normal and in no way affects steering. A replacement valve will also exhibit slight noise and is not always a cure for the objection. Investigate clearance around flexible coupling screws. Be sure steering shaft and gear are aligned so flexible coupling rotates in a flat plane and is not distorted as shaft rotates. Any metal to metal contacts through flexible coupling will transmit valve hiss into car.
Rattle	Pressure hose touching other parts of car	Adjust hose position
Rattle	Pump vanes not installed properly	Install properly
Rattle	Pump vanes sticking in rotor slots	Free up by removing burrs, varnish, or dirt
Gear noise (rattle or chuckle)	Loose pitman shaft adjustment **Note:** *A slight rattle may occur on turns because of increased clearance off the "high point." This is normal, and clearance must not be reduced below specified limits to eliminate this slight rattle.*	Adjust to specifications
Gear noise (rattle or chuckle)	Gear loose on frame	Check gear to frame mounting bolts. Tighten bolts to 70 ft-lb.
Rattle or chuckle	Steering linkage looseness.	Check linkage pivot points for wear. Replace if necessary
Groan	Low oil level	Fill reservoir to proper level
Groan	Air in the oil; poor pressure hose connection	Bleed system by operating steering from right to left, full turn
Growl	Excessive back pressure caused by hoses or steering gear	Locate restriction and correct. Replace part if necessary
Growl	Scored pump pressure plates, thrust plate, or rotor	Replace affected parts
Growl	Extreme wear of pump cam ring	Replace affected parts
Swish	Defective pump-flow control plunger	Replace part

CONDITION	POSSIBLE CAUSE	CORRECTION
Whine	Pump shaft bearing scored	Replace housing and shaft
Excessive wheel kickback or loose steering	Backlash in steering linkage	Adjust parts affected or replace worn parts
	Air in system	Add oil to pump reservoir and bleed by operating steering
	Excessive "on-center" lash	Adjust to specification
	Loose thrust bearing-preload adjustment	Adjust to specification
	Worn poppet valve	Replace poppet valve
	Steering gear loose on frame	Tighten attaching bolts to 70 ft-lb.
	Steering gear flexible coupling too loose on shaft or rubber-disc mounting screws loose	Tighten flange pinch bolts to 30 ft-lb. if serrations are not damaged. Tighten upper flange to coupling nuts to 20 ft-lb.
	Steering linkage ball joints worn enough to be loose	Replace loose components
	Front wheel bearings incorrectly adjusted or worn	Adjust bearings or replace with new parts as necessary
Poor return of steering	Tires underinflated	Inflate to specified pressure
	Lower coupling flange rubbing against steering gear adjuster plug	Loosen pinch bolt and assemble properly
	Steering wheel rubbing against directional-signal housing	Adjust steering jacket
	Frozen steering shaft bearings	Replace bearings
	Steering linkage binding	Replace pivots
	Steering gear to column misalignment	Align steering column
	Tie rod pivots not centralized	Adjust tie rod ends as required, to center pivots
	Lack of lubricant in suspension ball joints	Repack
	Steering gear adjustments too tight	Check adjustment with pitman arm disconnected. Readjust if necessary
	Sticky or plugged valve spool	Remove and clean, or replace valve
	Rubber spacer binding in shift tube	Make certain spacer is properly seated. Lubricate inside diameter with silicone
	Improper front wheel alignment	Check and adjust to specifications
	Tight steering shaft bearings	Replace bearings
Car leads to one side or the other (Keep in mind road condition and wind. Test car on flat road going in both directions.)	Front end misaligned	Adjust to specifications

CONDITION	POSSIBLE CAUSE	CORRECTION
	Steering shaft rubbing ID of shift tube	Align column
	Unbalanced or badly worn steering-gear valve	Replace valve
	Note: *If this is cause, steering effort will be very light in direction of lead and heavy in opposite direction.*	
	Steering linkage not level	Adjust as required
Steering wheel surges or jerks when turning with engine running, especially during parking	Loose pump belt	Adjust tension to specification
	Sticky flow control valve	Inspect for varnish or damage. Replace if necessary
	Insufficient pump pressure	Check pump pressure. (See pump pressure test.) Replace relief valve if defective
	Steering linkage hitting engine oil pan at full turn	Correct clearance
Momentary increase in effort when turning wheel fast to right or left	Pump belt slipping	Tighten or replace belt
	Low oil level in pump	Check oil level in pump reservoir
Hard steering or lack of assist	High internal leakage	Check pump pressure. (See pump pressure test.)
	Loose pump bolt	Adjust belt tension to specification
	Low oil level in reservoir	Fill to proper level. If excessively low, check all lines and joints for evidence of external leakage
	Lack of lubricant in suspension ball joints	Repack
	Tires not properly inflated	Inflate to recommended pressure
	Steering gear to column misalignment	Align steering column
	Improper front or rear wheel alignment	Check and adjust as necessary
	Steering gear adjusted too tight	Test steering system for binding with front wheels off floor. Adjust as necessary
	Excessive friction in steering linkage	Check tie rod pivot points for excessive friction. Replace the affected pivot
	Excessive caster or toe-in	Adjust caster and toe-in to specifications
	Incorrect camber	Check and adjust to specifications
	Suspension arms bent or twisted	Check wheel camber and caster. Replace bent arms with new ones

CONDITION	POSSIBLE CAUSE	CORRECTION
	Steering knuckle bent	Replace with new knuckle
	Frame bent	Check frame for proper alignment or cracking. Repair or replace as necessary
	Front springs weak and sagging	Check standing height. Weak or sagging springs should be replaced with new ones
	Insufficient oil pressure	If above checks do not reveal cause of hard steering, check pump pressure
Low oil pressure due to restriction in hoses	Check for kinks in hoses	Remove kink
	Foreign object stuck in hose	Remove hoses and remove restricting object or replace hose
Low oil pressure due to steering gear (See pump pressure test.)	Pressure loss in cylinder due to worn piston ring or scored housing bore	Remove gear from car for disassembly and inspection of ring and housing bore
	Leakage at valve rings, valve body due to worm seal	Remove gear from car for disassembly and replace seals
	Loose fit of spool in valve body, or leaky valve body	Replace valve
	Damaged poppet valve	Replace valve
Low oil pressure due to steering pump	Loose belt	Adjust tension to specification
	Low oil level	Fill reservoir to proper level
	Air in the oil	Locate source of leak and correct. Bleed system
	Defective hoses or steering gear	Correct as necessary
	Flow control valve stuck or inoperative	Remove burrs or dirt or replace
	Loose screw in end of flow control valve	Tighten
	Cracked or broken thrust or pressure plate	Replace part
	Pressure plate not flat against cam ring	Replace pressure plate
Steering System External Leakage	Extreme wear of cam ring	Replace parts
	Scored pressure plate, thrust plate, or rotor	Replace parts. (If rotor, replace with rotating group kit.)
	Vanes not installed properly	Install properly
	Vanes sticking in rotor slots	Free up by removing burrs, varnish, or dirt

Courtesy of Chevrolet Motor Division

JOB REPORT SHEET III-C2

DIAGNOSING POWER STEERING PROBLEMS ASSIGNMENT 1

Student name________________________________Date______________________

Make of car_______________Model_______________Year______________________

Time started______________Time finished______________Total time______________

Time recommended __________

A. Fluid Check

1. Recommended level of fluid is: ______________________________________

2. Recommended fluid is: ____

3. Fluid was checked at (normal) (cold) engine temperature.
4. Fluid at recommended level? (Yes) (No)

B. Fluid leakage check from power steering system

1. Power steering pump: ____

2. Power steering gear: ____

3. Power steering hose connections: __________________________________

C. Belt-tension check

1. Recommended belt tension: ______________________________________
2. Belt tension before adjustment: __________________________________
3. Belt tension after adjustment: ___________________________________

D. Loose component mounting check

1. Pump mounting: ____________________

____

____

2. Gear box mounting: ____________________

____

____

3. Hose brackets or clamps: ____________________

____

____

E. Pressure checks

1. Recommended pressure: No turn—Idle

Minimum: ______________ Maximum: ______________

2. Recommended pressure: Full turn against stop—Idle

Minimum: ______________ Maximum: ______________

3. Pressure checks show unit (OK) (Not OK)

F. The following problems need correction:

1. ____________________
2. ____________________
3. ____________________
4. ____________________
5. ____________________

DATE COMPLETED ______________ INSTRUCTOR'S SIGNATURE ______________

JOB REPORT SHEET III-C2

ASSIGNMENT 2

DIAGNOSING POWER STEERING PROBLEMS

Student name ________________________ Date ________________

Make of car ____________ Model ____________ Year ____________

Time started ____________ Time finished ____________ Total time ____________

Time recommended ____________

A. Fluid Check

1. Recommended level of fluid is: ____________________________

____

2. Recommended fluid is: ____________________________

____

3. Fluid was checked at (normal) (cold) engine temperature.

4. Fluid at recommended level? (Yes) (No)

B. Fluid leakage check from power steering system

1. Power steering pump: ____________________________

____

____

2. Power steering gear: ____________________________

____

____

3. Power steering hose connections: ____________________________

____

____

C. Belt tension check

1. Recommended belt tension: ____________________________

2. Belt tension before adjustment: ____________________________

3. Belt tension after adjustment: ____________________________

D. Loose component mounting check

1. Pump mounting: ______________________________

____

____

2. Gear box mounting: ______________________________

____

____

3. Hose brackets or clamps: ______________________________

____

____

E. Pressure checks

1. Recommended pressure: No turn—Idle

Minimum: ______________ Maximum: ______________

2. Recommended pressure: Full turn against stop—Idle

Minimum: ______________ Maximum: ______________

3. Pressure checks show unit (OK) (Not OK)

F. The following problems need correction:

1. ______________________________
2. ______________________________
3. ______________________________
4. ______________________________
5. ______________________________

DATE COMPLETED______________ INSTRUCTOR'S SIGNATURE ______________

JOB REPORT SHEET III-C2
ASSIGNMENT 3

DIAGNOSING POWER-STEERING PROBLEMS

Student name ______________________ Date ______________

Make of car ____________ Model ____________ Year ____________

Time started ____________ Time finished ____________ Total time ____________

Time recommended ____________

A. Fluid Check

1. Recommended level of fluid is: ______________________

____

2. Recommended fluid is: ______________________

____

3. Fluid was checked at (normal) (cold) engine temperature.
4. Fluid at recommended level? (Yes) (No)

B. Fluid leakage check from power-steering system

1. Power steering pump: ______________________

____

____

2. Power steering gear: ______________________

____

____

3. Power steering hose connections: ______________________

____

____

C. Belt tension check

1. Recommended belt tension: ______________________
2. Belt tension before adjustment: ______________________
3. Belt tension after adjustment: ______________________

D. Loose component mounting check

1. Pump mounting: ______________________________

____

____

2. Gear box mounting: ______________________________

____

____

3. Hose brackets or clamps: ______________________________

____

____

E. Pressure checks

1. Recommended pressure: No turn—Idle

Minimum: ______________ Maximum: ______________

2. Recommended pressure: Full turn against stop—Idle

Minimum: ______________ Maximum: ______________

3. Pressure checks show unit (OK) (Not OK)

F. The following problems need correction:

1. ______________________________

2. ______________________________

3. ______________________________

4. ______________________________

5. ______________________________

DATE COMPLETED ______________ INSTRUCTOR'S SIGNATURE ______________

Student name________________________________Date______________________

Directions: Select the best answer to each question and circle the correct letter at the left.

1. Hydraulic oil will transmit:
   I. Pressure
   II. Motion
      A. I only
      B. II only
      C. Both I and II
      D. Neither I nor II
2. A piston that is 1 inch in diameter moves 5 inches. Another piston 2 inches in diameter connected hydraulically by a tube will move:
      A. 2.5 inches
      B. 5 inches
      C. 10 inches
      D. 15 inches
3. A piston that is 1 inch in diameter moves 5 inches. Another piston ½ inch in diameter connected hydraulically by a tube will move:
      A. 2.5 inches
      B. 5 inches
      C. 10 inches
      D. 15 inches
4. Hydraulic oil transmits pressure pound for pound per:
   I. Cylinder
   II. Square inch
      A. I only
      B. II only
      C. Both I and II
      D. Neither I nor II
5. One hundred pounds of pressure is exerted against the applying piston. The applied piston exerts 200 pounds pressure. Which piston is the largest of the two?
   I. The applying piston
   II. The applied piston
      A. I only
      B. II only
      C. Both are the same size
6. The power steering pump used in the automobile is capable of producing pressures of more than:
   I. 1,000 psi
   II. 2,000 psi
      A. I only
      B. II only
      C. Both I and II
      D. Neither I nor II

7. Fifty pounds of pressure is exerted against the applying piston. The applied piston is the same diameter as the applying piston. What is the pressure exerted by the applied piston?
   I. 50 lb.
   II. 50 psi
      A. I only
      B. II only
      C. Both I and II
      D. Neither I nor II
8. The Saginaw pump uses________________to produce the pressure and motion in the hydraulic oil:
   I. Slippers
   II. Rollers
      A. I only
      B. II only
      C. Both I and II
      D. Neither I nor II
9. The rollers are placed in a star-shaped rotor inside a cam ring on the:
   I. Chrysler pump
   II. Saginaw pump
      A. I only
      B. II only
      C. Both I and II
      D. Neither I nor II
10. All power steering pumps must have a:
   I. Back-pressure valve
   II. Flow-control valve
      A. I only
      B. II only
      C. Both I and II
      D. Neither I nor II

DATE COMPLETED________________INSTRUCTOR'S SIGNATURE ________________

TEST III-C1

Student name________________________________Date____________________

Directions: Select the best answer to each question and circle the correct letter at the left.

Select the best answer to each question and circle the correct letter at the left.

1. In the power steering gearbox, the:
   I. Ball system has steel balls that act as roller threads between the worm and the ball nut
   II. Power piston and the ball nut are combined in the same piece
      A. I only
      B. II only
      C. Both I and II
      D. Neither I nor II

2. The________________power steering system is of the integral type:
   I. Chrysler's Constant Control
   II. Ford's Torsion Bar
      A. I only
      B. II only
      C. Both I and II
      D. Neither I nor II
3. In a left turn, the power piston always moves:
   I. Up
   II. Down
      A. I only
      B. II only
      C. Either I or II
      D. Neither I nor II
4. The teeth on the sector (pitman) shaft are meshed with:
   I. The worm gear
   II. The ball nut teeth
      A. I only
      B. II only
      C. Both I and II
      D. Neither I nor II
5. The reaction assembly of the Chrysler Constant Control power steering gear has a:
   I. Cylinder piston
   II. Right turn reaction ring
      A. I only
      B. II only
      C. Both I and II
      D. Neither I nor II
6. The back pressure valve for the Chrysler Constant Control power steering gear is placed in the:
   I. Return hose
   II. Pressure hose
      A. I only
      B. II only
      C. Both I and II
      D. Neither I nor II

7. In a left turn, the spool valve for the Chrysler Constant Control power steering gear allows fluid to flow to the:
   I. Upper side of the power piston
   II. Lower side of the power piston
      A. I only
      B. II only
      C. Both I and II
      D. Neither I nor II
8. In the Chrysler Constant Control power steering gear, the reaction ring is moved by the up-or-down movement of the:
   I. Worm shaft
   II. Power piston
      A. I only
      B. II only
      C. Both I and II
      D. Neither I nor II
9. During a "no turn" condition, the spool valve allows pressure to the:
   I. Top side of the piston only
   II. Bottom side of the piston only
      A. I only
      B. II only
      C. Both I and II
      D. Neither I nor II
10. In the Chrysler Constant Control power steering gear, the following things happen in a left turn:
   I. The worm shaft threads itself down
   II. The piston and ball nut move up
      A. I only
      B. II only
      C. Both I and II
      D. Neither I nor II

DATE COMPLETED____________________ INSTRUCTOR'S SIGNATURE ____________________

unit IV

Front End Alignment

In this unit on front end alignment, the five basic angles of front end geometry are discussed. Prealignment checks of the entire suspension system and alignment procedures are covered. Also included is the straight beam axle and rear axle housing alignment service as practiced by professional automotive mechanics.

OBJECTIVE IV-A (written test)

After studying the following preparation materials, you will be expected to answer a minimum of 14 out of 20 multiple-choice questions in a written test without reference materials on the following steering geometry angles:
1. Caster
2. Camber
3. Steering-axis inclination
4. Toe-in or toe-out
5. Turning radius

Preparation:

Study pages 299 through 307.

STEERING GEOMETRY

The term *steering geometry* is used as a description of the angles built into the front suspension and steering system. These angles provide easy handling, directional stability, and good tire life on properly designed cars.

There are five basic angles used in the design of the front suspension and steering system. They are *caster, camber, steering axis inclination, toe-in* or *toe-out,* and *turning radius.* Before you can successfully align the front wheels of an automobile, these angles and their functions must be completely understood.

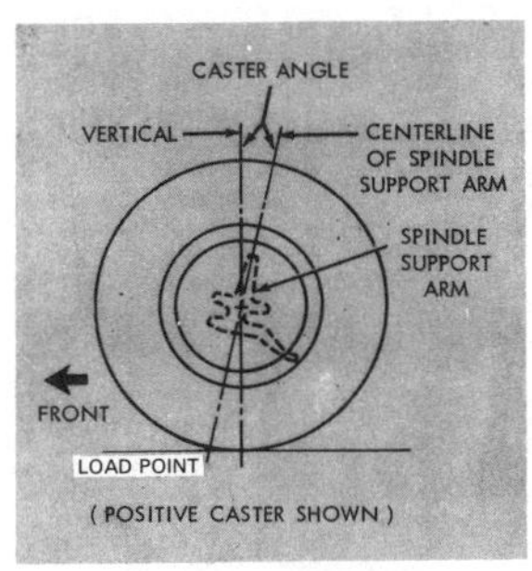

FIGURE IV -1 Positive caster is the backward tilt from true vertical of the spindle support arm (steering knuckle). (Courtesy of Moog Automotive, Inc.)

CASTER

Caster is defined as the forward or rearward tilt (from true vertical) of the steering knuckle at the top. *Positive caster* is the amount that the top of the king pin or steering knuckle tips from true vertical back toward the rear of the car (Fig. IV-1). *Negative caster* is the amount that the top of the king pin or steering knuckle tips forward from true vertical toward the front of the car.

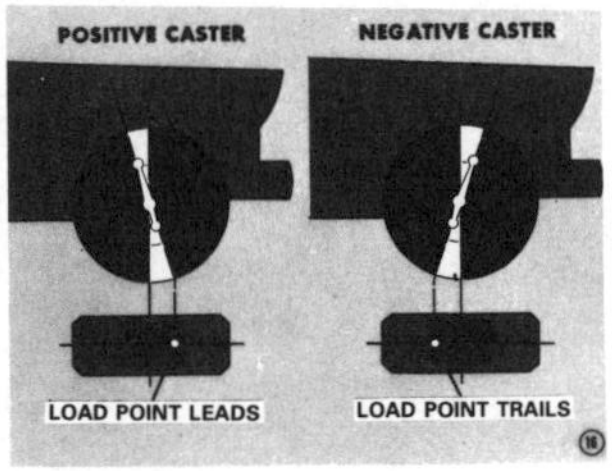

FIGURE IV-2 Positive and negative caster. (Courtesy of Chrysler Motors Corp.)

The main purpose of caster angle is to provide directional control of the vehicle. It is not a tire wearing angle, but it will create problems in steering if not corrected (Fig. IV-2).

In Fig. IV-1, there are two dotted lines, one representing true vertical and the other the center line of the king pin or steering knuckle. Caster is the angle formed by these two dotted lines. Therefore, we can say that caster is an angle and is measured in degrees of a circle. This angle can be read either positive (+) or negative (-). In Fig. IV-3, true vertical is listed as 0 (zero) caster. All of the readings to the right, or rear, are positive. Everything forward, or to the left in this illustration, is negative.

The tilting of the steering knuckle back, toward the driver (positive caster), gives the front wheels a tendency to run in a straight line and to return to a straight forward position after a turn. In Fig. IV-1, the true vertical line strikes the pavement in the center of the tire—the *wheel contact point.* If the imaginary steering-knuckle center line is extended to the pavement, where it touches the pavement becomes the *lead point.*

How wheel contact-point and lead point work is best illustrated by using the front wheel of a bicycle as an example (Fig. IV-4). Notice how the lead point in Fig. IV-4 is out in front of the wheel contact point. A bicycle may be ridden in a straight line without touching the handle bars because the tire has a tendency to follow the lead point. With positive caster, the forces (weight and speed) acting on the tire through its wheel contact point and lead point tend to straighten the wheel. The lead point is the straight ahead control of the caster angle.

The caster angle will also try to return the wheel to a straight ahead position after a turn. This is accomplished with the help of another angle called *steering axis inclination.*

In newer cars, because of changes in the design of the individual front suspension, wider tires, and the influence of steering axis inclination, caster has lost its supreme influence over directional control. Many new car makers now specify a negative caster setting. The change has come about because the early cars had large diameter tires with a very narrow tread area. As a rule, the smaller the tread area in comparison to the diameter, the harder it is to keep the wheel going in a straight line. Thus, narrow tires needed a heavy positive caster angle (Fig. IV-5). However, negative caster

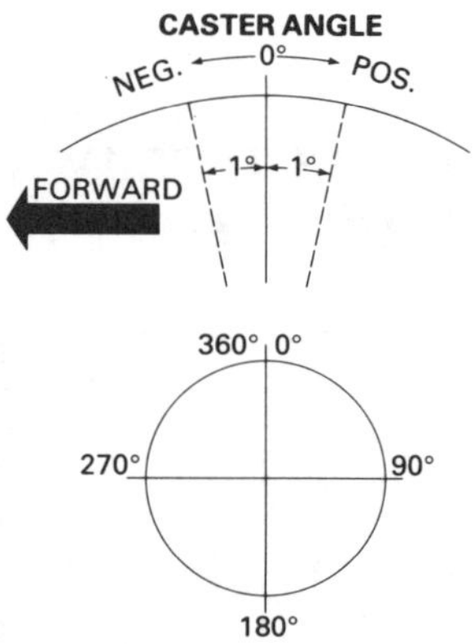

FIGURE IV-3 Caster is measured in degrees (+) or (−) from true vertical.

FIGURE IV-4 Caster angle in a bicycle front tire.

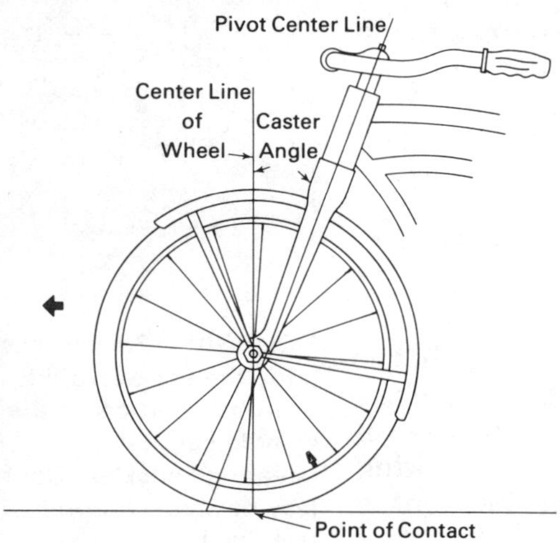

FIGURE IV-5 Two tires showing difference in the width of the tread area. The wider the tire the stronger the directional control and the harder it is to turn the wheel.

settings are easier to steer. Chrysler Corporation design is a good example of modern caster angle treatment. On their cars with power steering, the specifications call for a positive caster, while those without power steering call for a negative caster setting.

It should be remembered that steering effort (force) is directly related to caster setting. For example, if wider tires than recommended are mounted on an automobile, it may be necessary to adjust the caster to a less positive or even a negative setting to allow for easier steering.

Another steering problem stems from highways being built with a crown in the center and a slope to each side. This is fine for water runoff, but it presents a problem to automobile steering since the natural tendency is to go down the slope. To prevent this, the caster angle is set at a less positive or more negative setting on the left side (Fig. IV-6). **The rule is:** *The car will pull to the side that has the leading top ball joint.* Most manufacturers recommend that the right wheel have a more positive or less negative caster setting than the left wheel. For example, in Fig. IV-7, a center line is drawn through the ball joint on the right side and extends to the left side of the automobile. Notice that the center of the ball joint on the left side is just above or in front of the one on the right. The left side has a more negative or less positive setting on caster. Therefore, it will cause the car to pull to the left on flat pavement, but should go straight, on the highway with a crown.

The spread in caster must stay within the manufacturer's recommendation. For example, if the manufacturer recommends a positive one degree caster with a plus or minus one-half-degree tolerance, the adjustment may be made anywhere within a positive one-half degree to a positive one-and-one-half-degree setting. However, the difference in caster between the two sides of the automobile must not be more than one-quarter-degree under normal conditions. Following are some examples where caster settings affect the directional control of the automobile.

FIGURE IV-6 Car wants to pull to the side of the road or off the crown.

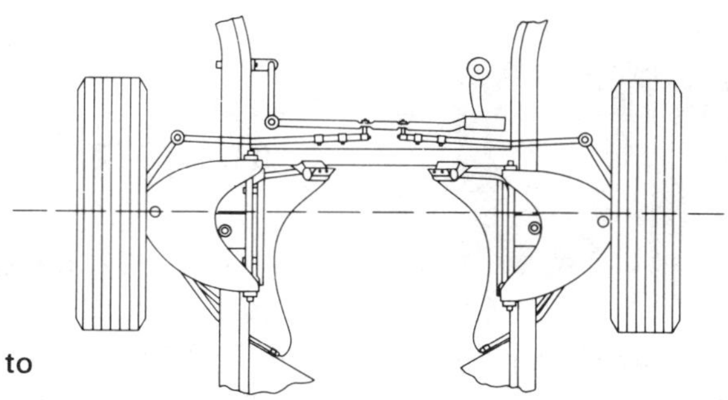
FIGURE IV-7 Car wants to pull to the side with the leading top ball joint.

These caster settings will result in a drift or pull to the right or off the crown in the road.

	Left Wheel		*Right Wheel*
a.	+1¼	↗	+1
b.	-1/4	↗	-1/2
c.	+1/2	↗	-1/4

(Arrows indicate direction of pull.)

The extremity of the pull will depend on the difference between the two sides. Example c above will pull harder to the right than example a or b because of the larger difference between the two settings.

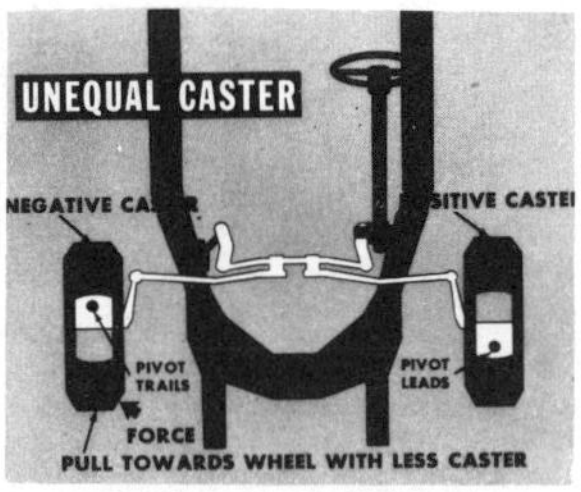

FIGURE IV-8 The negative caster on this car will make it pull to the right. (Courtesy of Chrysler Motors Corp.)

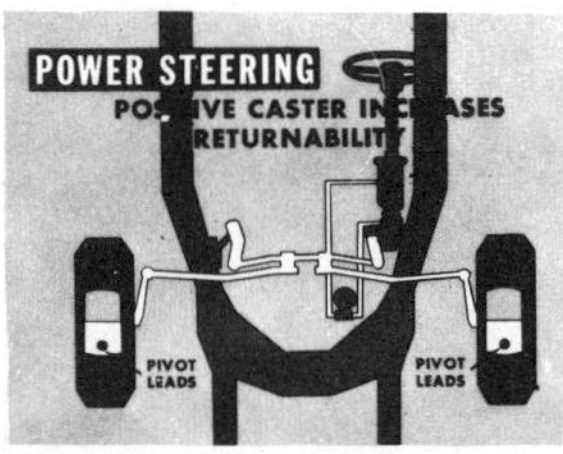

FIGURE IV-9 Equal but heavy positive caster will cause hard steering. (Courtesy of Chrysler Motors Corp.)

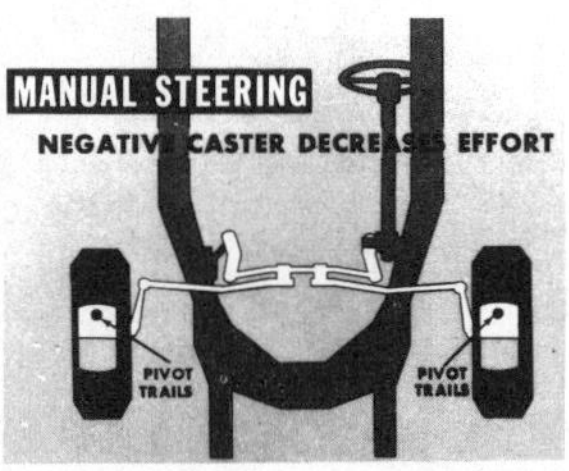

FIGURE IV-10 Equal but heavy negative caster will cause easier steering and may result in wander or weave. (Courtesy of Chrysler Motors Corp.)

In each of the following examples, the left ball joint is the leading ball joint. However, these caster settings will still result in a drift or pull to the left, because they exceed the allowable difference between the two wheels needed to control the crown of the road.

	Left Wheel		*Right Wheel*
a.	+1/2	↖	+1½
b.	-1/2	↖	+1/4
c.	-1¼	↖	-1/4

(Arrows indicate direction of pull.)

On a normal crown road, the following caster settings will give the car a tendency to hold a straight ahead position. Notice that the left ball joint is the leading ball joint and that the difference between the two wheels does not exceed one-quarter degree.

	Left Wheel		*Right Wheel*
a.	+1/4	↑	+1/2
b.	-1/2	↑	-1/4

(Arrows indicate direction of pull.)

Caster is not considered a tire wearing angle. It does, however, play a very important role in the directional stability of the automobile. Some of the harmful effects of an incorrect caster setting are:

1. Unequal caster causes the vehicle to pull toward the side with the most negative-caster or least positive caster setting (Fig. IV-8).
2. Too much negative or too little positive caster will cause wander and weave, with instability at high speeds (Fig. IV-9).
3. Too much positive caster causes hard steering, excessive road shock, and shimmy (Fig. IV-10).

CAMBER

Camber is defined as the amount that the center line of the tire leans either inward or outward from true vertical at the top. See Figure IV-11. *Positive camber* is the angle of outward tilt at the top of the tire from true vertical. The bottom illustration in Fig. IV-12 shows negative camber, which is the amount that the top of the tire leans inward from true vertical. In Fig. IV-11, the true vertical position is *zero camber*. The objective of the manufacturer is to have the camber angle as close to zero as possible under actual driving conditions.

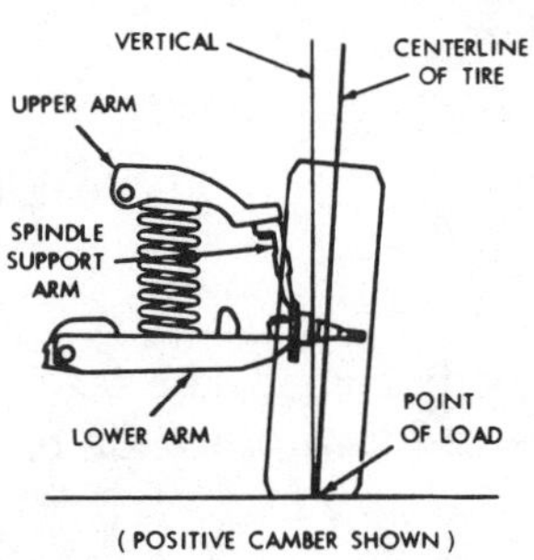

FIGURE IV-11 Positive camber is the outward tilt of the tire at the top. (Courtesy of Moog Automotive, Inc.)

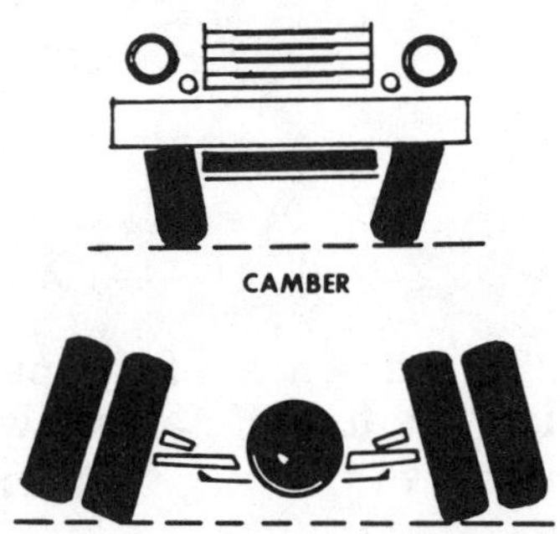

FIGURE IV-12 Positive camber is the outward tilt, and negative camber is the inward tilt at the top of the tire. (Courtesy of Chevrolet Motor Division)

Camber, like caster, is measured in degrees of a circle. The angle is formed by an imaginary true vertical line and the actual center line of the wheel assembly (Fig. IV-13). True vertical is 0 (zero) camber, while those settings that move the center line out are positive (+), such as +¼, +½, +1. When the center line of the tire moves in toward the center of the car, the settings are listed as negative (-), such as -½, -1.

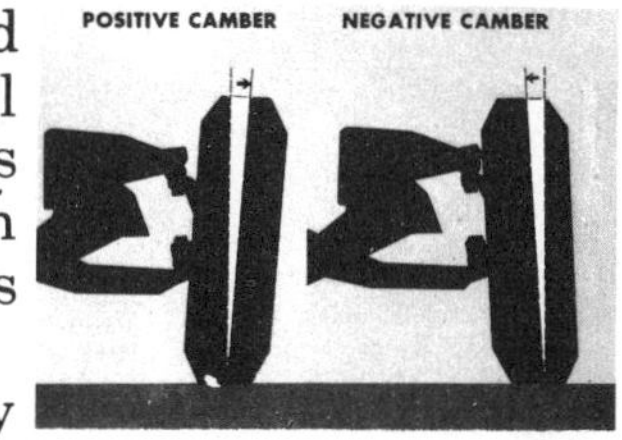

FIGURE IV-13 Positive and negative camber are compared in this illustration. (Courtesy of Chrysler Motors Corp.)

Camber, unlike caster, is a tire wearing angle. Therefore, we are not only concerned with the safety and comfort of the driver but also with the life of the tires. Re-examine pages 35–42 concerning tire wear problems. Figs. I-50 and I-51 show the results of camber wear.

The three main purposes of the camber angle are:

1. To provide easy steering by having the weight of the vehicle borne by the inner wheel bearing and spindle (see Fig. IV-14).
2. To bring the road contact of the tire more nearly under the point of load.
3. To prevent front tire wear by providing, as nearly as possible, total and even tread contact with the road surface under all driving conditions

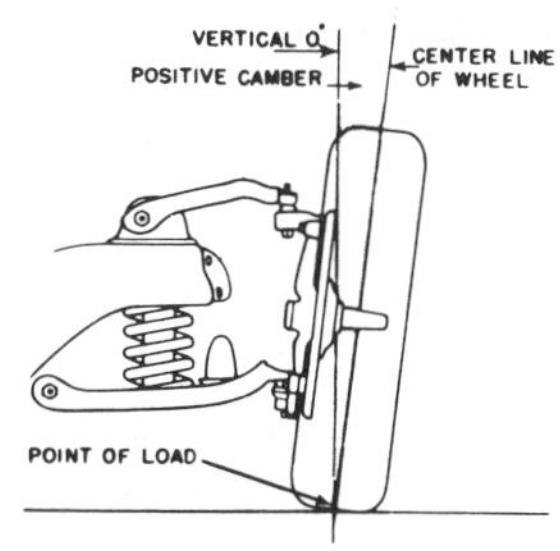

FIGURE IV-14 Positive camber allows the inner bearing to carry most of the vehicle weight. (Courtesy of Snap-on Tools Corp.)

Camber, like the other steering geometry angles, is under constant change when the automobile is in motion. If all roads were smooth and level with no curves or dips and the cars were always driven at the same speed with the same amount of load, it would be easy to design an automobile that would have zero running camber. Tires would last much longer and steering problems would be reduced to nothing. These ideal conditions do not exist, so it is necessary to adjust the camber angle to meet the average driving needs of the automobile. Most modern cars utilize a camber setting between zero and a positive (+) 1 degree. This small amount of positive (+) camber prevents the wheels from going to a negative camber condition when the vehicle is loaded.

Most cars have a front suspension design that gives a possible negative camber in jounce (when the wheel moves upward in relation to the body of the car), and a positive or very little camber change in rebound (when the tire moves down in relation to the body of the car) (Fig. IV-15).

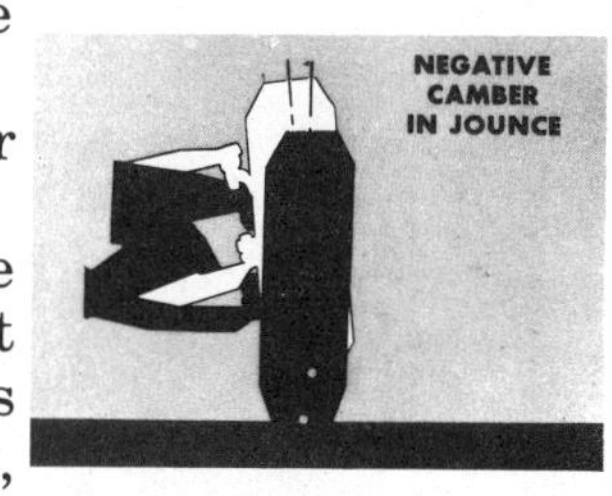

FIGURE IV-15 During jounce, the wheel moves up in relation to the body and usually causes a negative camber reading. (Courtesy of Chrysler Motors Corp.)

Others require a negative camber setting that will result in a zero camber setting under load and normal driving conditions.

When the car goes into a turn, the outside front tire is moved into a more negative camber setting. The centrifugal force causes a significant weight shift toward the outside wheel See Fig. IV-16. The car body and frame is lowered on the outside and raised on the inside. Because of this weight shift, the inside tire is moved into a more positive camber setting. Notice in the illustration how both tires have almost complete tread contact with the road surface.

The negative camber at the outside wheel has the effect of bracing the tire tread.

FIGURE IV-16 Camber settings in a hard turn are affected by vehicle weight shifts. (Courtesy of Chrysler Motors Corp.)

Camber has been used in some cases to control automobile response to the crown in the road. In Fig. IV-17, notice that the two lines drawn from the tire tread come to a common point. A tilted wheel will have a tendency to rotate or roll around that point (apex) of its cone. To demonstrate this, roll a tire away from yourself; when the speed slows to a point where the tire starts to lean one way or the other, the tire will roll in a circle to that side. The further the tire leans over, the smaller the circle. Because of tire characteristic, we can say that the car will pull to the side with the most positive camber angle.

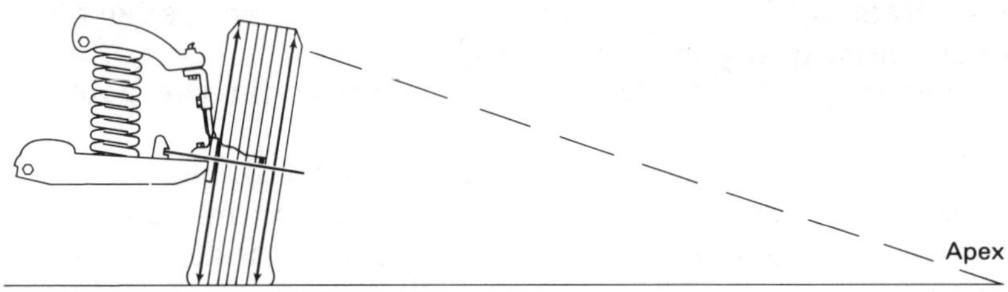

FIGURE IV-17 A wheel has a tendency to roll around the apex or the cone created by the leaning tire. Car will pull to the side with the most positive or least negative camber.

Chrysler Corporation has designed their cars to use the camber angle to control responses to the crown in the road. The camber angle specifications for most Chrysler Corporation cars read: left, +½; right, +¼.

It should be remembered that both caster and camber angles play an important part in keeping the car moving in a straight line.

Examine the sample caster and camber settings listed below. The arrows indicate the resultant force or pull on the vehicle.

Camber			Caster		
Left		*Right*	*Left*		*Right*
+1/2	↑	+1/2	+1/2	↑	+1/2
+1/2	↖	+1/4	+1/4	↖	+1/2
+1/4	↗	+1/2	+1/2	↗	+1/4
-1/2	↗	-1/4	-1/4	↗	-1/2

	Example 1			*Example 2*	
	Left		*Right*	*Left*	*Right*
Camber	+1/2	↑	+1/2	+1/2	+1/4
Caster	+1/4		+1/2	+1/2	+1/2
Crown in road		↗			

Notice that the camber angle in the first example has a straight ahead effect. Caster has a tendency to pull to the left and the road crown pulls to the right. Hopefully, if everything else, like tire pressure and size, is in order, the car will go down the road straight. The second example has the camber controlling the road crown, while the caster has a straight ahead effect. Any time two forces are acting against one force, the result is usually in favor of the two.

	Example 3			*Example 4*		
	Left		*Right*	*Left*		*Right*
Camber	+1/4	↗	+1/2	+1/2	↖	+1/4
Caster	-1/4	↖	0	-1/4	↖	0
Crown in road		↗			↗	

In Example 3, camber and road crown are both pulling to the right. Generally, the car will also pull that way. In Example 4, the camber and caster are pulling the car to the left, overcompensating for the road crown.

Unless the manufacturer specifies camber for road crown control, it is our recommendation that caster be used. Let's review the rules once more.

1. The car will pull to the side with the leading top ball joint, or the side with the least positive or most negative caster setting.
2. A tilted tire will roll around the point or apex of its cone; therefore, the car will pull to the side with the most positive or least negative camber setting.

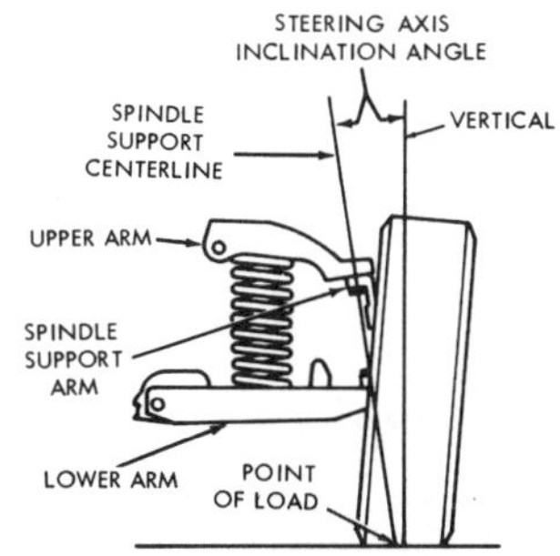

FIGURE IV-18 Steering axis inclination is the inward tilt at the top of the kingpin or ball joint assembly. (Courtesy of Moog Automotive, Inc.)

STEERING AXIS INCLINATION

Fig. IV-18 illustrates *steering axis inclination,* which may also be called *kingpin inclination* or *ball joint inclination.* Steering axis inclination is the inward tilt of the ball joint or the kingpin towards the center of the car. This angle is not adjustable but is measured in degrees. It is a constant angle designed and built into the spindle and steering knuckle and can only change if the spindle or steering knuckle is bent (Fig. IV-19). Likewise, it can only be corrected by replacement of bent or damaged parts.

The purpose of the steering inclination angle is to:

1. Reduce the need for excessive camber and caster angles
2. Distribute the weight of the vehicle more nearly under the road contact of the tire
3. Provide a pivot point about which the wheel will turn, thus producing easy steering

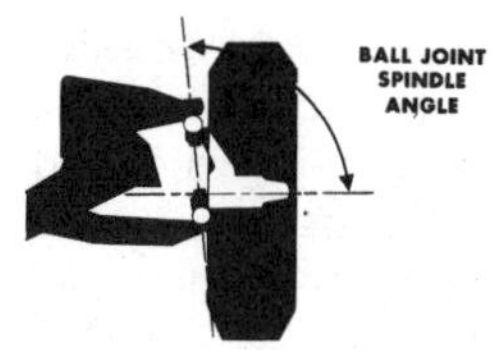

FIGURE IV-19 The angle formed by the steering knuckle and the spindle is greater than 90 degrees. The angle is not adjustable. (Courtesy of Chrysler Motors Corp.)

With the spindle steering knuckle tilted in at the top, the end of the spindle makes an arc as the wheels are turned. In Fig. IV-20 there are three pictures with the wheel assembly and backing plate removed showing only the spindle. In picture *A,* the spindle is in the straight out position. The distance from the end of the spindle to the floor can be measured at this point. Now, if we turn the spindle to the right, as in picture B, or left as in picture C, and recheck the distance to the floor, we will find that the distance is less than the straight out position. This means that the spindle moves in

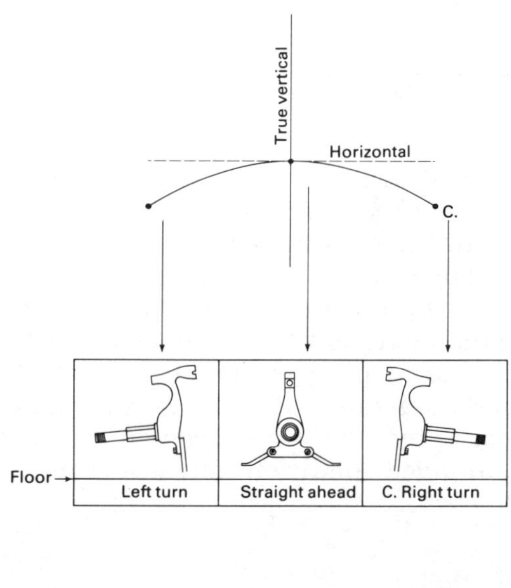

FIGURE IV-20 Distance from the end of the spindle to the floor changes as the position is changed from a right-to-left turn through the straight ahead position.

an arc. It is obvious that with the wheel assembly in place, the spindle cannot move down, so the body of the car must move up. When you let go of the steering wheel, the weight of the vehicle pushes the wheel back to its straight ahead position.

TOE-IN OR TOE-OUT

Toe-in or toe-out is the difference in distance between the extreme front and the extreme rear of the tires at spindle height. Toe-in is measured in fractions of an inch and is usually about ⅛ inch (Fig. IV-21). One method to check toe-in, is to scribe a center mark around the circumference of each front tire (Fig. IV-22). Measure from the center mark in the front, about halfway up on the tire, and compare that measurement to the one at the rear of the tire. If the distance at the front is less than the rear, the tires have a toe-in condition. If the distance at the front is greater than the one at the rear, the tires have a toe-out condition.

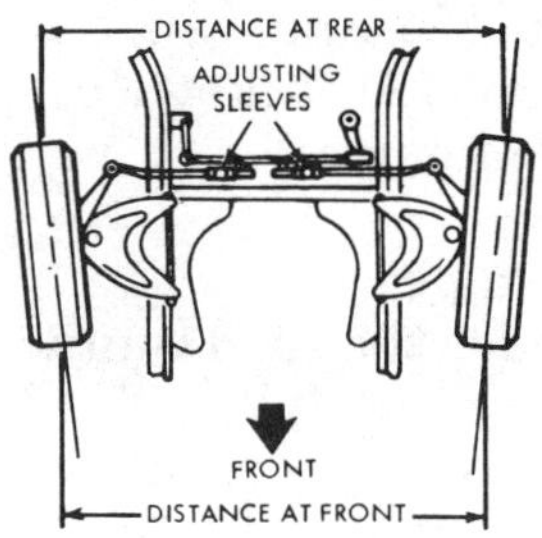

FIGURE IV-21 Toe-in is usually about ⅛ inch. The drawing is exaggerated. (Courtesy of Moog Automotive, Inc.)

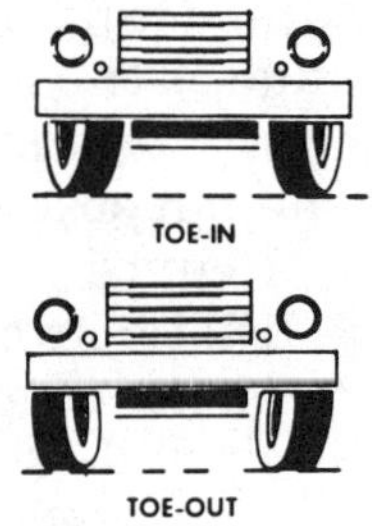

FIGURE IV-22 Toe-in and toe-out in small amounts are normal to most cars. (Courtesy of Chevrolet Motor Division)

Toe-in helps to counteract the tendency of the wheels to toe out due to friction between the tire and the road. Play in the tie-rod ends, idler arm, and other steering joints makes toe-in settings more critical because tires must be in a straight ahead or zero toe-in position during normal driving and load conditions.

Toe-in or toe-out is the most severe tire-wearing angle of all front suspension angles. For examples of tire wear, refer to page 37, Figs. I-48 and I-49 for toe-in and toe-out wear.

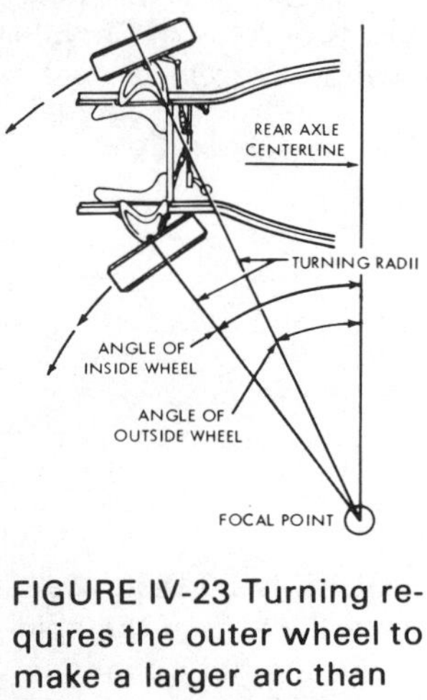

FIGURE IV-23 Turning requires the outer wheel to make a larger arc than the inner wheel. (Courtesy of Moog Automotive, Inc.)

TURNING RADIUS

In a turn, the outer wheel must travel in a much larger arc than the inside wheel (Fig. IV-23). Therefore the inside wheel must turn at a sharper angle than the outside wheel.

The *turning radius angle* is that part of the steering geometry that determines the relative position of the front wheels to each other during a turn. This angle is sometimes referred to as toe-out on turns.

The turning radius angle, even though it can cause serious tire wear, is not adjustable. This angle is designed into the steering arms and can only be corrected by the replacement of the steering arms or other bent parts whose damaged condition may affect it.

As illustrated by the accompanying pictures, when the front wheels are straight ahead, angle A is the same as angle B (Fig. IV-24).

As the wheels are turned to the left (Fig. IV-25), the steering arm on the outside wheel starts to close angle A, while the steering arm on the inside wheel starts to open angle B.

Angle A must first come to 90 degrees and then close to a smaller angle, while angle B is already past 90 degrees and continues to open wider (Fig. IV-23). As a result, the change is greater on the inside wheel, and therefore that wheel is capable of a much sharper turn.

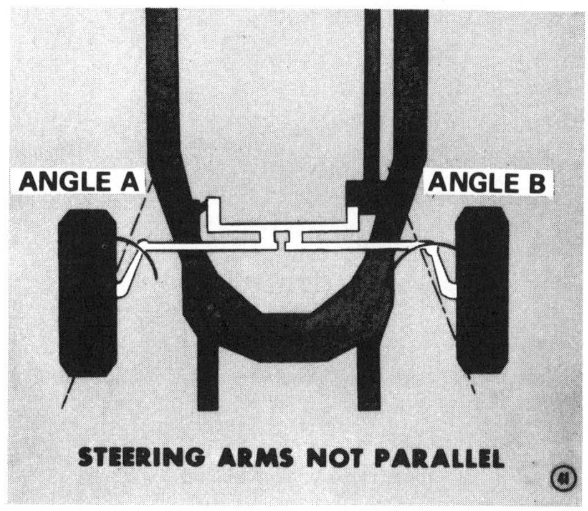

FIGURE IV-24 Steering arms have predetermined bends for control of turning radius. (Courtesy of Chrysler Motors Corp.)

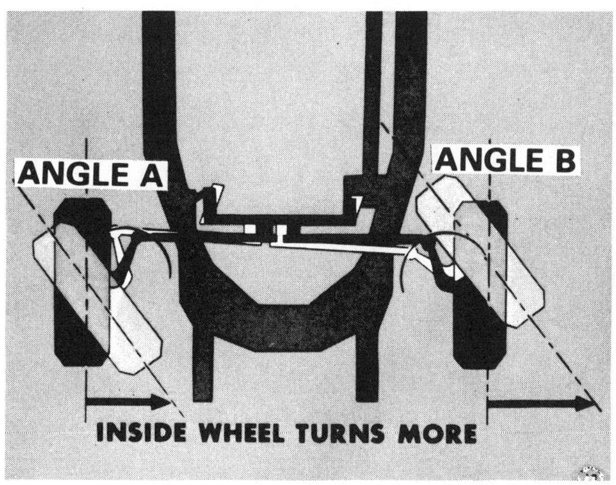

FIGURE IV-25 Inside wheel always turns at a larger angle than the outside wheel, thus the inside wheel must turn more sharply. (Courtesy of Chrysler Motors Corp.)

TEST IV-A

Student name________________________Date________________

Directions: Select the best answer to each question and circle the correct letter at the left.

1. The forward or backward tilt from true vertical of the top of the kingpin or ball joint assembly is called:
   I. Caster
   II. Camber
   A. I only
   B. II only
   C. Either I or II
   D. Neither I nor II
2. The forward tilt from true vertical of the top of the kingpin or ball joint assembly is called:
   I. Negative caster
   II. Negative camber
   A. I only
   B. II only
   C. Either I or II
   D. Neither I nor II
3. Caster and camber are measured in:
   I. Inches
   II. Degrees
   A. I only
   B. II only
   C. Either I or II
   D. Neither I nor II
4. The inward or outward tilt from true vertical of the top of the wheel is called:
   I. Caster
   II. Camber
   A. I only
   B. II only
   C. Either I or II
   D. Neither I nor II
5. The backward tilt from true vertical of the top of the kingpin or ball joint assembly is called:
   I. Positive caster
   II. Positive camber
   A. I only
   B. II only
   C. Both I and II
   D. Neither I nor II
6. Which of the following two angles will cause abnormal tire wear?
   I. Positive caster
   II. Negative caster
   A. I only
   B. II only
   C. Both I and II
   D. Neither I nor II

7. Which of the following two angles will cause abnormal tire wear?
   I. Positive camber
   II. Negative camber
      A. I only
      B. II only
      C. Both I and II
      D. Neither I nor II
8. The inward tilt from true vertical of the top of the kingpin or steering knuckle is called:
   I. Turning radius
   II. Steering axis inclination
      A. I only
      B. II only
      C. Both I and II
      D. Neither I nor II
9. Which of the two angles listed below will reduce scuffing of the tires?
   I. Turning radius
   II. Steering axis inclination
      A. I only
      B. II only
      C. Both I and II
      D. Neither I nor II
10. Toe-in or toe-out is measured in:
   I. Inches
   II. Degrees
      A. I only
      B. II only
      C. Both I and II
      D. Neither I nor II
11. Which of the following two angles can be used to control the crown in the road?
   I. Caster
   II. Camber
      A. I only
      B. II only
      C. Both I and II
      D. Neither I nor II
12. Wide tires will reduce the need for:
   I. Positive caster
   II. Negative caster
      A. I only
      B. II only
      C. Both I and II
      D. Neither I nor II
13. The car will pull to the side with the most:
   I. Positive camber
   II. Negative caster
      A. I only
      B. II only
      C. Both I and II
      D. Neither I nor II
14. Which of the following two angles will best produce directional control?
   I. Positive caster
   II. Positive camber
      A. I only
      B. II only
      C. Both I and II
      D. Neither I nor II

15. Turning radius is controlled by:
    I. Steering arms
    II. Steering knuckles
        A. I only
        B. II only
        C. Both I and II
        D. Neither I nor II
16. Which angle will cause the most severe abnormal tire wear?
    I. Camber
    II. Toe-in
        A. I only
        B. II only
        C. Both I and II
        D. Neither I nor II
17. The purpose of the steering axis inclination angle is to:
    I. Reduce the need for excessive caster and camber
    II. Assist the return of the front wheels to the straight-ahead position
        A. I only
        B. II only
        C. Both I and II
        D. Neither I nor II
18. Which of the two angles listed below is not adjustable?
    I. Caster
    II. Camber
        A. I only
        B. II only
        C. Both I and II
        D. Neither I nor II
19. Which of the following settings will possibly cause a car to pull to the right?
    I. Caster left +¼ right -¼
    II. Camber left +½ right +¾
        A. I only
        B. II only
        C. Both I and II
        D. Neither I nor II
20. Which of the following settings will possibly cause a car to pull to the left?
    I. Caster left +¼ right +1
    II. Camber left -¼ right 0
        A. I only
        B. II only
        C. Both I and II
        D. Neither I nor II

DATE COMPLETED____________________ INSTRUCTOR'S SIGNATURE ____________________

OBJECTIVE IV-B (written assignment)

Working as a team with one other student, you will inspect a minimum of three late model instructor assigned automobiles in preparation for front end alignment.

The following steps will be followed:

1. Road test.
2. Inspect tires.
3. Inspect front wheel bearings.
4. Inspect suspension.
5. Inspect steering.

Recommended time is a maximum of 30 minutes per automobile.

PREALIGNMENT INSPECTION

Preparation:

1. Review Units I, II, and III.
2. Review Objective IV-A, "Steering Geometry Angles," pages 299 through 307
3. Study pages 313 through 317.

Materials and Tools Needed:

1. Jack
2. Air hose and tire inflator
3. Air gauge
4. Dial indicator and stand

The main purpose of the steering and suspension system is to allow maximum control of the vehicle and to permit all wheels to roll straight and even, under all driving conditions. For this to be accomplished, it is necessary for the front and rear wheels to be in proper alignment. The first step in any alignment job is to make certain that the suspension and steering system of the automobile will accept an alignment.

If worn or loose parts are present, they must be replaced before the wheels can maintain a true alignment position. In many cases, a loose suspension or steering part may not be noticed in the alignment process. This does not mean that the specified adjustment cannot be made. But it does mean that when the car is back on the road and subjected to normal driving conditions, the worn parts will cause excessive variations in the alignment settings.

Units I, II, and III of this text have explained the procedure for checking each of the components in the wheel, suspension, and steering assemblies. In this unit you are expected to utilize that information in checking all assemblies as they relate to the front end alignment.

Use the following chart, excluding no parts, in the prealignment check.

JOB REPORT SHEET IV-B

PREALIGNMENT INSPECTION

ASSIGNMENT 1

Student name________________________________Date____________________

Make_______________Model_______________Year____________________

Time started_______________Time finished_______________Total time_______________

Recommended time _______________

	Lt. Front		Rt. Front		Lt. Rear		Rt. Rear	
	OK	Not OK	OK	Not OK	OK	Not OK	OK	Not OK
WHEEL ASSEMBLY								
Tire Air Pressure ( )								
Wheel Rim Size ( )								
Run Out								
Wheel Balance								
Excessive Tire Wear								
Tire Size ( )								

WHEEL BEARINGS	Lt. Front OK	Lt. Front Not OK	Rt. Front OK	Rt. Front Not OK	Lt. Rear OK	Lt. Rear Not OK	Rt. Rear OK	Rt. Rear Not OK
Front Wheel Bearing Adj.								
Wheel bearing Condition								
Axle Bearing Condition								

SUSPENSION	Lt. Front OK	Lt. Front Not OK	Rt. Front OK	Rt. Front Not OK	Lt. Rear OK	Lt. Rear Not OK	Rt. Rear OK	Rt. Rear Not OK
Trim Height (Torsion Bar)								
Shock Absorbers								
Spring Shackles								
Control Arm Bushings								
Ball Joints								
Strut Rod Bushings								

Stabilizer Bar OK (   ) Not OK (   )

Track Bar OK (   ) Not OK (   )

STEERING	Lt. Front OK	Lt. Front Not OK	Rt. Front OK	Rt. Front Not OK
Tie Rod Ends—Outer				
Tie Rod Ends—Inner				
Adjusting Sleeves				
Center Link and Joints				
Idler Arm				
Pitman Arm				
Steering Arms				

STEERING GEAR OK (   ) Not OK (   )

ROAD TEST DESCRIPTION:____

____

DATE COMPLETED_______________INSTRUCTOR'S SIGNATURE ____________________

JOB REPORT SHEET IV-B

PREALIGNMENT INSPECTION

ASSIGNMENT 2

Student name ______________________ Date ______________

Make ____________ Model ____________ Year ______________

Time started ____________ Time finished ____________ Total time ____________

Recommended time ____________

	Lt. Front		Rt. Front		Lt. Rear		Rt. Rear	
	OK	Not OK	OK	Not OK	OK	Not OK	OK	Not OK
WHEEL ASSEMBLY								
Tire Air Pressure ( )								
Wheel Rim Size ( )								
Run Out								
Wheel Balance								
Excessive Tire Wear								
Tire Size ( )								
WHEEL BEARINGS								
Front Wheel Bearing Adj.								
Wheel bearing Condition								
Axle Bearing Condition								
SUSPENSION								
Trim Height (Torsion Bar)								
Shock Absorbers								
Spring Shackles								
Control Arm Bushings								
Ball Joints								
Strut Rod Bushings								

Stabilizer Bar OK (   ) Not OK (   )

Track Bar OK (   ) Not OK (   )

STEERING	Lt. Front OK	Lt. Front Not OK	Rt. Front OK	Rt. Front Not OK
Tie Rod Ends—Outer				
Tie Rod Ends—Inner				
Adjusting Sleeves				
Center Link and Joints				
Idler Arm				
Pitman Arm				
Steering Arms				

STEERING GEAR OK (   ) Not OK (   )

ROAD TEST DESCRIPTION: ______________________________

DATE COMPLETED ____________ INSTRUCTOR'S SIGNATURE ____________

JOB REPORT SHEET IV-B

PREALIGNMENT INSPECTION

ASSIGNMENT 3

Student name________________________________ Date____________________

Make______________ Model______________ Year____________________

Time started______________ Time finished______________ Total time______________

Recommended time ____________

	Lt. Front		Rt. Front		Lt. Rear		Rt. Rear	
	OK	Not OK	OK	Not OK	OK	Not OK	OK	Not OK
WHEEL ASSEMBLY								
Tire Air Pressure ( )								
Wheel Rim Size ( )								
Run Out								
Wheel Balance								
Excessive Tire Wear								
Tire Size ( )								
WHEEL BEARINGS								
Front Wheel Bearing Adj.								
Wheel bearing Condition								
Axle Bearing Condition								
SUSPENSION								
Trim Height (Torsion Bar)								
Shock Absorbers								
Spring Shackles								
Control Arm Bushings								
Ball Joints								
Strut Rod Bushings								

Stabilizer Bar ____________ OK ( ) Not OK ( )

Track Bar ____________ OK ( ) Not OK ( )

STEERING				
Tie Rod Ends—Outer				
Tie Rod Ends—Inner				
Adjusting Sleeves				
Center Link and Joints				
Idler Arm				
Pitman Arm				
Steering Arms				

STEERING GEAR ____________ OK ( ) Not OK ( )

ROAD TEST DESCRIPTION:____

____

DATE COMPLETED______________________ INSTRUCTOR'S SIGNATURE ______________________

After studying the preparation assignment you will be expected to correctly answer a minimum of 21 out of 30 multiple choice questions in a written test without reference materials on front wheel-alignment topics that use the following controls:
1. Shim type
2. Eccentric bushing type
3. Slot type

OBJECTIVE IV-C (written test)

Preparation:

1. Review all material for Objective IV-A, pages 299 through 307.
2. Study pages 319 through 348.

FRONT END ALIGNMENT

Alignment procedures vary considerably with each make and model of car. The basic principles, however, do not change. Caster refers to the same angle in a small sports car as it does in a large diesel truck. This objective material will be devoted to the problems connected with the adjusting of caster, camber, and toe-in on all makes and models. Manufacturers have designed different ways of controlling the adjustments on the front suspension. Fig. IV-26 shows some of those methods that are in use on American designed automobiles. All methods of controlling front end alignment are usually variations of one of the following:
1. Shims at upper or lower inner control arm
2. Eccentric (some with a strut bar for caster adjustments)
3. Slotted holes in the upper control arm or slotted holes in the frame bracket with both parts being serrated for a lock fit.

Changes in caster and camber will always affect the toe-in setting; therefore, caster and camber must be changed before toe-in.

The method used by the manufacturer to control the adjustment of caster and camber will determine which one of these angles should be adjusted first. There will be times when both caster and camber may be adjusted simultaneously. However, accuracy in doing this usually comes only with experience.

CASTER/CAMBER ADJUSTMENT CHART

Make and Model	*No.**	*Year*	*Manner of Caster Adjustment*	*Manner of Camber Adjustment*
American Motors All models except Pacer	7	82-64	Lengthen strut to increase positive caster, shorten strut to decrease positive caster.	Rotate cam to achieve desired measurement.
Pacer	4	80-75	Rotate rear cam to achieve desired measurement.	Rotate both eccentric cams to achieve desired measurements.
Chrysler Corporation				
Full-sized Chrysler, Dodge, and Plymouth Cordoba, Mirada LeBaron, Diplomat	 2 2 2	 82-74 82-75 82-77	Increase positive caster by sliding front adjusting bolt away from engine and rear adjusting bolt toward engine.	Increase positive camber by moving both adjusting bolts away from the engine.
Volare, Aspen, Caravelle Charger, Coronet, Magnum Satellite, Monaco, Fury St. Regis	2 2 2 2	82-76 82-73 78-73 80-79	Decrease positive caster by sliding front adjusting bolt toward engine and rear adjusting bolt away from engine.	Decrease positive camber by moving both adjusting bolts toward the engine.

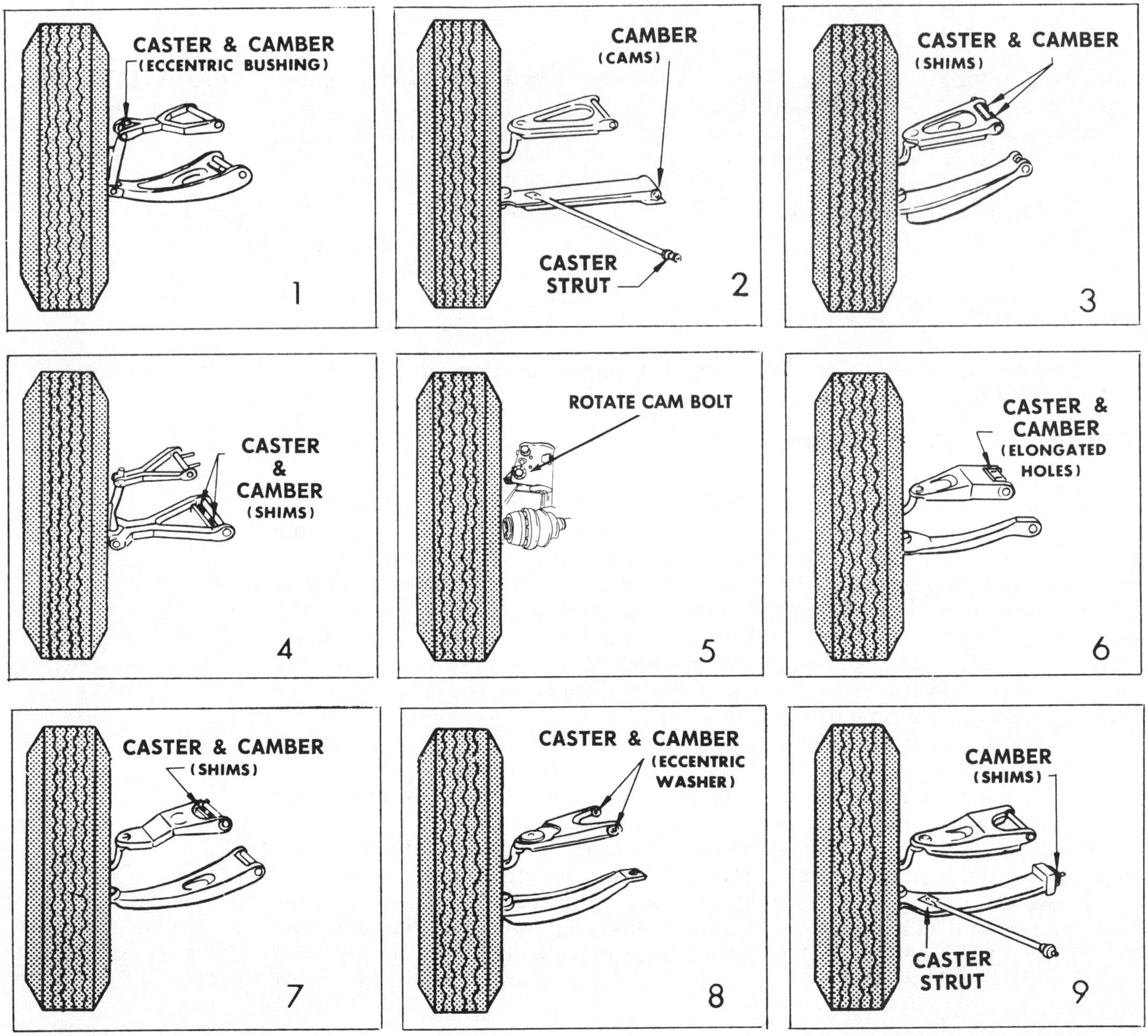

FIGURE IV-26 Caster and camber adjustments are shown for different model cars by determining the model on one chart and checking the adjustment on the diagram chart. (Courtesy of Snap-on Tools Corp.)

CASTER/CAMBER ADJUSTMENT CHART

Make and Model	No.*	Year	Manner of Caster Adjustment	Manner of Camber Adjustment
Chrysler Corporation				
Full-sized Chrysler, Dodge, and Plymouth	4	73-71	Increase positive caster by rotating front cam in the proper direction to move the front of the control arm away from the engine and rotate rear cam to move rear of the control arm toward the engine. Decrease positive caster by reversing the above procedure.	Increase positive camber by rotating both cams to move the control arm away from engine. Decrease positive camber by rotating both cams to move the control arm toward the engine.
Challenger, Barracuda	4	74-71		
Dart, Valiant	4	76-71		
Charger, Satellite, Coronet	4	72-71		
Chrysler Corporation (McPherson strut vehicles)				
Omni, 024, Horizon, TC3		82-78	Caster is not adjustable.	Turn eccentric cam on the strut.
Ford Motor Co.				
Zephyr, Fairmont		82-78	Not adjustable.	Not adjustable.
Capri, Mustang		82-79		
Cougar, Thunderbird		82-80		
Ford Motor Co. (spring on upper control arm)				
Fairlane, Torino, Montego	7	71-66	Increase positive caster by shortening the lower control arm strut. Decrease positive caster by lengthening the lower control arm strut.	Increase positive camber by rotating lower control arm adjusting cam to move the arm inward. Decrease positive camber by reversing the above procedure.
Maverick, Comet	7	77-66		
Mustang, Cougar	7	73-71		
Granada, Monarch	7	82-75		
Versailles	7	80-77		
Ford Motor Co. (slotted upper control arm adjustment)				
Full-sized Ford, Mercury and Lincoln models	2	82-71	Increase positive caster by moving front adjusting bolt away from engine, rear adjusting bolt toward engine. Decrease positive caster by moving rear adjusting bolt away from engine, front adjusting bolt toward engine.	Increase positive camber by moving both adjusting bolts away from engine. Decrease positive camber by moving both adjusting bolts toward engine.
Torino, Montego	2	76-72		
Thunderbird, Meteor	2	79-71		
Pinto, Bobcat	2	80-71		
Mustang II	2	78-74		
Cougar, LTD II	2	79-74		
General Motors Corporation				
Buick Motor Division (shim pack adjustment)				
All models except Skyhawk, 1980-79 Riviera and 1980 F.W.D. Skylark	3	82-71	Increase positive caster by adding shims to rear shim pack and removing shims from front shim pack. Decrease positive caster by removing shims from rear shim pack and adding shims to front shim pack.	Increase positive camber by removing shims equally from front and rear shim packs. Decrease positive camber by adding shims equally to front and rear shim pack.
Buick Motor Division (McPherson strut)				
Skylark (F.W.D.)	9	82-80	Caster is not adjustable.	Turn eccentric cam on strut mounting bracket as necessary.
Buick Motor Division (cam-type adjustment)				
Skyhawk	8	80-75	Turn rear cam as necessary	Turn front cam as necessary
Buick Motor Division (upper control arm cam adjustment)				
Riviera (F.W.D.)	8	80-79	Turn front and rear cams as necessary.	Turn front and rear cams as necessary
Cadillac Motor Division (shim pack adjustment)				
All models except Eldorado and 1980-82 Seville	3	82-77	Increase positive caster by adding shims to rear shim pack and removing shims from front shim pack. Decrease positive caster by removing shims from rear shim pack and adding shims to front shim pack.	Increase positive camber by removing shims equally from front and rear shim packs. Decrease positive camber by adding shims equally to both front and rear shim packs.
Seville	3	79-76		
Cadillac Motor Division (cam-type adjustment)				
Eldorado (F.W.D.)	8	82-71	Turn front and rear cams as necessary	Turn front and rear cams as necessary.
Seville (F.W.D.)	8	82-80		
Cadillac Motor Division (eccentric on upper ball joint)				
All models except Eldorado and Seville	6	76-65	Increase positive caster by shortening lower control arm strut. Decrease positive caster by lengthening lower control arm strut.	Turn ball joint eccentric as necessary.

CASTER/CAMBER ADJUSTMENT CHART

Make and Model	No.*	Year	Manner of Caster Adjustment	Manner of Camber Adjustment
Chevrolet Motor Division (shim pack adjustment)				
All models except Vega, Monza and 1980 Citation	3	80-71	Increase positive caster by adding shims to rear shim pack and removing shims from front shim pack. Decrease positive caster by removing shims from rear shim pack and adding shims to front shim pack.	Increase positive camber by removing shims equally from front and rear shim packs. Decrease positive camber by adding shims equally to front and rear shim packs.
Chevrolet Motor Division				
Vega, Monza	8	80-72	Turn rear cam as necessary.	Turn front cam as necessary
Chevette	9	82-79	Rearrange washers at upper control arm.	Rotate upper ball joint flange.
Chevrolet Motor Division (McPherson strut)				
Citation (F.W.D.)	5	82-80	Caster is not adjustable.	Turn eccentric cam on strut mounting bracket as necessary.
Olsmobile Division (shim pack adjustment)				
All models except Starfire, Toronado and 1980 Omega (F.W.D.)	3	80-71	Increase positive caster by adding shims to rear shim pack and removing shims from front shim pack. Decrease positive caster by removing shims from rear shim pack and adding shims to front shim pack.	Increase positive camber by removing shims equally from front and rear shim packs. Decrease positive camber by adding shims equally to front and rear shim packs.
Oldsmobile Division (cam-type adjustment)				
Starfire	4	80-75	Turn rear cam as necessary.	Turn front cam as necessary.
Oldsmobile Division (McPherson strut)				
Omega (F.W.D.)	5	82-80	Caster is not adjustable.	Turn eccentric cam on strut mounting bracket as necessary.
Oldsmobile Division				
Toronado	4	82-72	Turn front and rear cams as necessary.	Turn front and rear cams as necessary.
Pontiac Motor Division (shim pack adjustment)				
All models except Astre, Sunbird and 1980 Phoenix (F.W.D.)	3	80-71	Increase positive caster by adding shims to rear shim pack and removing shims from front shim pack. Decrease positive caster by removing shims from rear shim pack and adding shims to front shim pack.	Increase positive camber by removing shims equally from front and rear shim packs. Decrease positive camber by adding shims equally to front and rear shim packs.
Pontiac Motor Division (McPherson strut)				
Phoenix (F.W.D.)	5	82-80	Caster is not adjustable.	Turn eccentric cam on strut mounting bracket as necessary.
Pontiac Motor Division (cam-type adjustment)				
Astre, Sunbird	8	80-73	Turn rear cam as necessary.	Turn front cam as necessary.

*Refer to **Figure IV-26** for location of adjustments on front wheel drive vehicles.

OBJECTIVE IV-C (manipulative)

Working as a team with one other student, you will align the front wheels on at least one instructor assigned automobile with shim adjustment controls. The alignment will include checking and adjusting, where necessary, the caster, camber, toe-in, steering axis inclination, turning radius, and centering of the steering wheel. Time recommended: Twice the Flat-rate time allowed in the Motors or Chilton Flat Rate manual.

Preparation:

1. Study pages 323 through 330
2. Study manufacturer's procedure and specifications on the specific automobile to which you are assigned.
3. Carefully observe instructor's demonstration of front end alignment.

Materials and Tools Needed:

1. Alignment equipment
2. Jacks
3. Air hose and tire inflator
4. Air gauge
5. Special tools for front-end alignment. (See manufacturer's instructions.)
6. Set of box/open-end combination wrenches
7. Set of sockets and ratchet

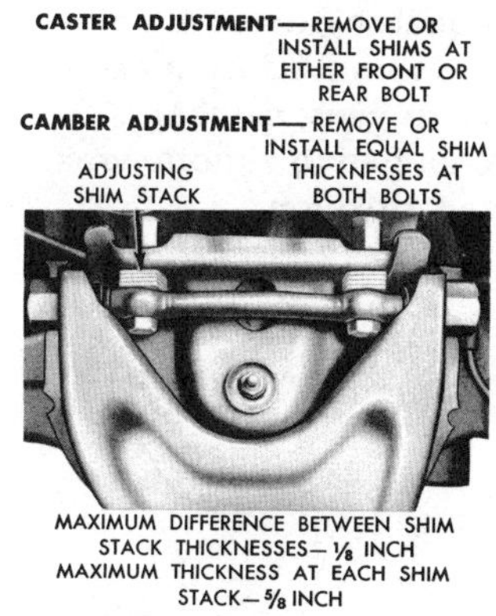

FIGURE IV-27 Ford shim adjustment for camber and caster. (Courtesy of Ford Customer Service Division)

THE SHIM ADJUSTMENT

The shim adjustment has had many variations, with the two most popular being the methods used by Ford Motor Company and General Motors Corporation. Other manufacturers such as Chrysler Corporation and American Motors have used other variations of the shim type adjustment.

FORD SHIM ADJUSTMENT

The Ford front suspension shown in Fig. IV-27 has the upper inner control arm shaft located on the outside of the frame bracket. Both caster and camber are adjusted by removing or adding shims to both shim packs. For example, if it becomes necessary to change the camber to a more positive position, an equal number of shims are added to both front and rear shim packs.

In Fig. IV-28, notice that the upper ball joint has moved straight out, thus changing the camber angle but not the caster angle.

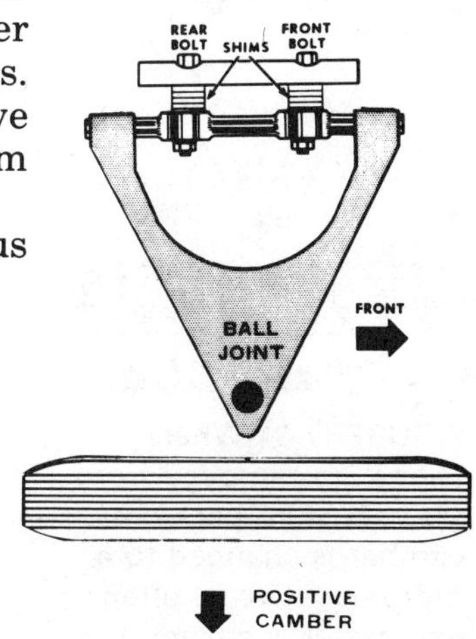

FIGURE IV-28 When an equal thickness of shims are added to each shim pack, camber is changed to a more positive position. (Courtesy of Snap-on Tools Corporation)

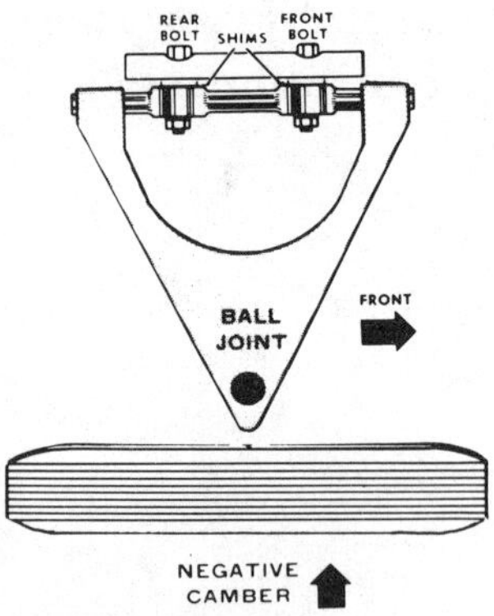

FIGURE IV-29 When an equal thickness of shims are removed from each shim pack, camber is changed to a more negative position.

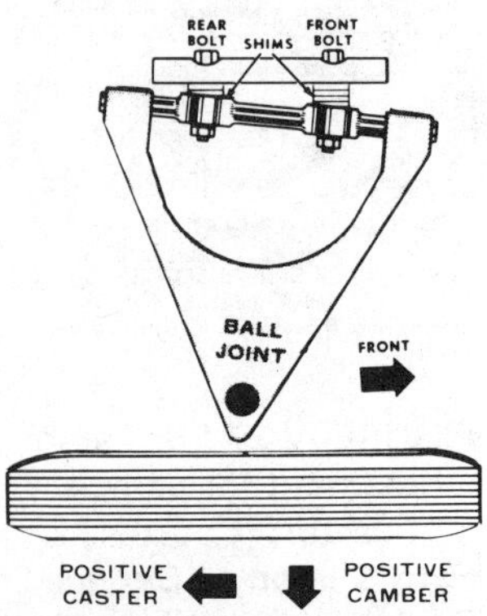

FIGURE IV-30 When shims are added to the front shim pack only, camber and caster both are changed to a more positive position.

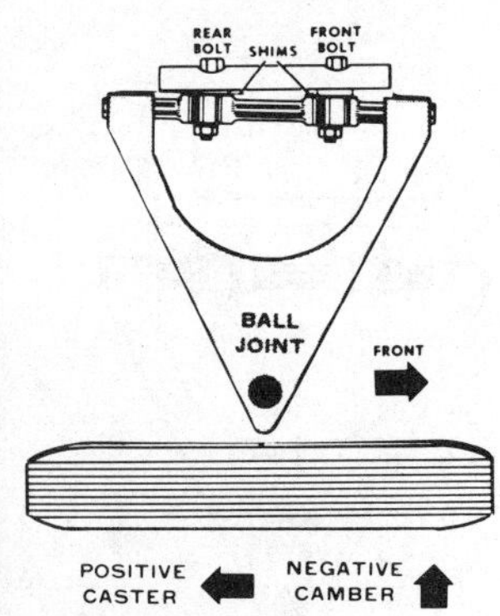

FIGURE IV-31 When shims are removed from the rear shim pack only, camber is changed to a more negative position and caster is changed to a more positive position.

Courtesy of Snap-on Tools Corporation

To change the camber angle to a more negative position, remove an equal number of shims from both front and rear shim packs (Fig. IV-29). This will move the top ball joint to a more negative camber angle without changing the caster angle.

Caster is adjusted by removing or adding shims to either the front or the rear shim pack. If caster needs to be adjusted to a more positive position, shims could be added either to the front shim pack or removed from the rear shim pack.

Fig. IV-30 shows shims being added to the front shim pack. Notice how this has moved the upper ball joint to the outside and to the rear of the automobile.

Fig. IV-31 shows that the upper ball joint has moved in and to the back of the automobile because shims were removed from the rear shim pack.

Both methods changed the caster to a more positive position. By adding shims to the front shim pack, the camber was also changed to a more positive position. However, the removal of shims from the rear shim pack changed the camber to a more negative position. The method used for changing the caster should be controlled by the needs of the camber angle.

When both caster and camber need to be adjusted, the automotive technician should study the situation before making any adjustments. The technician can change both caster and camber with one operation, simply by knowing the direction that the upper ball joint will move when shims are either added or subtracted from the shim pack. If the front suspension needs a more negative camber angle and a more positive caster angle (Fig. IV-31), the automotive technician can change both caster and camber at the same time by removing shims from the rear shim pack. This moves the upper ball joint to a more negative camber angle and, at the same time, back to a more positive caster angle.

When it is not possible to adjust the caster and camber angle at the same time, the caster angle on the Ford shim type must be adjusted first.

The moving of the caster angle will always affect the camber angle. But changing of the camber angle with an equal thickness of shims in both front and rear shim packs will probably not change the caster angle.

The following check list will assist in making the right changes in both caster and camber on the Ford type shim adjustment:

1. *To change camber to a more positive angle without changing caster:* Add an equal number of shims to both front and rear shim pack.
2. *To change camber to a more negative angle without changing caster:* Remove an equal number of shims from both front and rear shim pack.
3. *To change caster to a more positive angle without changing the camber excessively:* Remove shims from the rear shim pack and add an equal thickness to the front shim pack.
4. *To change caster to a more negative angle without changing camber excessively:* Remove shims from the front shim pack and add an equal thickness to the rear shim pack.
5. *To change caster and camber to more positive angles simultaneously:* Add shims to the front shim pack only.
6. *To change caster to a more negative angle and camber to a more positive angle simultaneously:* Add shims to the rear shim pack only.

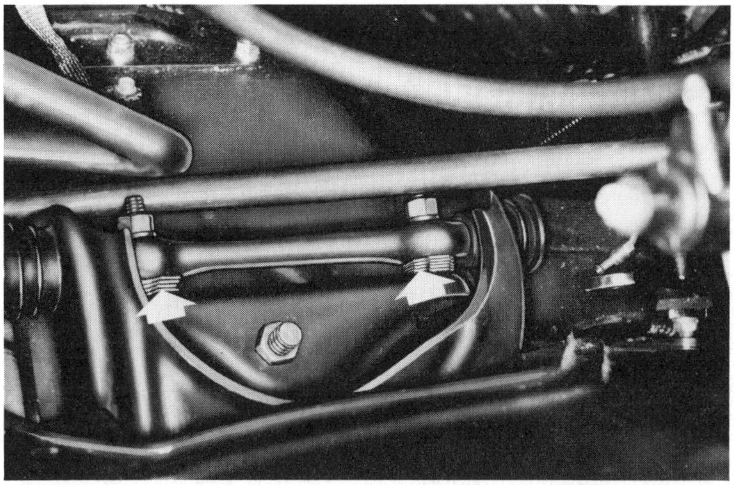

FIGURE IV-32 Arrows show where to make shim adjustments on General Motors cars. (Courtesy of Chevrolet Motor Division)

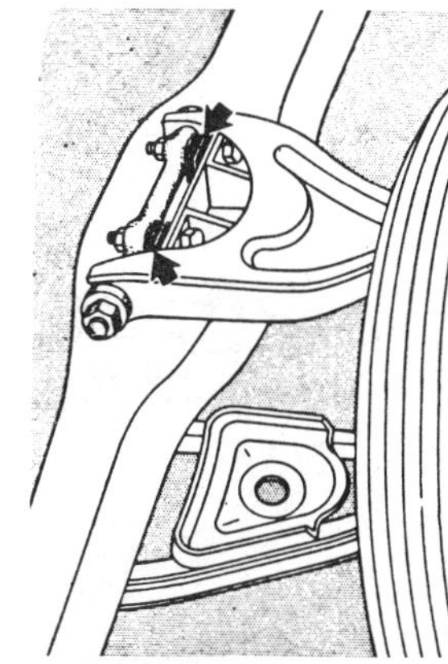

FIGURE IV-33 Arrows show where to make shim adjustments on General Motors cars. (Courtesy of Snap-on Tools Corp.)

GENERAL MOTORS SHIM ADJUSTMENT

The only difference between Ford and General Motors shim adjustments is that on General Motors cars, the shim pack is on the inside of the frame bracket (Fig. IV-32).

All adjustments on the General Motors shim pack are just opposite of those made on the Ford Automobiles where the shim pack is on the outside of the frame bracket (Figs. IV-33 and 34). It should be noted that the normal wear in the front suspension will cause the camber on both types to gradually move towards a negative setting. This means that, with age, the shim pack on the Ford automobiles will grow larger, while the shim pack on the General Motors automobiles will diminish in size.

To adjust caster and camber on the General Motors automobiles, shims are either removed from or added to the front or the rear shim pack.

Where the camber angle needs to be more positive, shims would be removed in equal amounts from the front and rear shim packs (Fig. IV-35). This moves the top ball joint out to a more positive position.

When shims are added in equal amounts to both front and rear shim pack, the top ball joint will move in and the camber will be more negative or less positive (Fig. IV-36).

Caster is adjusted to a more positive setting by adding shims to the rear shim pack or by removing shims from the front shim pack.

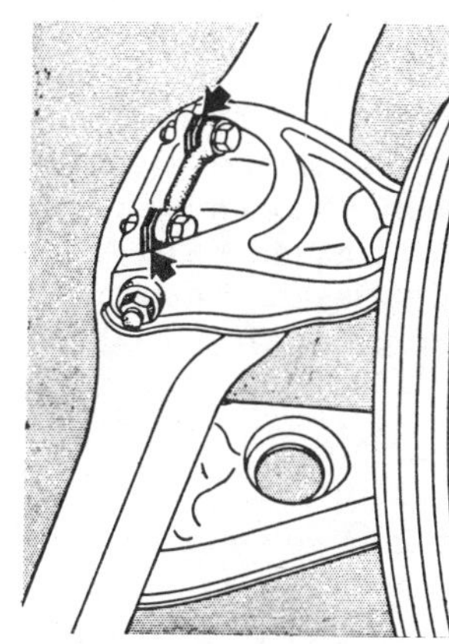

FIGURE IV-34 Ford Shim Adjustment. (Courtesy of Snap-on Tools Corp.)

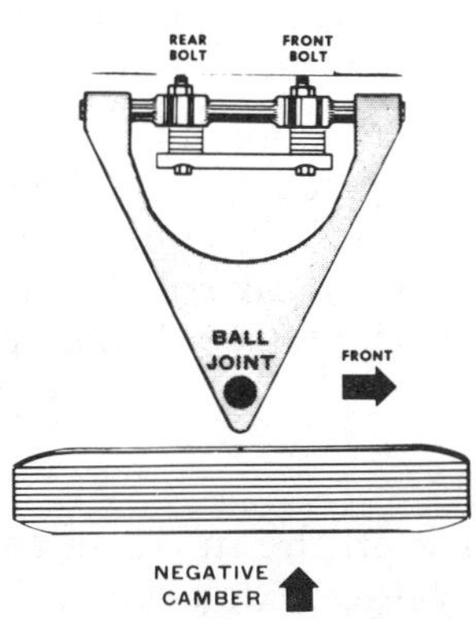

FIGURE IV-36 When an equal thickness of shims are added to each shim pack, camber is changed to a more negative position.
Courtesy of Snap-on Tools Corporation

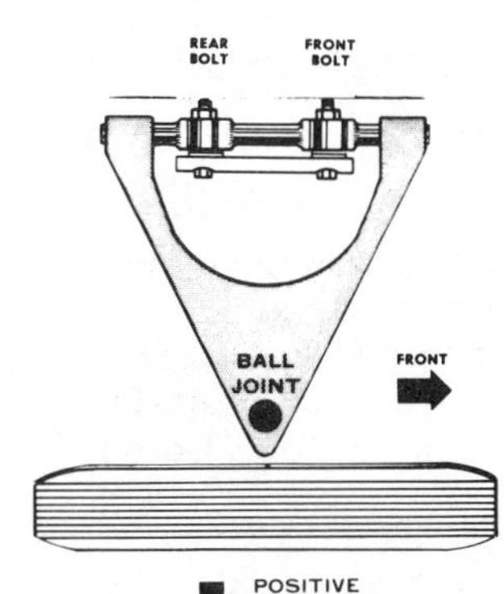

FIGURE IV-35 When an equal thickness of shims are removed from each shim pack, camber is changed to a more positive position.
Courtesy of Snap-on Tools Corporation

Caster is adjusted to a more negative setting by adding to the front shim pack or removing from the rear shim pack (Fig. IV-37). Caster is adjusted more positive by adding shims to the rear shim pack (Fig. IV-38).

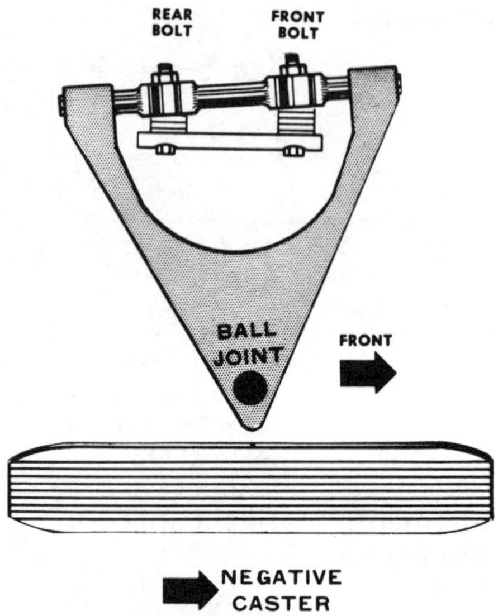

FIGURE IV-37 When shims are added to the front shim pack only, caster and camber are changed to a more negative position. (Courtesy of Snap-on Tools Corp.)

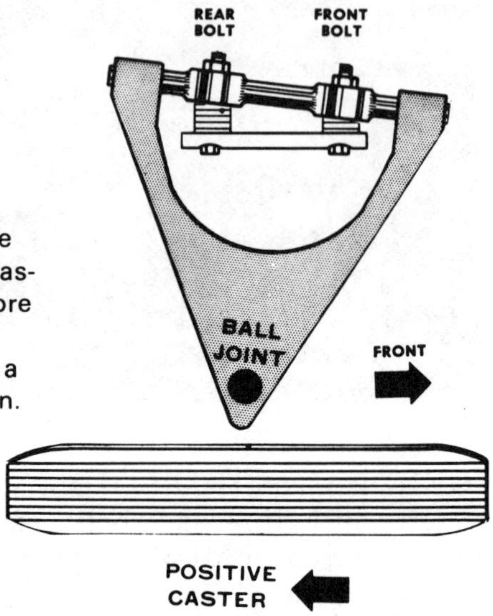

FIGURE IV-38 When shims are added to the rear shim pack only, caster is changed to a more positive position and camber is changed to a more negative position. (Courtesy of Snap-on Tools Corp.)

The following check list will help in making the right changes to both caster and camber on the General Motors shim pack adjustment:

1. *To change camber to a more positive angle without changing caster:* Remove equal thickness of shims from both front and rear shim packs.
2. *To change camber to a more negative angle without changing caster:* Add an equal thickness of shims to both front and rear shim packs.
3. *To change caster to a more positive angle without changing camber excessively:* Remove shims from the front shim pack and add an equal thickness to the rear shim pack.
4. *To change caster to a more negative angle without changing camber excessively:* Remove shims from the rear shim pack and add an equal thickness to the front shim pack.
5. *To change caster and camber to more positive angles simultaneously:* Remove shims from the front shim pack only.
6. *To change caster to a more negative angle and camber to a more positive angle simultaneously:* Remove shims from the rear shim pack only.

When caster and camber adjustments must be made separately, caster must be adjusted before camber.

Each division of General Motors will vary a little in the design of the front suspension for their automobiles. However, the shim adjustment has been used at one time or another by all divisions. One thing that will change the thickness of shims required to adjust caster is the distance between the two shim packs. The closer the shim packs are to each other the fewer shims are necessary to make the necessary caster changes.

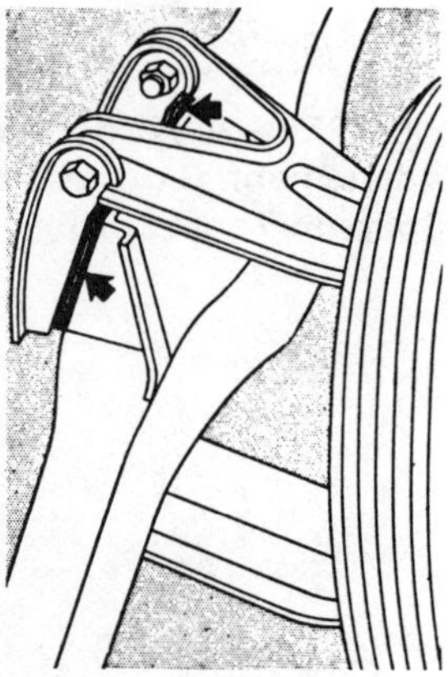

FIGURE IV-39 Shims were used at the upper-inner control arms of some Chrysler Corporation vehicles. (Courtesy of Snap-on Tools Corp.)

CHRYSLER CORPORATION SHIM ADJUSTMENT

Chrysler products that use shims have the shim pack on the inside of the frame. This means that the adjustments for those Chrysler models would be the same as for the General Motors automobiles (Fig. IV-39). However, a different size and shape of shim is used in Chrysler Corporation products.

AMERICAN MOTORS SHIM ADJUSTMENT

American Motors uses still another variation to the shim adjustment on the Rambler American. The shim pack was placed between the lower inner control arm and the inside of the frame bracket (Fig. IV-40). The adding or removing of shims moved the lower trunnion instead of the top trunnion. Both caster and camber were adjusted with shims. For an example of how the adjustment worked, examine the following problem.

The wheel needs a more negative caster setting and a more positive camber setting.

By adding shims to the rear shim pack, the lower trunnion will be pulled in toward the center of the car and back towards the rear, thus giving the tire a more positive camber, and a more negative caster setting.

It is important to remember that the lower trunnion is being moved, not the top one (Fig. IV-41). The caster and camber angles, therefore, change just opposite to those of the General Motors type, with the shim pack at the inside of the upper frame bracket.

FIGURE IV-40 American Motors uses shims at the lower-inner control arm shaft on the Rambler American. (Courtesy of American Motors Corp.)

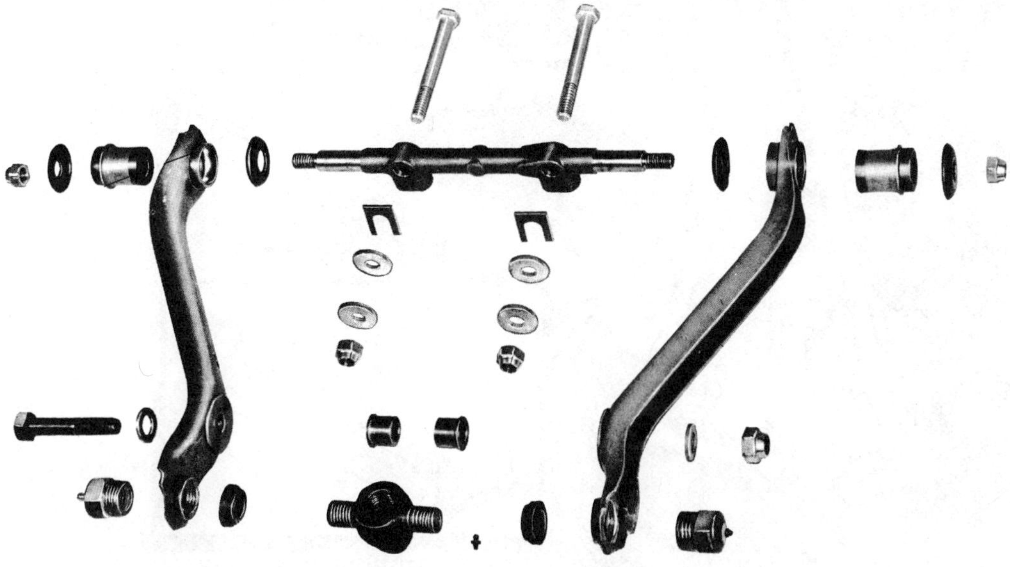

FIGURE IV-41 Exploded view of lower control arm and shim placement. (Courtesy of American Motors Corp.)

FORD THUNDERBIRD SHIM

The Thunderbird front suspension is illustrated in Fig. IV-42. The shims for adjusting the camber are placed between the pivot bracket and the frame mounted bracket. The caster is adjusted by shortening or lengthening the strut rod (Fig. IV-42). With this type of adjustment, camber must always be made before the caster adjustment. However, the camber angle should be rechecked after changing the caster. An extreme amount of movement in the strut rod will also affect the camber angle.

The following check list will help in making caster and camber adjustments on the Ford Thunderbird:

1. *To change the camber to a more positive position:* Remove the necessary thickness of shims from the shim pack.
2. *To change the camber to a more negative position:* Add the necessary thickness of shims to the shim pack.
3. *To change the caster to a more positive position:* Loosen the strut retaining bolts and nuts. Then shorten the strut rod by moving the lower control arm forward. This moves the lower ball joint toward the front of the car, giving a more positive caster setting.
4. *To change the caster to a more negative position:* Loosen the strut retaining bolts and nuts. Then lengthen the strut rod by moving the lower control arm back. This moves the lower ball joint toward the rear of the car, giving a more negative caster setting.

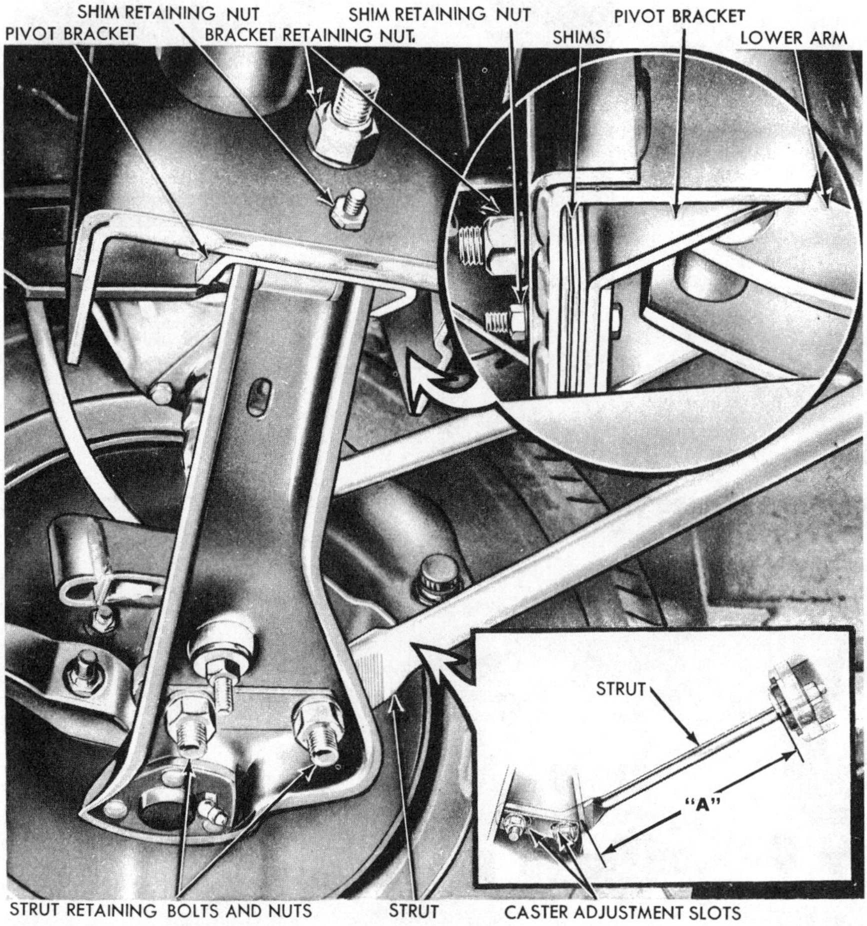

FIGURE IV-42 Lower control arm camber shim placement on the older Thunderbirds is shown in expanded circle. Caster is controlled by shortening or lengthening the strut rod. (Courtesy of Ford Customer Service Division)

JOB REPORT SHEET IV-C1

SHIM PACK ALIGNMENT

Student name____________________ Date____________________

Make of car____________ Model____________ Year____________

Time started____________ Time finished____________ Total time____________

Flat-rate time ____________

A. Caster Setting
1. Check caster setting and specifications.
   (Note: In some cases, caster and camber may be set at the same time.)
   a. Caster setting before adjusting:

      Left wheel:____________ Right wheel:____________
   b. Factory caster specifications:

      Left wheel:____________ Right wheel:____________
2. Adjust caster by removing or adding shims to the front or the rear shim pack or by adjusting the strut rod.
   a. Caster setting after adjustment

      Left wheel:____________ Right wheel:____________

Notes:____

____

____

B. Camber Setting
1. Check camber setting and specifications.

   a. Camber setting before adjustment:

      Left wheel:____________ Right wheel:____________
   b. Factory camber specifications:

      Left wheel:____________ Right wheel:____________
2. Adjust camber by removing or adding shims to both shim packs.
   Camber setting after adjustment:

      Left wheel:____________ Right wheel:____________

Notes:____

____

____

C. Toe-in Setting

1. Check toe-in setting

   a. Toe-in or toe-out setting before adjustment: ______________________________

   b. Toe-in factory specifications: ______________________________

2. Set toe-in by turning tie rod end adjusting sleeves on one or both tie rod ends.

   Toe-in setting after adjustment: ______________________________

3. Center steering wheel.

Notes: ______________________________

____

____

D. Turning Radius Check (20-degree turn)

1. Factory specifications:

   Inside wheel: ______________ Outside wheel: ______________

2. Turning radius settings:

   Inside wheel: ______________ Outside wheel: ______________

Report: ______________________________

____

____

E. Road Test

Report: ______________________________

____

____

DATE COMPLETED ______________ INSTRUCTOR'S SIGNATURE ______________

OBJECTIVE IV-C2 (manipulative)

Working as a team with one other student you will align the front wheels on at least one instructor assigned automobile with eccentric adjustments. The alignment will include checking and adjusting where necessary the caster, camber, toe-in, steering axis inclination, turning radius, and centering of the steering wheel. Time recommended: twice the flat-rate time allowed in the Motors or Chilton Flat Rate manuals.

Preparation:

1. Study pages 331 through 340.
2. Study the manufacturer's procedure and specifications for the specific automobile to which you are assigned.
3. Carefully observe instructor's demonstration of front end alignment.

Materials and Tools Needed:

1. Alignment equipment
2. Jacks
3. Air hose and tire inflator
4. Air gauge
5. Special tools for front end alignment. (See Manufacturer's instructions.)
6. Set of box/open end combination wrenches
7. Set of sockets and ratchet

ECCENTRIC ADJUSTMENTS

Before ball joints came into use, some form of an eccentric adjustment was used in almost every car. This was usually an eccentric bolt in the connection between the upper control arm and the steering knuckle. The ball joint has replaced that bushing, and now many auto manufacturers have converted production to include an eccentric washer and bolt assembly at the upper or lower inner control arm bushing (Fig. IV-43).

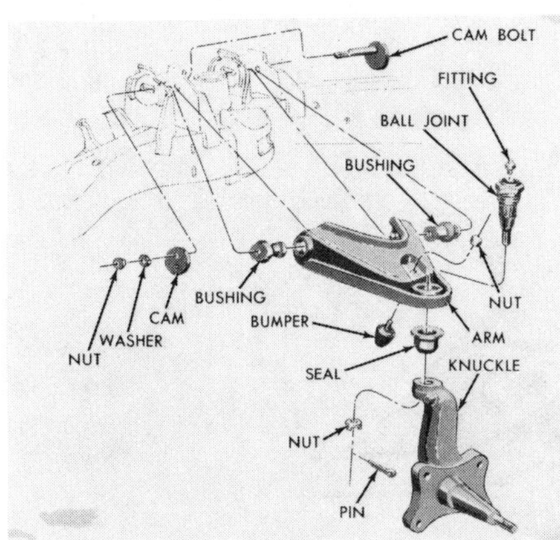

FIGURE IV-43 Exploded view of upper control arm with camber and caster controls in the cam washer and bushings. (Courtesy of Chrysler Motors Corp.)

CAM TYPE WASHERS

Eccentric washers are nothing more than large washers with the hole for the bushing bolt drilled off center (Fig. IV-44). When mounted properly, the washers are encaged in a bracket attached to the frame. The washer that is on the bolt head side of the bushing is mounted either in a square hole or welded solidly to the bolt head (Fig. IV-45). This means that the washers move in the cage when the bolt is turned. As the washer turns, the bushing center is moved either in or out (Fig. IV-46). This is what changes the camber and sometimes the caster adjustments.

FIGURE IV-44 Exploded view of lower-inner control arm eccentric for camber adjustment. (Courtesy of American Motors Corp.)

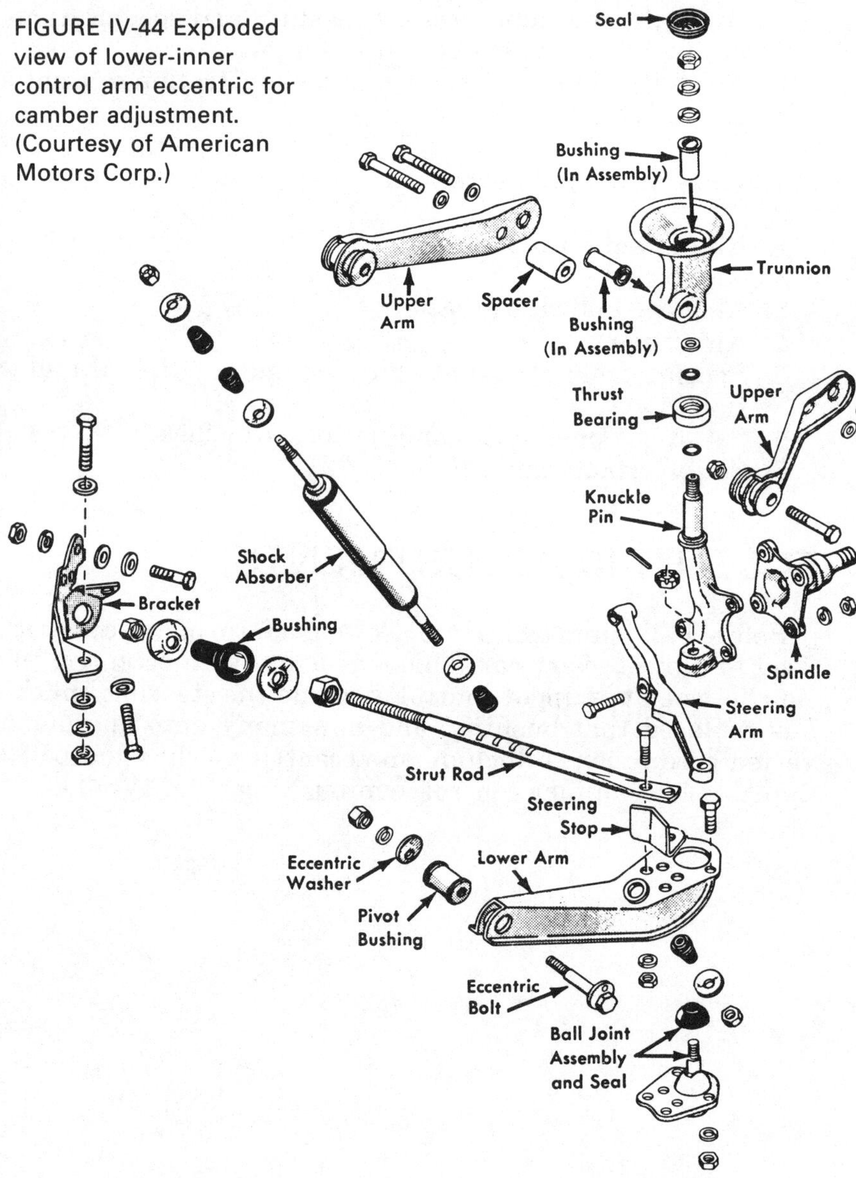

FIGURE IV-45 Lower-inner control arm eccentric controls camber only. (Courtesy of Chevrolet Motor Division)

FIGURE IV-46 Notice how the eccentric or cam-like washer has the bolt off center to move for adjustment. (Courtesy Snap-on Tools Corp.)

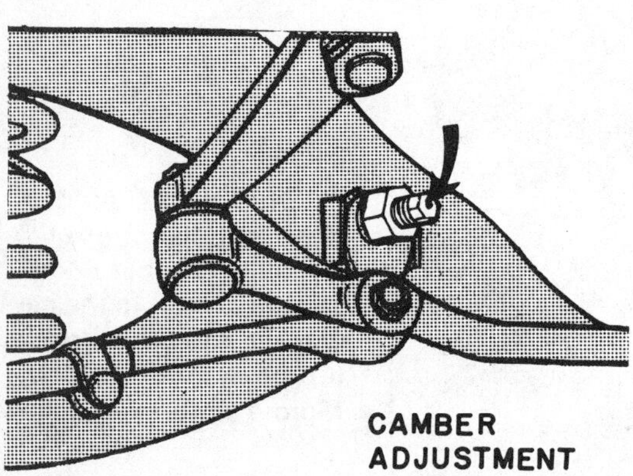

UPPER INNER CONTROL ARM ECCENTRICS

Late model Chrysler products and some Ramblers have used the upper inner eccentric bushing or washer cams (Fig. IV-47).

Adjustments are made on eccentric bushings or washer cams by first loosening the locknut. Then, with the appropriate size wrench or socket, turn the bushing center bolt by moving the high side of the cam either to the inside or the outside of the car. By moving the high side of the cam to the outside, the upper ball joint is also moved out (Fig. IV-48). By moving the high side of the cam in, the upper ball joint also moves in. Both caster and camber are adjusted with the eccentric method when there are two eccentric bushings at the upper inner control arm shaft.

The following list will help in making the right changes in both caster and camber when there are two eccentrics at the upper inner control arm bushings.

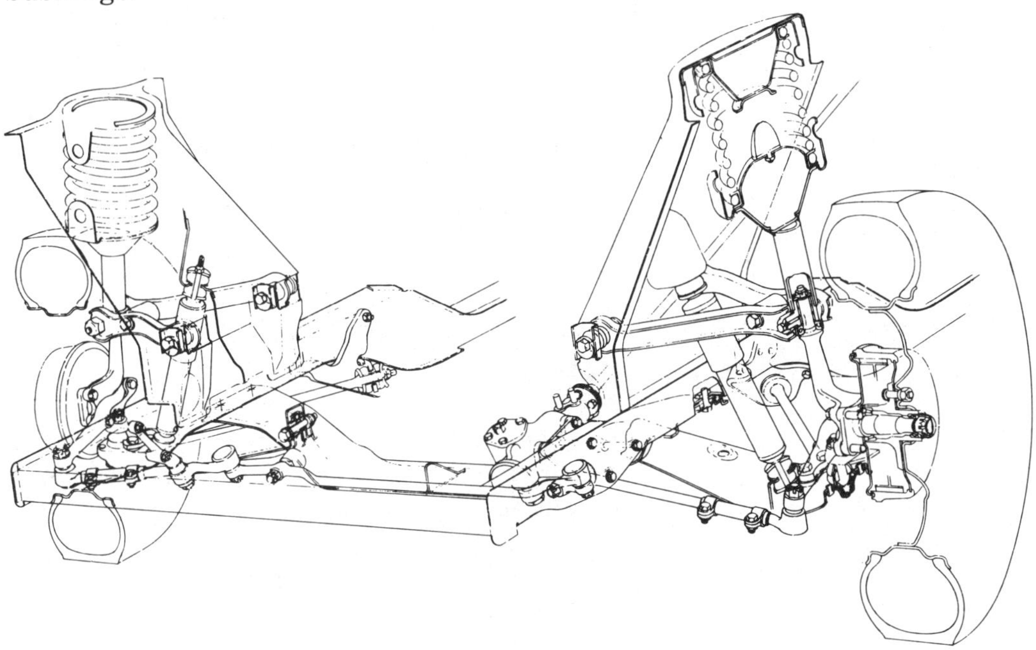

FIGURE IV-47 Upper-inner control arm eccentric bushings used by Rambler. (Courtesy of American Motors Corp.)

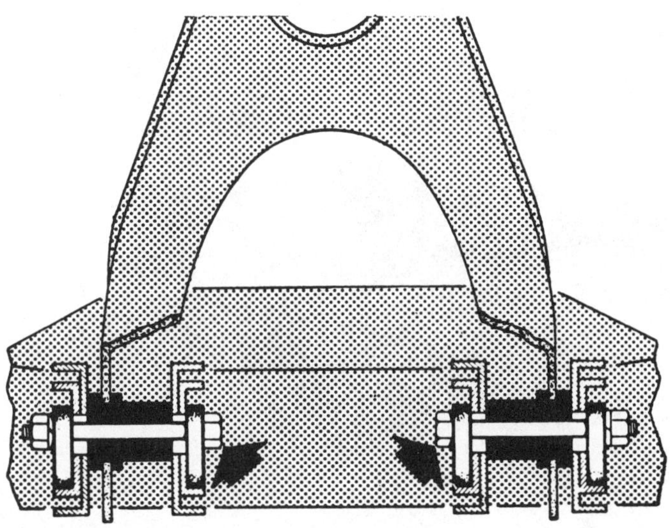

FIGURE IV-48 Eccentric adjustments for both caster and camber indicated by arrows. (Courtesy of Snap-on Tool Corp.)

1. *To change camber to a more positive angle without changing caster:* Turn both eccentrics out an equal amount. This will move the top ball joint outward.
2. *To change camber to a more negative angle without changing caster:* Turn each eccentric in an equal amount. This will move the top ball joint inward.
3. *To change caster to a more positive angle without changing camber excessively:* Turn the front eccentric out one-half the required caster change. Turn the rear eccentric in the remainder of the required caster change.
4. *To change caster to a more negative angle without changing camber excessively:* Turn the rear eccentric out one half of the required caster change. Then turn the front eccentric in the remainder of the required caster change.
5. *To change caster and camber to more positive angles simultaneously:* Turn the front eccentric out. This will move the top ball joint out to a more positive camber setting and backward to a more positive caster setting.
6. *To change caster and camber to more negative angles simultaneously:* Turn the front eccentric in. This will move the top ball joint in to a more negative camber setting and forward to a more negative caster setting.

When caster and camber adjustments must be made separately on this type of eccentric adjustment, you must adjust caster and then camber.

CAM WASHERS AT THE LOWER INNER CONTROL ARM BUSHINGS

The design of the lower control arm can be in the shape of an A. This design always requires two eccentrics while the single arm, as used on some Chevrolets, requires only one eccentric (Fig. IV-49).

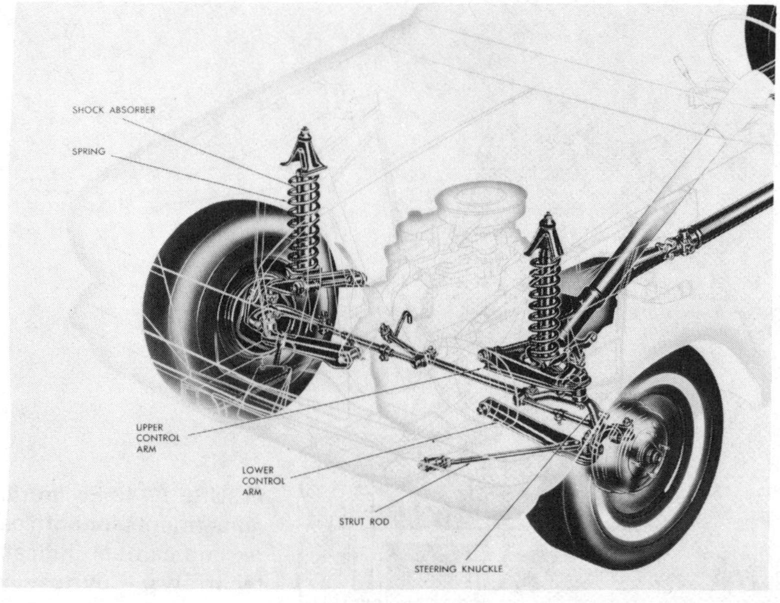

FIGURE IV-49 The strut rod is used to control caster. (Courtesy of Chevrolet Motor Division)

SINGLE CAM WASHER ADJUSTMENT AT THE LOWER CONTROL ARM

Manufacturers have used the eccentric adjustment at the lower inner control-arm bushing in some models, such as the Chevrolet shown in Fig. IV-50. When only one eccentric adjustment is placed at the lower inner control arm bushing, it is used to adjust camber only. Camber is adjusted by loosening the locknut and turning the center bolt of the bushing. This will move the bushing and the lower ball joint either in or out. Caster is adjusted by shortening or lengthening a strut rod. See Fig. IV-51.

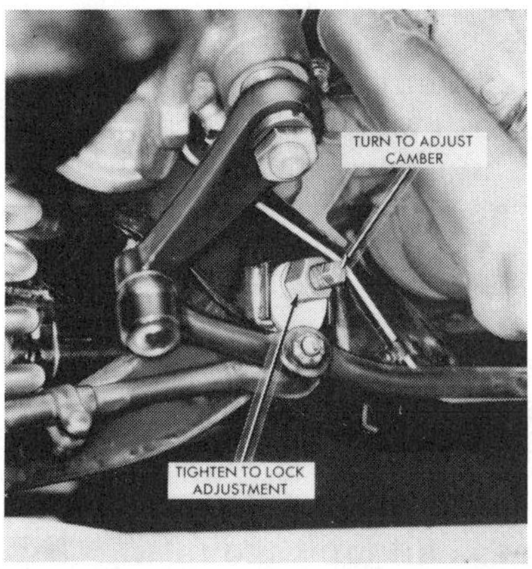

FIGURE IV-50 Camber adjustment is accomplished by turning the center bolt. (Courtesy of Chevrolet Motor Division)

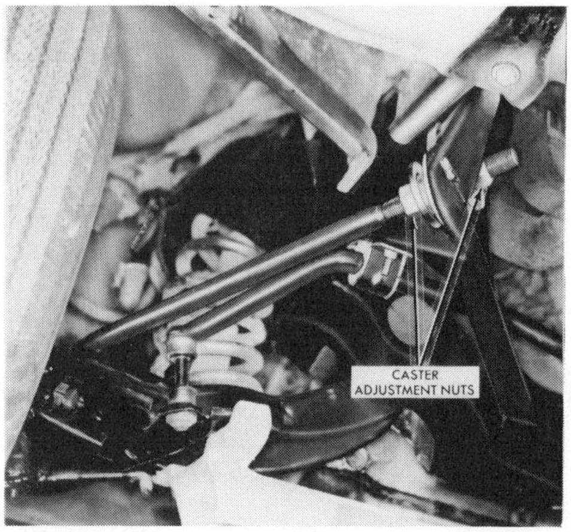

FIGURE IV-51 Lengthening or shortening the strut rod adjusts caster settings. (Courtesy of Chevrolet Motor Division)

The following will be helpful in making the right caster and camber adjustments on this type of eccentric:

1. *To change camber to a more positive position:* Turn the eccentric in. This moves the lower ball joint in and gives the tire a more positive camber setting.
2. *To change camber to a more negative position:* Turn the eccentric out. This moves the lower ball joint out and gives the tire a more negative camber setting.
3. *To change caster to a more positive position:* This adjustment will depend on the position of the strut rod. When the strut rod is mounted from the lower control arm to the front of the car, it will be necessary to shorten the strut rod. If the strut rod is mounted from the rear of the lower control arm to the frame, the strut rod must be lengthened (Fig. IV-52). The strut rod is shortened or lengthened by loosening the locking nuts and turning them in the desired direction.

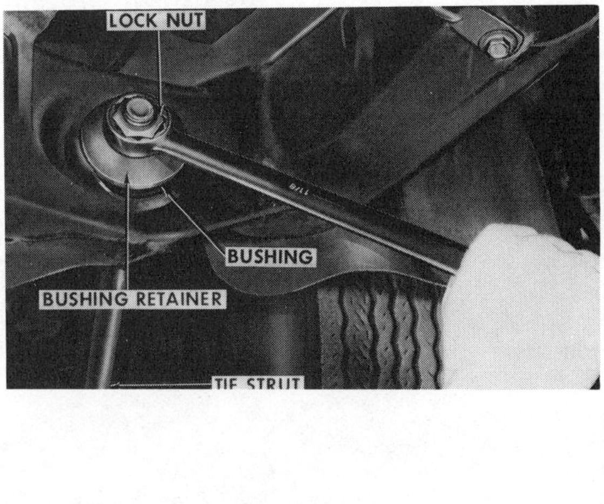

FIGURE IV-52 Strut rod control of caster is managed by loosening and turning the lock nuts. (Courtesy of Cadillac Motor Car Division)

4. *To change caster to a more negative position:* When the strut rod is mounted in front of the lower control arm, it must be lengthened. When it is mounted in the rear of the lower control arm, it must be shortened.

Note: *Not all strut rods are used for caster adjustments. For example, there are two upper control-arm eccentrics in Chrysler products.*

TWO CAM WASHER ADJUSTMENTS AT THE LOWER CONTROL ARM

American Motors used this method on some models such as the 1961 Classic and Ambassador. The two eccentric adjustments were placed on the lower inner control arm bushings and were used to adjust both caster and camber. The method used in adjusting caster and camber would be just opposite to that used for the two eccentrics at the upper inner control arm bushings.

ECCENTRIC UPPER BALL JOINT MOUNTING

Cadillac uses an eccentric bushing at the upper ball joint mounting (Fig. IV-53). This bushing allows for camber adjustment only. Caster is adjusted with a strut rod mounted to the lower control arm.

When adjusting camber on Cadillacs with eccentric bushings, the final position of the upper ball joint stud should be in the rear half of the upper steering knuckle as shown in Fig. IV-54A or Fig. IV-54B. The reason for the above rule is this: As the eccentric bushing is rotated around the axis of the upper ball joint stud, the upper knuckle on the steering knuckle not only moves in and out for camber adjustments but also moves forward and backward. This *forward-backward* movement (which does not affect caster) is accompanied by a simultaneous lifting or lowering movement of the steering arm. In turn, the steering arm movement lifts or lowers the pivot point of the outer tie rod end. This up-and-down movement of the tie rod end pivot point has a direct effect on the amount of toe-in and toe-out during jounce-rebound and turning movements of the wheel. Therefore, it is important that camber be properly adjusted (Fig. IV-54A or Fig. IV-54B) in order to maintain the outer tie rod end in its proper position according to factory specifications. If camber is not properly adjusted (Fig. IV-54C), the outer tie rod end will not be in its proper position. Unpredictable and uncalculated toe changes will then occur in the steering and suspension system, which may cause customer dissatisfaction over poor vehicle performance.

Fig. IV-54A depicts a steering knuckle halfway between positive and full negative camber. At this point the upper knuckle is at its maximum forward travel, the upper ball joint stud is in the rear half of the upper knuckle, and the pivot point of the outer tie-rod end is lifted to its maximum acceptable height denoted by dimension A.

Fig. IV-54B depicts a steering knuckle at maximum positive or maximum negative camber. When compared to Fig. IV-54A, rotation of the eccentric

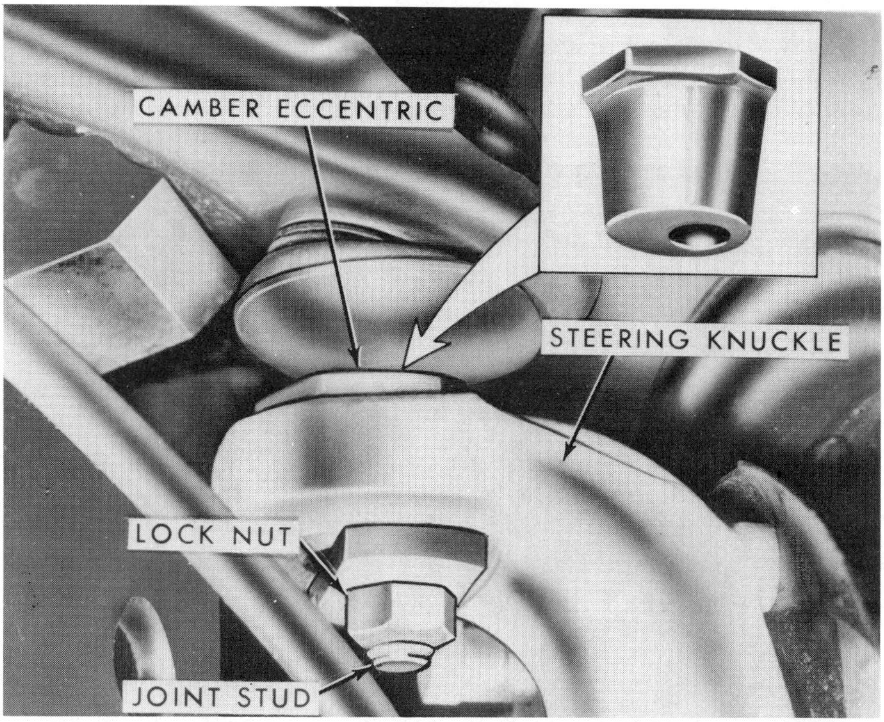

FIGURE IV-53 Camber is controlled by eccentric bushing at the upper ball joint stud. (Courtesy of Cadillac Motor Car Division)

bushing has caused the upper knuckle to move in a rearward direction. Maximum camber (positive or negative) is attained when the center lines of the eccentric bushing knuckle and upper ball joint stud are in line. Further rotation of the eccentric bushing will not yield additional camber. The rearward movement of the upper knuckle in turn lowers the pivot point of the outer tie-rod end to a maximum acceptable depth denoted by dimension B.

Fig. IV-54C depicts the undesirable situation where the upper ball joint stud is in the front half of the upper knuckle. It is a sure sign that the tie rod end is improperly located. Note that the pivot point of the outer tie rod end is now below the maximum dimension B. This situation must be avoided.

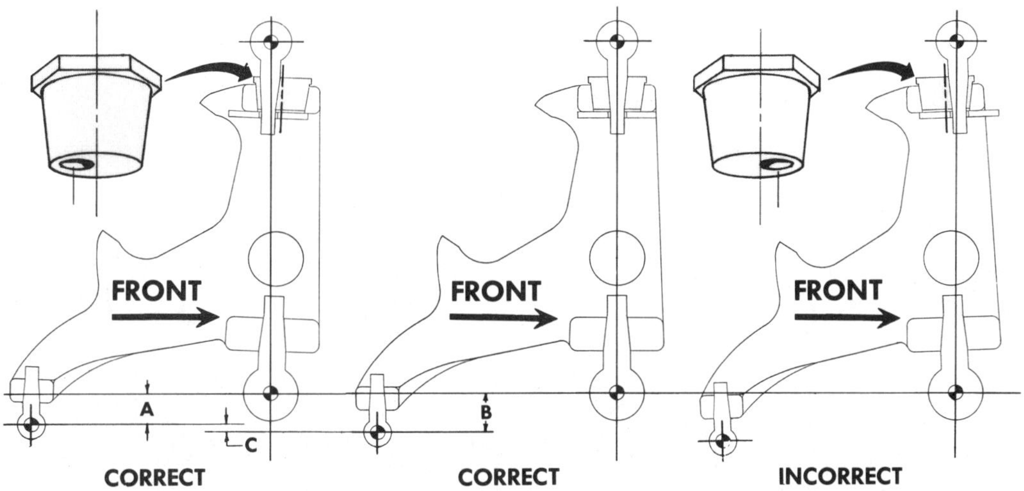

FIGURE IV-54 A, B, C (Courtesy of Moog Automotive, Inc.)

If camber is adjusted within the extremes depicted in Figs. IV-54A and IV-54B, the alignment specialist can be assured that the pivot point of the outer tie rod end will be properly located and within the acceptable range denoted by dimension C, as per factory specifications (Fig. IV-55).

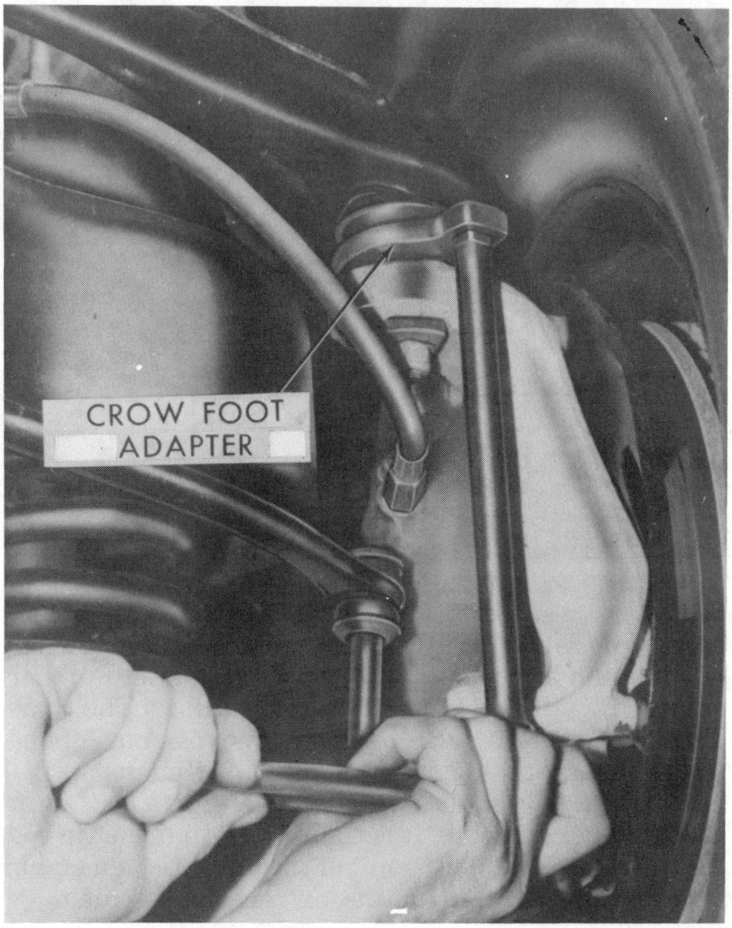

FIGURE IV-55 A crow foot adapter is used to turn the upper ball joint stud eccentric. (Courtesy of Cadillac Motor Car Division)

JOB REPORT SHEET IV-C2

ECCENTRIC ALIGNMENT

Student name____________________Date____________________

Make of car_______________Model_______________Year_______________

Time started_______________Time finished_______________Total time_______________

Flat-rate time _______________

A. Camber Setting

1. Check camber setting and specifications
   a. Camber setting before adjustment:

      Left wheel:_______________Right wheel:_______________

   b. Factory camber specifications

      Left wheel_______________Right wheel_______________
2. Adjust camber by turning the front and rear eccentrics out or in.
   a. Camber setting after adjustment:

      Left wheel:_______________Right wheel:_______________

Notes:____

____

____

(**Note:** *In some cases, caster and camber may be set at the same time.*)

B. Caster Setting

1. Check caster setting and specifications.
   a. Caster setting before adjusting:

      Left wheel:_______________Right wheel:_______________

   b. Factory caster specifications:

      Left wheel:_______________Right wheel:_______________
2. Adjust caster by turning the front or the rear eccentric in or out, or by the strut rod.
   a Caster setting after adjustment:

      Left wheel:_______________Right wheel:_______________

Notes:____

____

____

C. Toe-in Setting

1. Check toe-in setting.
   a. Toe-in or toe-out setting before adjustment: ____________________

   b. Toe-in factory specifications:____________________

2. Set toe-in by turning tie-rod-end adjusting sleeves on one or both tie-rod ends.

   a. Toe-in setting after adjustment: ______________________

3. Center steering wheel:

Notes: ______________________

____

____

D. Turning Radius Check (20 degree turn)

1. Factory specifications:

   Inside wheel: __________ Outside wheel: __________

2. Turning-radius settings:

   Inside wheel: __________ Outside wheel: __________

Report: ______________________

____

____

E. Road Test

Report: ______________________

____

____

DATE COMPLETED __________ INSTRUCTOR'S SIGNATURE __________

OBJECTIVE IV-C3 (manipulative)

Working as a team with one other student, you will align the front wheels on an instructor assigned automobile with slotted adjustments. The alignment will include checking and adjusting where necessary the caster, camber, toe-in, steering axis inclination, turning radius, and centering of the steering wheel. Time recommended: twice the flat-rate time allowed in the Motors or Chilton Flat Rate manuals.

Preparation:

1. Study pages 341 through 348.
2. Study the manufacturer's procedure and specifications for the specific automobile to which you are assigned.
3. Carefully observe instructor's demonstration of front end alignment.

Materials and Tools Needed:

1. Alignment equipment
2. Jacks
3. Air hose and tire inflator
4. Air gauge
5. Special tools for front end alignment. (See Manufacturer's instructions.)
6. Set of box/open-end combination wrenches
7. Set of sockets and ratchet

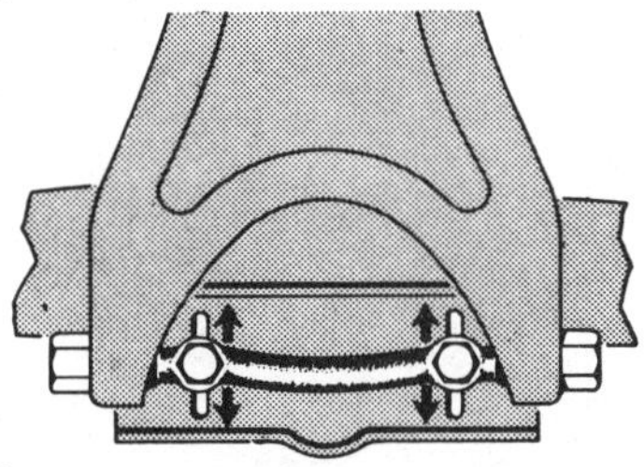

FIGURE IV-56 Slotted control for caster and camber is used by some Ford models. (Courtesy of Snap-on Tools Corp.)

SLOTTED UPPER CONTROL ARM OR SLOTTED FRAME BRACKET BOLT HOLES

This type of adjustment has been used quite frequently in later model cars. A good example of this adjustment is the Ford suspension shown in Fig. IV-56.

Caster and camber adjustments are made by loosening the two bolts that hold the upper inner arm shaft to the frame (Fig. IV-56). On this Ford suspension, the upper inner arm shaft has two holes for the bolts, and the holes in the frame are slotted. The slotted holes in the frame allow the upper arm and ball joint to move in or out to the front or back, thus allowing for the adjustment of caster and camber.

The two special tools shown in Figs. IV-57 and IV-58 are used as aids in adjusting the caster and camber on slotted controls (Fig. IV-35). The following is an example of how to make the actual adjustment on this type of control on a 1973 Ford.

1. Install the special tools. See Fig. IV-57.
2. Loosen the upper arm inner shaft bolts two or three complete turns. For a positive camber, tighten both clamps. This will pull the upper inner shaft out. For a more negative camber, loosen both clamps. This will allow the upper inner shaft to move in.

The following list will help in making the proper caster and camber adjustments:

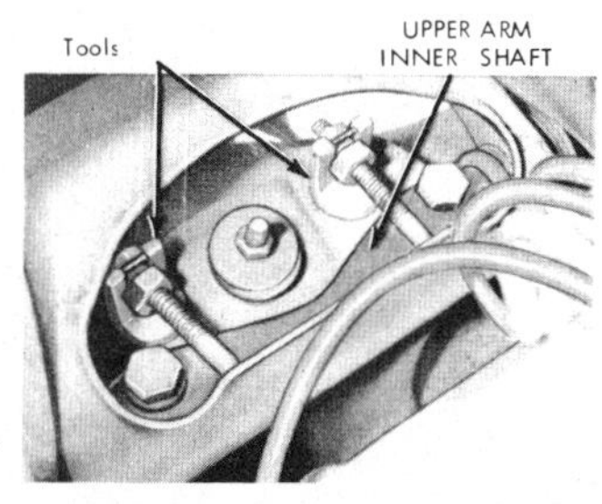

FIGURE IV-57 One type of designated tool to hold the upper control arm in the correct position while caster and camber adjustments are made. (Courtesy of Ford Customer Service Division)

1. *To adjust the camber angle to a more positive position:* Tighten both tool clamps an equal amount.
2. *To adjust the camber angle to a more negative position:* Loosen both tool clamps an equal amount.
3. *To adjust the caster angle to a more positive position and without changing the camber angle excessively:* Loosen the rear tool clamp and tighten the front tool clamp an equal number of turns.
4. *To adjust the caster angle to a more negative position without changing the camber angle excessively:* Loosen the front tool clamp and tighten the rear tool clamp an equal number of turns.
5. *To adjust both caster and camber to a more positive position simultaneously:* Tighten front tool clamp only.
6. *To adjust both caster and camber to a more negative position simultaneously:* Turn the rear tool bolt counterclockwise. Loosen the front tool clamp only.

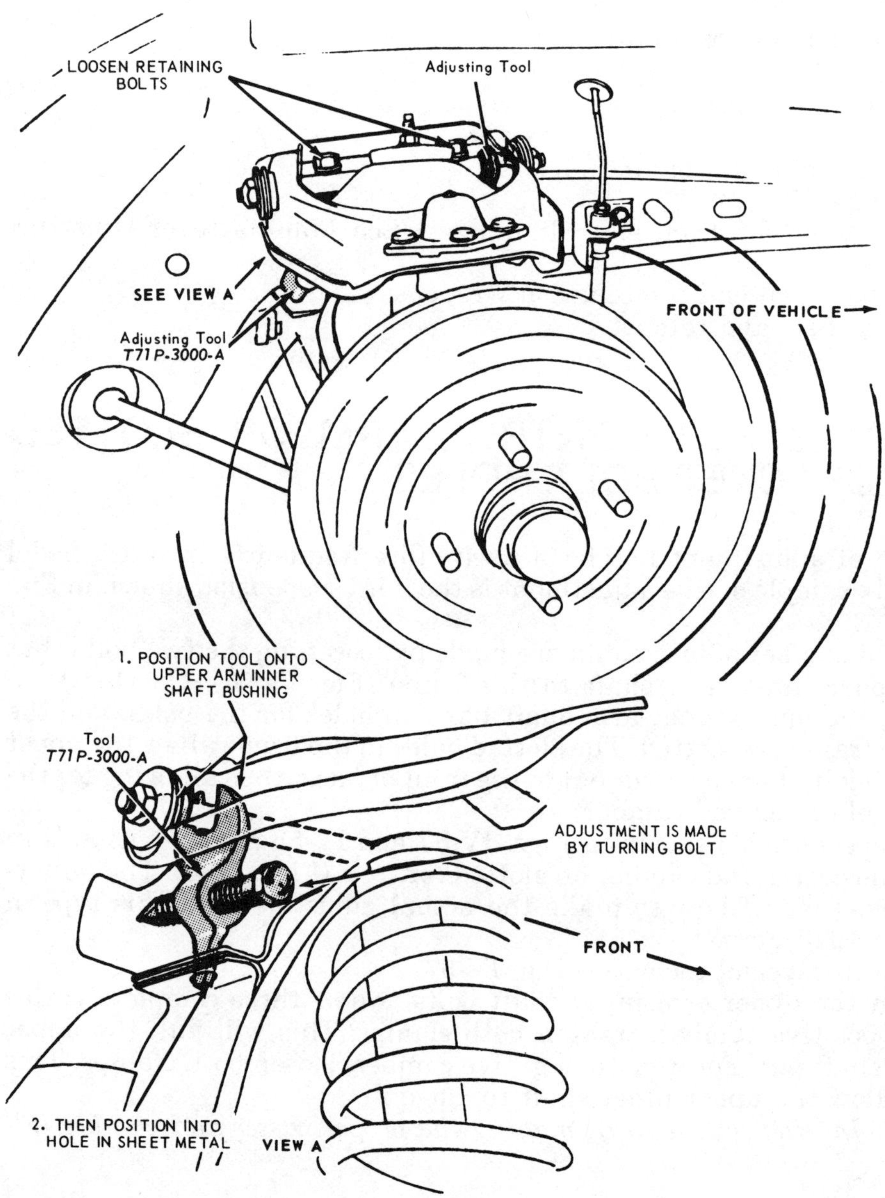

FIGURE IV-58 This special tool is used to help position upper control arm during caster and camber adjustment on slotted controls. (Courtesy of Ford Customer Service Division)

TOE-IN ADJUSTMENTS

Toe-in should always be the last adjustment made on any wheel alignment job. The reason for this is that both caster and camber angle changes will affect the toe-in setting (Fig. IV-59).

The toe-in adjustment is also the most critical adjustment when it comes to tire wear. Toe-in adjustments, unlike caster and camber adjustments, are all made in the same way—by shortening or lengthening the tie-rod ends. The only difference between automobiles is in the amount of toe-in suggested by the manufacturer and the placement or number of tie-rod ends.

The tie rod ends may be found either in the front of the front-suspension system or in the back of the front suspension system. It is obvious that when the tie rod ends are in front of the front suspension system, the toe-in setting is increased by shortening the tie rod ends. When the tie rod ends are in back of the front-suspension system, toe-in is increased by lengthening the tie rod ends.

Note: *Bending is never recommended as a method of adjusting toe-in.*

An important part of adjusting toe-in is the centering of the steering wheel. When the steering wheel is not centered, it is not only annoying to the driver, but it is also dangerous in that it reduces the effectiveness of the steering gear.

To adjust the toe-in on any make or model of automobile, follow these steps:

1. Place the front wheels and steering wheel in a straight ahead position. (If the steering wheel is not level or in a straight ahead position when driving, it is possible to position it during the toe-in adjustment.)
2. Loosen the two clamp bolts on each tie-rod sleeve.
3. If the steering wheel is in the straight ahead position, adjust the toe-in by shortening or lengthening both tie rod end sleeves an equal distance. Any adjustment that requires less than a one-half turn of the tie rod end sleeve will usually not affect the steering wheel position and can be made by turning only one sleeve.

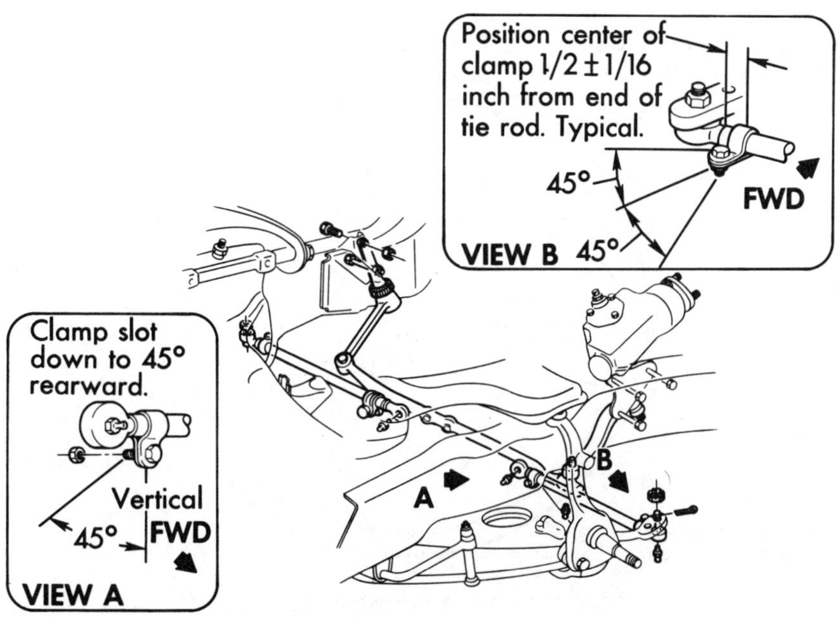

FIGURE IV-59 Positions are shown for tie rod end adjusting sleeve and clamp after toe-in adjustment. (Courtesy of Chevrolet Motor Division)

4. When the steering wheel is not centered, additional steps are needed in the adjusting of the toe-in.
   a. To determine the center position, turn the steering wheel as far right as it will go, then count the revolutions while turning the steering wheel all the way to the left. Then turn the steering wheel one-half of the total number of revolutions to the right. This should be the steering wheel's center position. This will also be the high point of the steering gear. If the steering wheel spokes are not centered, you should remove the steering wheel from the steering shaft and place it in the properly centered position. (See Unit V, "Steering Columns.")
   b. After the steering wheel is properly centered with the steering gear, adjustment of toe-in and proper positioning of the steering wheel with relation to the front wheels can be made on Ford automobiles by following the examples shown in Fig. IV-60.

After toe-in and the steering wheel have been adjusted to their proper position, tighten the adjusting sleeve clamps. On some cars, the clamp bolts must be in a certain position for proper clearance of other steering parts. See Fig. IV-61, IV-62 and IV-63.

This chapter has given a brief description of each type of adjustment controls. You should not assume that all the technical information needed for each automobile is in this unit. Manufacturers' service manuals, technical bulletins, and automotive repair manuals give more detailed procedures for servicing specific makes and models of automobiles. You should never try to align the front end on any car until you know the recommended procedure for that particular make and model. Above all, never assume that your alignment is completed until you have road tested the car.

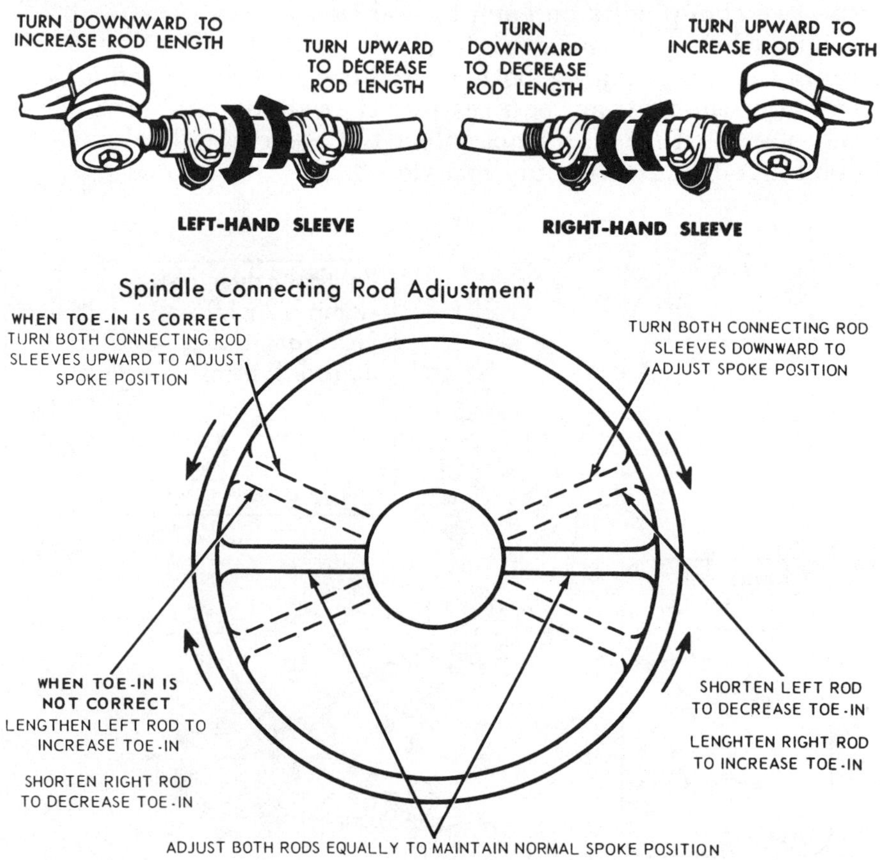

FIGURE IV-60 Toe-in and steering wheel spoke alignment adjustments are shown for most Fords. (Courtesy of Ford Customer Service Division)

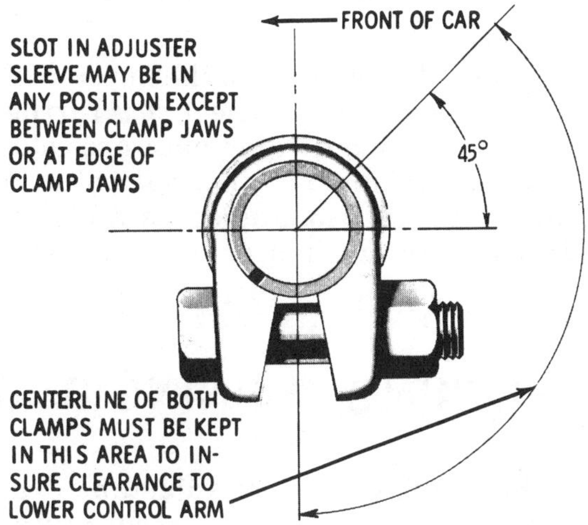

FIGURE IV-61 Position for adjusting sleeve clamp after toe-in adjustment. (Courtesy of Oldsmobile Division)

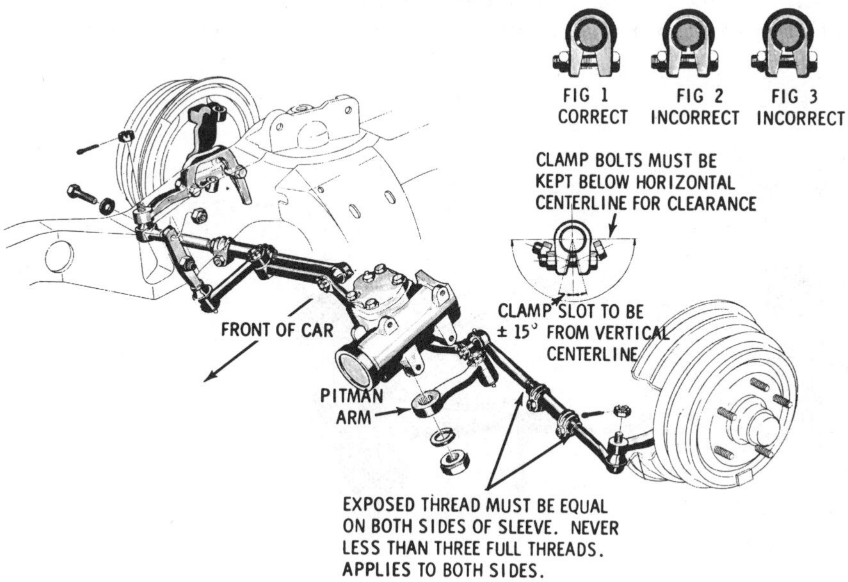

FIGURE IV-62 Typical steering linkage showing position of adjusting sleeve clamps. (Courtesy of Oldsmobile Division)

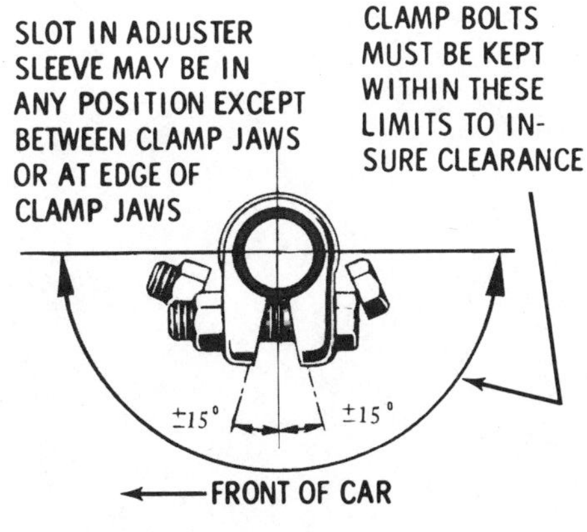

FIGURE IV-63 Direction of adjusting sleeve clamp bolt is important on some car models. (Courtesy of Oldsmobile Division)

CHECKING STEERING AXIS INCLINATION

This angle is built into the automobile and has no possible adjustment. An explanation of how to check it is further complicated because each manufacturer of alignment equipment uses a different method. Because of the direct effect of steering axis inclination on caster and camber, it can be assumed that if these two angles can be adjusted to within specifications, the steering axis inclination angle is OK.

CHECKING TURNING RADIUS

This angle is also not adjustable. However, it can be checked quite easily with turntable pads that are calibrated in degrees. Turn the inside wheel to a 20-degree turn and read the setting on the outside wheel (Fig. IV-64). This reading should match the specifications listed by the manufacturer. It will usually be somewhere between 18 and 19 degrees on the inside wheel. Correction can only be accomplished by replacing the steering arms.

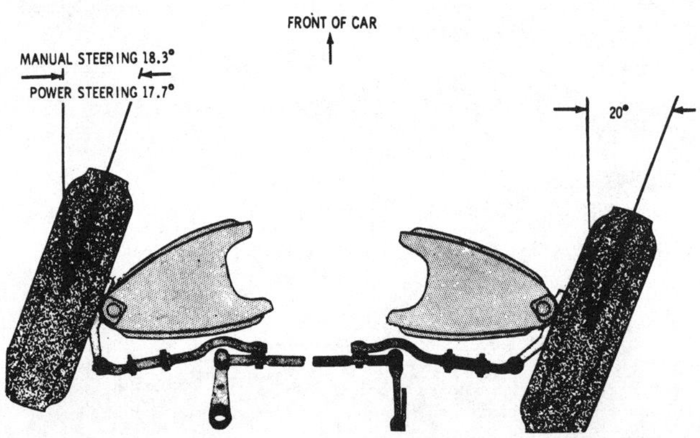

FIGURE IV-64 Checking for correct turning radius angle is accomplished with turn pads. (Courtesy of Oldsmobile Division)

JOB REPORT SHEET IV-C3

SLOTTED ALIGNMENT

Student name________________________________Date____________________

Make of car______________Model______________Year____________________

Time started______________Time finished______________Total time______________

Flat-rate time ______________

A. Caster Setting
   1. Check caster setting and specifications. (**Note:** *In some cases, caster and camber may be set at the same time.)*
      a. Caster setting before adjusting:

         Left wheel:______________Right wheel:______________
      b. Factory caster specifications:

         Left wheel:______________Right wheel:______________
   2. Adjust caster by moving upper control arm forward or backward at the ball joint.
      a. Caster setting after adjustment:

         Left wheel:______________Right wheel:______________

   Notes:____

   ____

   ____

B. Camber Setting
   1. Check camber setting and specifications.
      a. Camber setting before adjustment:

         Left wheel:______________Right wheel:______________
      b. Factory camber specifications:

         Left wheel:______________Right wheel:______________
   2. Adjust camber by moving the upper control arm in or out.
      a. Camber setting after adjustment:

         Left wheel:______________Right wheel:______________

   Notes:____

   ____

   ____

C. Toe-in Setting
   1. Check toe-in setting.
      a. Toe-in or toe-out setting before adjustment: ______________

b. Toe-in Factory specifications ______________________________

2. Set toe-in by turning tie rod end adjusting sleeves on one or both tie rod ends.

   a. Toe-in setting after adjustment ______________________________

3. Center steering wheel

Notes: ______________________________

____

____

D. Turning Radius Check (20-degree turn)

1. Factory specifications:

   Inside wheel: ______________ Outside wheel: ______________

2. Turning radius settings:

   Inside wheel: ______________ Outside wheel: ______________

Report: ______________________________

____

____

E. Road Test

Report: ______________________________

____

____

DATE COMPLETED ______________ INSTRUCTOR'S SIGNATURE ______________

TEST IV-C

Student name________________________________Date____________________

Directions: Select the best answer to each question and circle the correct letter at the left.

1. Ford shim adjustment at the upper inner control arm has the shim pack:
   I. Outside the frame bracket
   II. Inside the frame bracket
      A. I only
      B. II only
      C. Both I and II
      D. Neither I nor II
2. To change the camber to a more positive position on the Ford shim type:
   I. Remove shims from both shim packs
   II. Add shims to both shim packs
      A. I only
      B. II only
      C. Both I and II
      D. Neither I nor II
3. To change the caster to a more negative position on the Ford shim type:
   I. Add shims to the front shim pack
   II. Remove shims from the rear shim pack
      A. I only
      B. II only
      C. Both I and II
      D. Neither I nor II
4. To change the caster and camber to more positive positions at the same time on the Ford shim type:
   I. Add shims to front shim pack
   II. Add shims to rear shim pack
      A. I only
      B. II only
      C. Both I and II
      D. Neither I nor II
5. To change caster to a more positive position without changing camber on the Ford shim type:
   I. Remove shims from rear shim pack and add an equal thickness to the front
   II. Remove shims from front shim pack and add an equal thickness to the rear
      A. I only
      B. II only
      C. Both I and II
      D. Neither I nor II

6. To adjust camber to a more positive position on the General Motors type of shim pack:
   I. Remove shims from both shim packs
   II. Add shims to both shim packs
      A. I only
      B. II only
      C. Both I and II
      D. Neither I nor II
7. To adjust caster and camber to more positive positions at the same time on the General Motors shim type:
   I. Remove shims from the front shim pack
   II. Add shims to the rear shim pack
      A. I only
      B. II only
      C. Both I and II
      D. Neither I nor II
8. Camber only (not caster,) is adjusted by shims on the:
   I. Ford Thunderbird shim type
   II. Rambler lower control arm shim type
      A. I only
      B. II only
      C. Both I and II
      D. Neither I nor II
9. When caster is controlled by a strut rod:
   I. Camber is adjusted first
   II. Caster is adjusted first
      A. I only
      B. II only
      C. Both I and II
      D. Neither I nor II
10. Upper inner control arm eccentrics control:
   I. Caster
   II. Camber
      A. I only
      B. II only
      C. Both I and II
      D. Neither I nor II
11. When both eccentrics at the upper inner control arm bushing are moved out an equal amount:
   I. Camber is moved more positive
   II. Caster is not changed
      A. I only
      B. II only
      C. Both I and II
      D. Neither I nor II
12. To change the camber to a more positive position on a lower inner control arm with eccentric-type adjustments:
   I. Move the lower ball joint outward
   II. Move the lower ball joint inward
      A. I only
      B. II only
      C. Both I and II
      D. Neither I nor II

13. When the strut rod is mounted to the front of the suspension system, caster is adjusted to a more positive position by:
    I. Shortening the strut rod
    II. Lengthening the strut rod
        A. I only
        B. II only
        C. Both I and II
        D. Neither I nor II
14. The upper ball joint eccentric bushing in the Cadillac is used to adjust:
    I. Caster
    II. Camber
        A. I only
        B. II only
        C. Both I and II
        D. Neither I nor II
15. To maintain proper tie rod-end height on a Cadillac with upper ball joint eccentric bushing, the eccentric bushing must be positioned so that the upper ball joint stud is:
    I. In the front half of the steering knuckle hole
    II. In the rear half of the steering knuckle hole
        A. I only
        B. II only
        C. Both I and II
        D. Neither I nor II
16. Slotted upper control arms are used to adjust:
    I. Caster
    II. Camber
        A. I only
        B. II only
        C. Both I and II
        D. Neither I nor II
17. Adjustments for turning radius is made at the:
    I. Upper control arm
    II. Tie rod end
        A. I only
        B. II only
        C. Both I and II
        D. Neither I nor II
18. Toe-in is adjusted on the:
    I. Upper control arm
    II. Tie rod end
        A. I only
        B. II only
        C. Both I and II
        D. Neither I nor II
19. Toe-in adjustment is made with the wheels:
    I. Turned to 20 degrees
    II. In a straight ahead position
        A. I only
        B. II only
        C. Both I and II
        D. Neither I nor II

20. Turning radius is checked while the wheels are:
    I. Turned to 20 degrees
    II. In a straight ahead position
        A. I only
        B. II only
        C. Both I and II
        D. Neither I nor II

DATE COMPLETED____________________ INSTRUCTOR'S SIGNATURE ____________________________

After studying the preparation materials, you will be expected to correctly answer a minimum of seven out of ten multiple-choice questions in a written test without reference materials on the following topics related to front end alignment procedures in pickups and light trucks:

1. I-beam bending for camber
2. I-beam alignment for caster (shims)
3. I-beam bending for caster
4. Toe-in adjustment

OBJECTIVE IV-D (written test)

Preparation:

Study pages 353 through 360.

Truck manufacturers, as well as passenger car manufacturers, are putting more and more stress on the importance of maintaining correct wheel alignment. The major difference between truck and passenger car alignment has been in the suspension system used. In the past, the single I-beam axle has prevailed as the most practical suspension system for trucks and pickups. In new light trucks and pickups, however, manufacturers are rapidly replacing single I-beam axle suspension systems.

SINGLE I-BEAM AXLE

Although the single I-beam axle has been known for strength and durability, it still needs periodic checks and changes in setting of camber, caster, and toe-in (Fig. IV-65).

As with passenger cars, certain prealignment checks should be made. These checks should be made with the vehicle level and with its normal load such as campers, tool boxes, etc. Look for the following:

1. Loose or defective wheel bearing and proper torque on spindle nut
2. Play or excessive wear in kingpins or bushings
3. Loose steering connections, such as tie rod ends
4. Loose spring shackles
5. Loose spring U bolts
6. A loose or sheared spring center bolt
7. Sagging or broken front springs
8. Improper tire inflation
9. Unbalanced or improperly mounted tires
10. Different tire size on one or more wheels
11. Broken motor mounts
12. Brakes dragging
13. Lug bolts improperly torqued
14. Loose or defective shock absorber
15. Bent or worn spindle

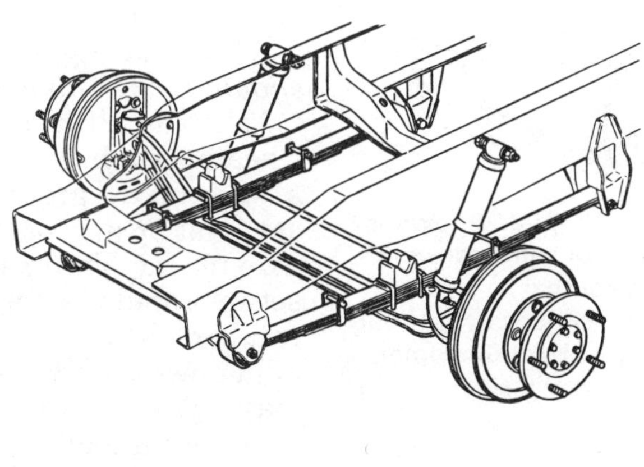

FIGURE IV-65 This is a typical single I-Beam Suspension used in trucks. (Courtesy of Chevrolet Motor Division)

OBJECTIVE IV-D1 (manipulative)

Working as a team with one other student, you will align the front wheels on at least one instructor assigned pickup or light truck with an I-beam axle. The alignment will include checking and adjusting where necessary the caster, camber, toe-in, steering axis inclination, turning radius, and centering of the steering wheels. Time recommended: Twice the flat-rate time allowed in the Motors or Chilton Flat Rate manual.

Preparation:

1. Study pages 354 through 360.
2. Study manufacturer's procedure and specifications on the specific truck to which you are assigned.
3. Carefully observe instructor's demonstration of front wheel alignment on single I-beam axle.

Materials and Tools Needed:

1. Alignment equipment
2. Jacks
3. Air hose and tire inflator
4. Air gauge
5. Bending equipment
6. Set of box/open-end combination wrenches
7. Set of sockets and ratchet, ½ inch drive
8. Assortment of tapered shims

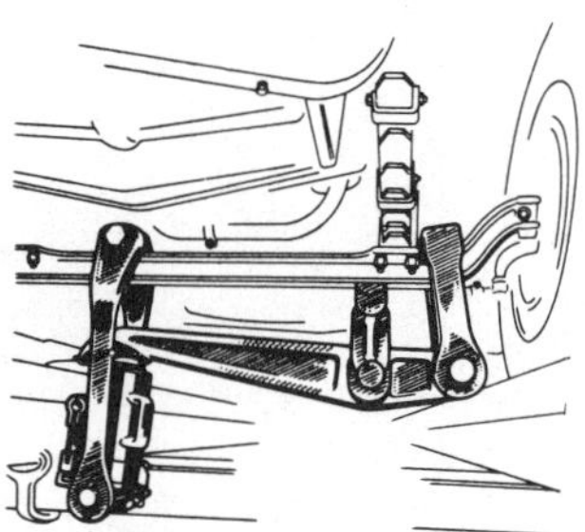

FIGURE IV-66 Shown is the tool setup for bending more positive camber on one wheel only. (Courtesy of Ford Customer Service Division)

CAMBER ADJUSTMENTS

The camber angle adjustment is always the first alignment operation on the single I-beam axle. This must be done by accurately bending the axle to meet specifications. When the axle has been bent or twisted excessively from its original shape, it should be replaced with a new axle. As a general rule, when an extreme bend or twist condition exists, barely visible minute fractures will occur. These fractures frequently cause eventual failure under normal operating conditions.

The straightening process should always be done with mechanical means such as hydraulic jacks and proper tie downs. *Do not use heat.*

Always straighten forgings with cold pressure—Under no circumstances should heat be applied.

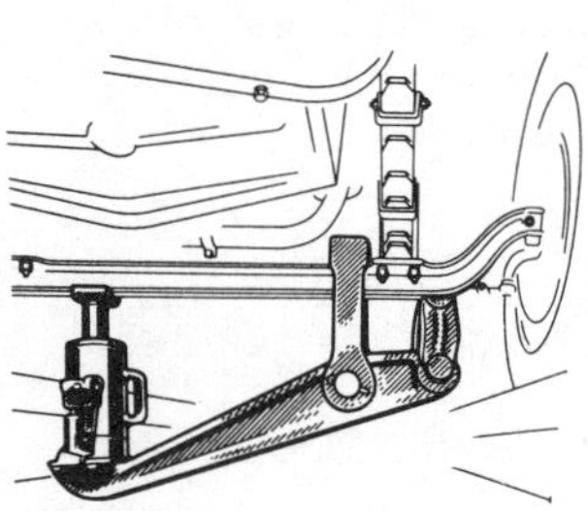

FIGURE IV-67 Shown is the tool setup for bending camber to a more negative position. (Courtesy of Ford Customer Service Division)

When handling camber on single I-beam, it should be remembered that the I-beam axle has a lot of spring. When bending for camber, the I-beam will always take a greater amount of bend than is actually desired. This compensates for spring back. However, caution must be used to avoid bending the axle too far. The examples shown in Figs. IV-66 through IV-70 illustrate the setup procedures for bending to achieve more positive camber. Normally, the camber on a single I-beam axle will move to the negative when out of adjustment. However, other factors, such as severe blows during an accident, may make it necessary to straighten or unbend the axle to remove excessive positive camber.

FIGURE IV-68 The tool setup shown is for bending both wheels to a more negative position. (Courtesy of Bear Manufacturing Co.)

FIGURE IV-69 The tool setup shown is for bending both wheels to a more positive position. (Courtesy of Bear Manufacturing Co.)

FIGURE IV-70 This tool setup is to bend the driver's side to a more negative and the passenger's side to a more positive position. (Courtesy of Bear Manufacturing Co.)

CASTER ADJUSTMENTS

Caster adjustments on the single I-beam axle are accomplished by one of two methods (Fig. IV-71). The first is by using a tapered shim placed between the spring and the I-beam axle. These caster shims come in various widths and degrees of angle taper (Fig. IV-72). Some of the shims are 1¾ or 2 inches wide, and angles of taper vary from ½ to 3 or 4 degrees. The tapered shim will actually place a twist in the axle, giving the desired caster angles. **Caution:** *The center bolt of the leaf spring must extend into the axle. Therefore, do not use a shim that is thicker than the end of the center bolt of the leaf spring is long.* If caster is out of adjustment so far that it cannot be compensated for by using only one shim, caster should be adjusted by bending the axle. Fig. IV-73 illustrates bending the caster to a positive position.

FIGURE IV-71 Caster shim for single I-beam caster adjustment. (Courtesy of Bear Manufacturing Co.)

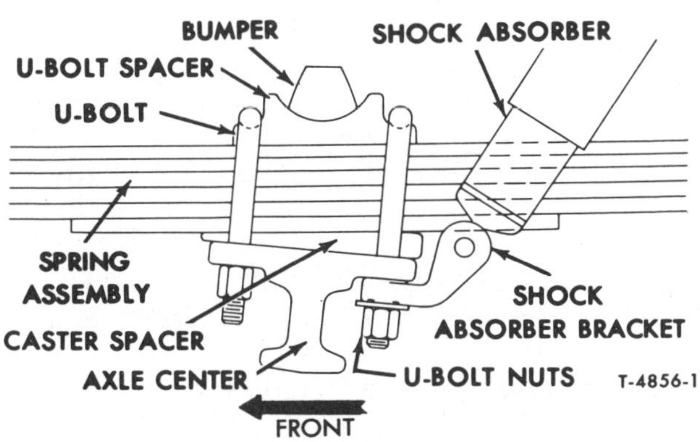

FIGURE IV-72 Tilting axle for caster adjustment on single I-beam. (Courtesy of Chevrolet Motor Division)

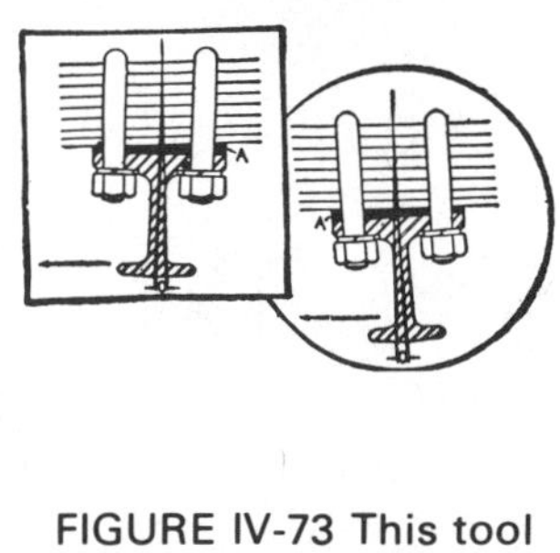

FIGURE IV-73 This tool setup is for bending more positive caster into the single I-beam axle. (Courtesy of Bear Manufacturing Co.)

TOE-IN ADJUSTMENTS

Toe-in on single I-beam axles is adjusted by lengthening or shortening the tie rod. In preparation for adjusting toe-in, center the steering gear on its high point and adjust it to eliminate excessive play (Fig. IV-74). With the wheels in the straight ahead position, check and adjust toe-in. Adjustment is made by loosening the tie rod end clamp bolts, then turning the tie rod until wheels have proper toe-in.

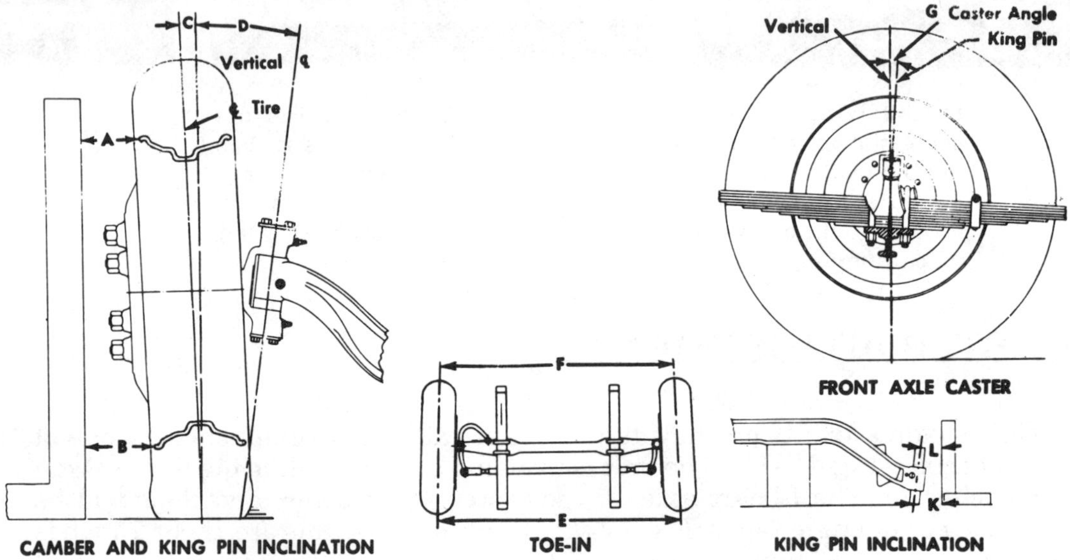

B MINUS A .... Camber (Inches)
C .... Camber (Degrees Positive)
D .... King Pin Inclination (Degrees)
E MINUS F .... Toe-in (Inches)
G .... Caster (Degrees Positive)
K MINUS L .... King Pin Inclination (Inches)

Refer to "Specifications" for correct dimensions.

FIGURE IV-74 Toe-in adjustment on the single I-beam suspension is controlled by a single adjusting sleeve on the tie rod assembly. (Courtesy of Chevrolet Truck Division)

FORD TWIN I-BEAM ADJUSTMENT

The Ford twin I-beam suspension gives the sturdiness of the single I-beam and the comfort of the independent suspension. It consists of two I-beams set parallel (Fig. IV-75). They are pivoted from the frame, then extend to the other side where the coil spring, kingpin, and wheel assembly are mounted.

CAMBER ADJUSTMENT

Camber is adjusted on the twin I-beam by bending the I-beams. Examples are shown in Figs. IV-76 and IV-77.

CASTER ADJUSTMENT

Caster is also adjusted on the twin I-beam by bending. Because of the design, it is the hardest of all angles to set. Figs. IV-78 and IV-79 show an example of bending caster on the twin I-beam.

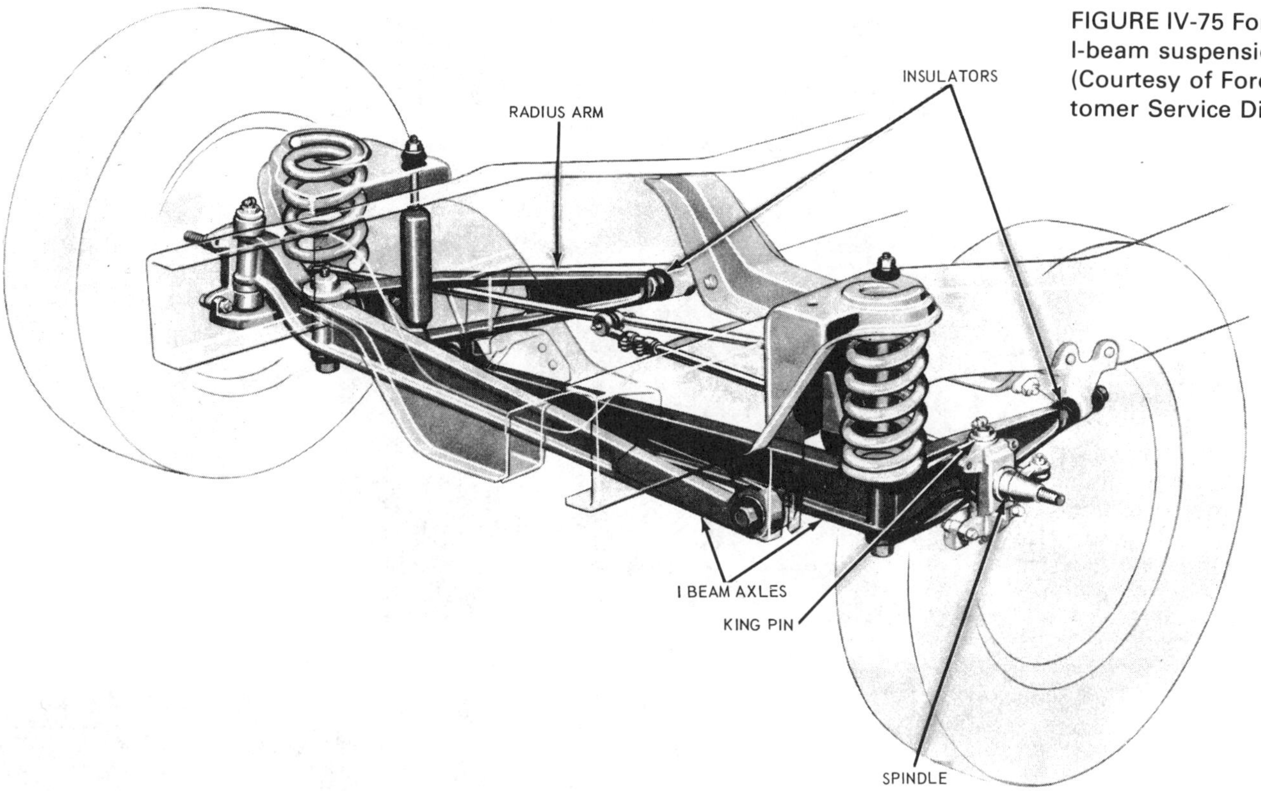

FIGURE IV-75 Ford's twin I-beam suspension. (Courtesy of Ford Customer Service Division)

FIGURE IV-76 This tool setup is for bending camber to a more positive position on the twin I-beam suspension. (Courtesy of Ford Customer Service Division)

FIGURE IV-77 This tool setup is to decrease camber on the twin I-beam suspension. (Courtesy of Ford Customer Service Division)

FIGURE IV-78 Caster bending shown here is to achieve a more positive position on the Ford twin I-beam suspension. (Courtesy of Ford Customer Service Division)

FIGURE IV-79 Caster bending is to accomplish a more negative position on the Ford twin I-beam suspension. (Courtesy of Ford Customer Service Division)

I-BEAM AXLE ALIGNMENT

Student name____________________ Date__________

Make of car__________ Model__________ Year__________

Time started__________ Time finished__________ Total time__________

Flat-rate time __________

A. Camber Setting
   1. Check camber setting and specifications.
      a. Camber setting before adjusting:

         Left wheel:__________ Right wheel:__________
      b. Factory camber specifications:

         Left wheel:__________ Right wheel:__________
   2. Adjust camber by bending axle.
      a. Camber setting after adjustment:

         Left wheel:__________ Right wheel:__________

B. Caster Setting
   1. Check caster setting and specifications.
      a. Caster settings before adjustments:

         Left wheel:__________ Right wheel:__________
      b. Factory caster specifications:

         Left wheel:__________ Right wheel:__________
   2. Adjust caster by using shims or bending.
      a. Size of shim used:

         Left side:__________ Right side:__________
      b. Caster setting after adjustment:

         Left wheel:__________ Right wheel:__________

C. Toe-in Setting
   1. Check toe-in setting.
      1. Toe-in setting before adjustment: __________

      b. Factory toe-in specifications:__________

   2. Set toe-in.
      a. Toe-in setting after adjustment: __________

   Notes:__________

D. Turning Radius Check

1. Factory specifications:

   Inside wheel:________________ Outside wheel:________________

2. Turning radius settings:

   Inside wheel:________________ Outside wheel:________________

Report: ____

____

____

E. Road Test

Report: ____

____

____

DATE COMPLETED____________________ INSTRUCTOR'S SIGNATURE ____________________

TEST IV-D

Student name______________________________Date____________________

Directions: Select the best answer to each question and circle the correct letter at the left.

Select the best answer for each question and circle the correct letter at the left.

1. On a single I-beam axle, always adjust:
   I. Caster first
   II. Camber first
   A. I only
   B. II only
   C. Either I or II
   D. Neither I nor II
2. When bending camber to a more positive position on a single I-beam, it is possible to bend both sides at the same time by placing:
   I. Two tie downs next to the kingpins and one jack in the center
   II. One tie down in the center and two jacks close to the kingpins
   A. I only
   B. II only
   C. Either I or II
   D. Neither I nor II
3. When bending camber to a more positive position on one side only and holding the other side:
   I. Use two tie downs and two jacks
   II. Use one tie down and one jack
   A. I only
   B. II only
   C. Either I or II
   D. Neither I nor II
4. When using shims to adjust caster to a more positive position on the single I-beam:
   I. Place the thick end of the shim to the front
   II. Place the thick end of the shim to the back
   A. I only
   B. II only
   C. Either I or II
   D. Neither I nor II
5. When bending the single I-beam for a more positive caster adjustment:
   I. Twist the axle to move the top of the kingpin to the front
   II. Twist the axle to move the top of the kingpin to the back:
   A. I only
   B. II only
   C. Either I or II
   D. Neither I nor II
6. Caster shims are tapered and come in sizes of:
   I. ½ to 3 degrees
   II. ⅛ to ¾ inch thick
   A. I only
   B. II only
   C. Either I or II
   D. Neither I nor II

7. Toe-in adjustment can be made by:
    I. Turning the tie rod adjusting sleeves
    II. Bending the tie rod
        A. I only
        B. II only
        C. Either I or II
        D. Neither I nor II
8. The Ford twin I-beam suspension has:
    I. Leaf springs
    II. Torsion bars
        A. I only
        B. II only
        C. Either I or II
        D. Neither I nor II
9. When bending for camber on the twin I-beam:
    I. Both sides can be bent at the same time
    II. One side at a time should be done
        A. I only
        B. II only
        C. Either I or II
        D. Neither I nor II
10. On the twin I-beam axle, caster is adjusted by:
    I. Shims
    II. Bending
        A. I only
        B. II only
        C. Either I or II
        D. Neither I nor II

DATE COMPLETED____________________ INSTRUCTOR'S SIGNATURE ______________________________

OBJECTIVE IV-E (manipulative)

Working as a team with one other student, you will align the rear axle housing on an instructor-assigned automobile. The alignment will include checking and adjusting camber and toe-in. Time will vary on this job, but it will require approximately three hours.

Preparation:

1. Study pages 363 through 367.
2. Carefully observe instructor's demonstration on rear axle alignment.

Materials and Tools Needed

1. Alignment equipment
2. Jacks
3. Bending equipment
4. Air hose and tire inflator
5. Air gauge
6. Air blow gun
7. Oxyacetyline welding equipment with large heating tip
8. Fire extinguisher
9. Drain pan with water and large rags

REAR AXLE ALIGNMENT

The rear axle must carry a lot of weight and withstand extreme road shock (Fig. IV-80). The natural tendency is for the rear-axle housing to bend into a negative camber setting at the wheels. The torque and pressure of heavy weight will also bend the rear-axle housing into a toe-out condition. Camber should be set slightly positive (less than ⅛°) and toe-in should be set with a slight amount of toe-in (less than ⅛").

Before attempting to align the rear-axle housing, conduct the same prealignment checks that are called for on the front-suspension alignment.

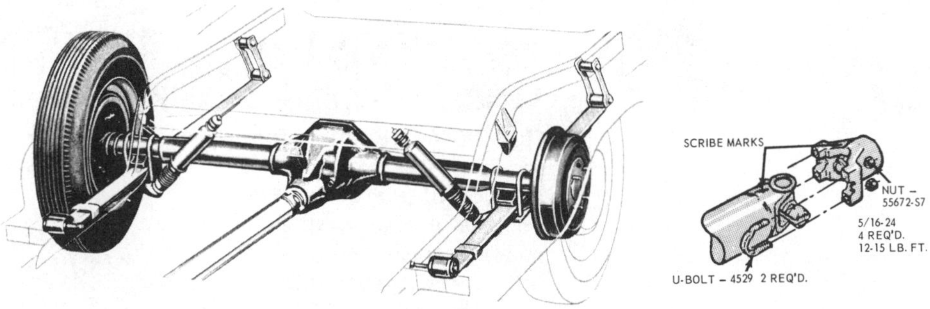

FIGURE IV-80 Typical rear assembly (Camber and toe-in are the only two adjustments possible). (Courtesy of Ford Customer Service Division)

Any alignment rack can be used for rear-axle alignment by backing the vehicle on to it so the rear wheels stop on the alignment pads (Fig. IV-81).

Camber and toe-in are the only two angles to be considered on rear axle alignment.

FIGURE IV-81 Back the car onto the alignment rack.

ADJUSTMENT OF CAMBER

The rear axle housing is a tubular housing, and extreme care must be taken to prevent the housing from collapsing during camber adjustment. There are two ways to align the rear axle housing:

1. Pressure bending with hydraulic jacks and tie downs like those used on the I-beam front axle.
2. Heat shrinking.

PRESSURE BENDING

To adjust both wheels from a negative to a positive camber setting, the tie downs should be placed at the outside edge of the axle. The hydraulic jacks should be toward the center, where they can apply the necessary pressure. (Fig. IV-82 and 83).

FIGURE IV-82 Bending for camber on the rear axle housing of a large truck. Camber alignment can be done without removing the axle from the vehicle. (Courtesy of Bear Manufacturing Co.)

FIGURE IV-83 Bending for toe-in on the rear axle housing of a large truck. For this operation on a passenger car, the housing must be removed. (Courtesy of Bear Manufacturing Co.)

HEAT SHRINKING FOR CAMBER

There is one basic principle that must be understood before attempting to align the rear axle with heat. Metal always expands when heated. The hotter it gets, the more it expands. It also shrinks when cooled. As a matter of fact, it shrinks more than it expands.

To straighten an axle that has a negative camber setting, you should follow these basic steps:

1. Fill axle housing to proper level.
2. Have the following available:
   a. Fire extinguisher
   b. Air hose
   c. Air blow gun
   d. Drain pan with water
   e. Rags
   f. Oxy-acetylene welding setup with large welding tip
3. Inspect the axle housing area for combustible materials such as gas leaks, oil leaks, and brake fluid leaks. (**Do not** *use heat on any axle that may result in a fire.)*
4. Install mirrors or gauges on rear wheels.
5. Support axle housing with a hydraulic jack so that the weight of the vehicle is off the wheels.
6. Heat axle to a bright red on the bottom of the axle (Fig. IV-84).
7. The camber setting will first move to a more negative position. When it has moved to a negative position which is about twice the amount that you want it to move positive, remove the heat.

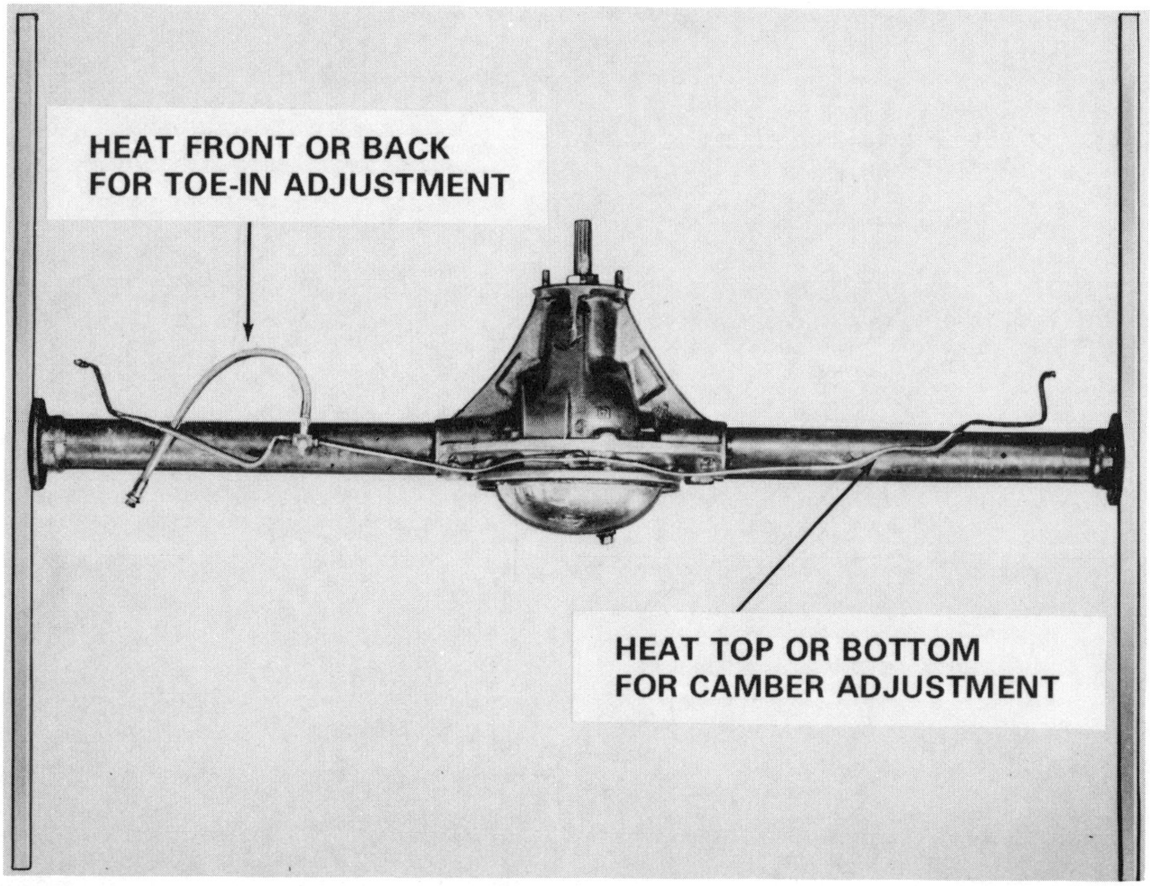

FIGURE IV-84 Heat on the top or the bottom of the axle for camber adjustment. Heat either the front or the back for toe-in adjustment. (Courtesy of American Motors Corp.)

8. Cool axle housing as fast as possible using rags and water. Compressed air blown over the axle housing will also help the cooling process.
9. When the axle is cool enough to place your hands on it, lower the axle housing and recheck the camber.
10. The vehicle should stand with supports holding most of the weight until the axle housing has completely cooled.

HEAT SHRINKING FOR TOE-IN

The process used for toe-in is the same as for camber, the only difference being the position of the heating spots. To increase toe-in, apply heat to the front of the axle. To reduce toe-in, apply heat to the back side of the axle.

JOB REPORT SHEET IV-E

REAR AXLE ALIGNMENT

Student name____________________Date____________________

Make of car____________Model____________Year____________

Time started____________Time finished____________Total time____________

Flat-rate time ____________

A. Camber Setting
   1. Check camber setting before adjustment.

      Left wheel:____________Right wheel:____________(Camber should be as close to "o" as possible.)
   2. Change camber by heating axle housing or by bending with pressure. (Use caution not to collapse axle housing when bending with pressure.)
      a. Setting after adjustment:

         Left wheel:____________Right wheel:____________

B. Toe-in Setting
   1. Check toe-in setting before adjusting:____________(Toe-in should be as close to "o" as possible.)
   2. Change toe-in by heating the front or rear of the axle housing.

      a. Toe-in setting after adjustment:____________

   Caution: *Gas tank is close to the rear axle housing and brake lines must be moved before heating.*

C. Cool Axle
   If possible, leave car supported on jacks and let cool until axle housing is cool to the touch.

D. Road Test

   Report: ____________________________________

   ____

DATE COMPLETED____________INSTRUCTOR'S SIGNATURE ____________

unit V

Steering Columns

This unit discusses steering columns and their modern safety and convenience features. General diagnosis and service procedures are covered.

OBJECTIVE V-A (written test)

After studying the preparation materials, you will be expected to correctly answer a minimum of 14 out of 20 multiple-choice questions on the following topics in a written test without reference materials.

1. Collapsible steering columns
2. Movable steering columns
3. Column ignition switches and locks
4. Hazard warning switch
5. Turn-signal switches
6. Horn switches

Preparation:

Study pages 369 through 379.

In the earliest years of the automobile, steering wheels, steering columns, and their gearboxes served only to provide a rigid mechanism through which the driver could steer the car. The original form of the column consisted of a shaft attached, at the lower end, to the steering gear and, at the upper end, to a steering wheel. On the outside of this steering shaft was a jacket of sheet metal in the shape of a tube. This provided a shield to cover the shaft and a means of holding the column and wheel in place within the passenger compartment.

As automobiles became more sophisticated, the steering column served as a holding device for throttle and spark controls, horn switches, light controls, transmission-shift linkages, and turn-signal switches. Eventually tilting steering wheels and telescoping steering wheels were installed on some cars. In this text these will be referred to as *movable steering columns*.

Most of these additions to the steering column were made to provide more convenience for the driver. Collapsible steering columns, however, were designed and installed for safety purposes. Another feature of the modern steering column is the locking mechanism. This anti-theft device incorporates the normal ignition switch and additional parts to lock the steering wheel in the general position in which it was turned by the driver (Fig. V-1). In most designs, the transmission linkage is also locked into "Park" or "Reverse" position when the key is removed.

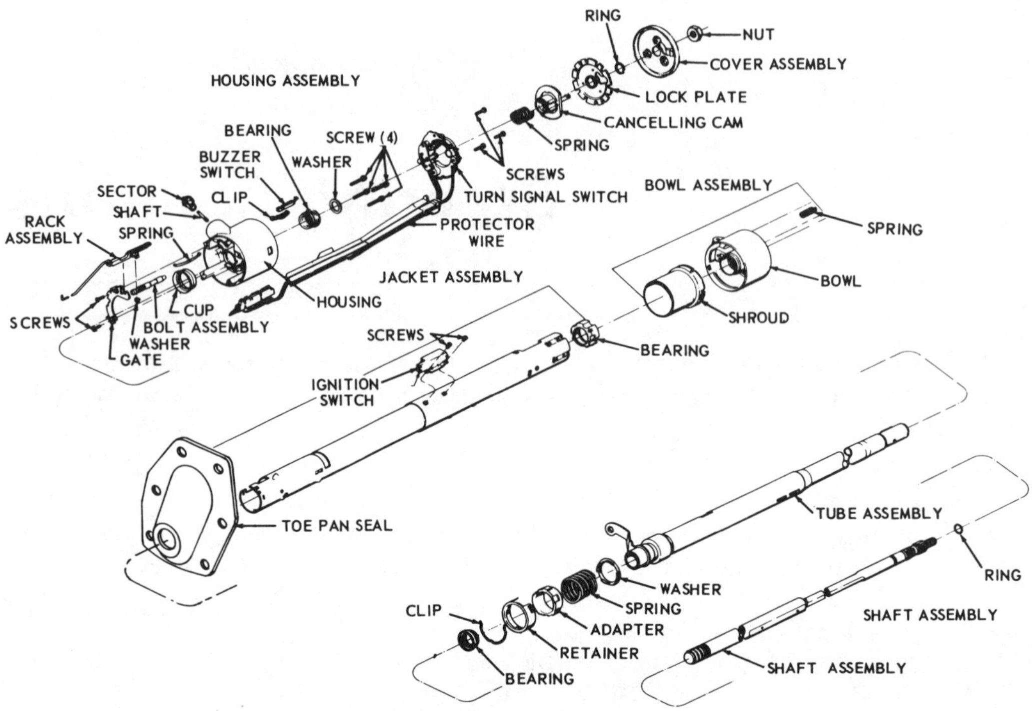

FIGURE V-1 This is a typical standard steering column, disassembled. (Courtesy of Cadillac Motor Car Division)

COLLAPSIBLE STEERING COLUMNS

In 1968, the Federal Highway Safety Act became law. This act specified that certain safety features be designed into steering columns. Among the new safety requirements were collapsible or energy-absorbing steering columns for all domestic automobiles.

When an automobile is involved in a frontal collision, the forward movement (primary force) is quickly halted, while the driver (secondary force) continues in a forward direction. A severe collision normally involves both primary and secondary forces (Fig. V-2). The secondary impact occurs when the driver is thrust forward against the steering wheel and column.

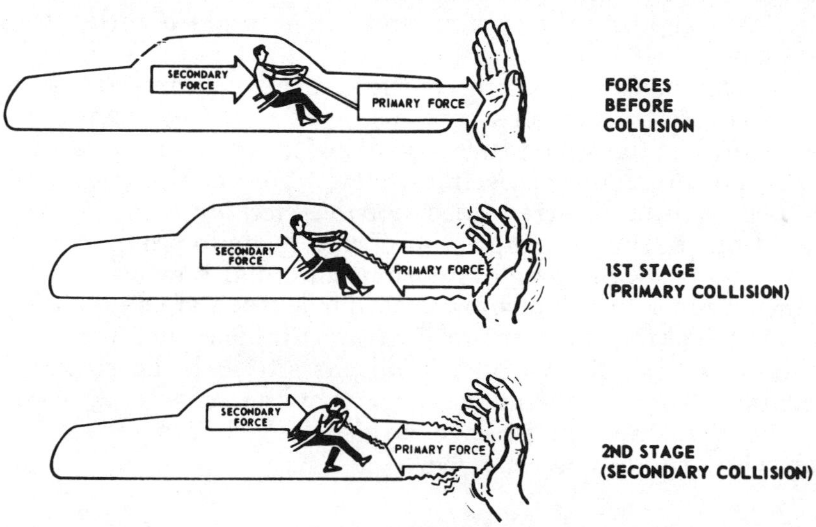

FIGURE V-2 The drawing illustrates collision forces in action during a frontal collision. (Courtesy of Chevrolet Motor Division)

FIGURE V-3 Collapsible steering column—mesh type. (Courtesy of Cadillac Motor Car Division)

The collapsible steering column is designed to absorb both forces, thereby reducing the secondary or driver impact. During a collision, the steering column compresses, reducing its tendency to move further into the passenger compartment. A split second later, when the driver is thrown forward, his energy is partially absorbed by the compressible parts of the column.

There are four basic types of collapsible steering columns in use today on domestic passenger cars. They are:

1. A steering column using a jacket with a collapsible mesh section (Fig. V-3).
2. A column with a jacket having a collapsible bellow or accordian section (Fig. V-4).

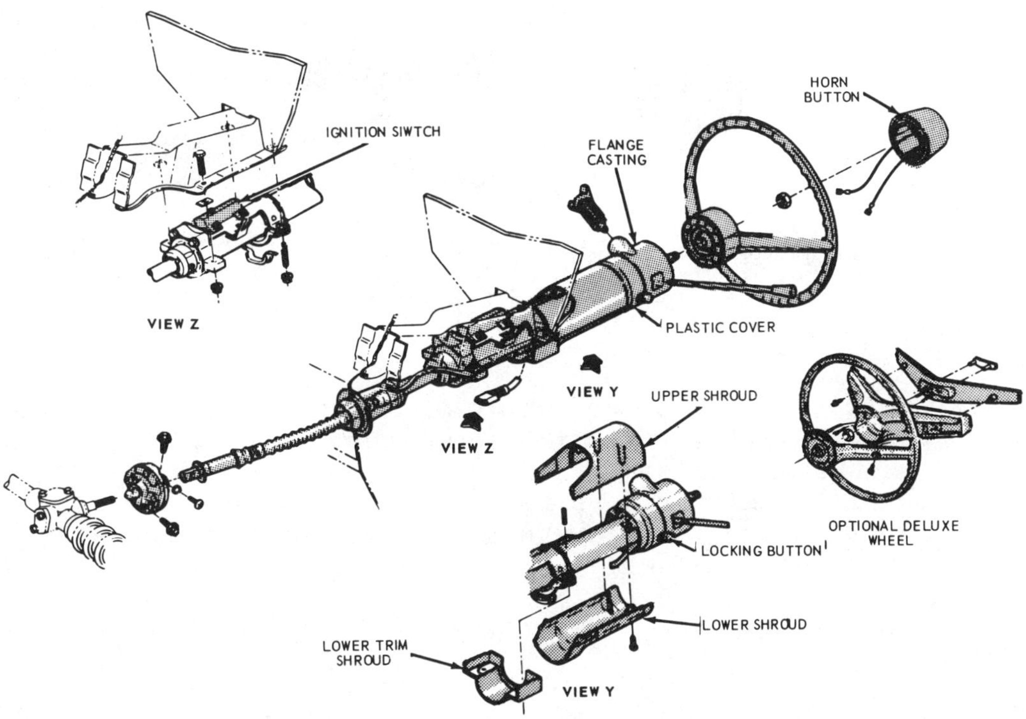

FIGURE V-4 Collapsible steering column—accordian type. (Courtesy of Ford Customer Service Division)

3. A steering column consisting of an outer jacket composed of two tubes of different sizes joined together by two rows of ball bearings pressed into position (Fig. V-5).
4. A corrugated cylinder fastened to the steering wheel and bearing housing (Fig. V-6).

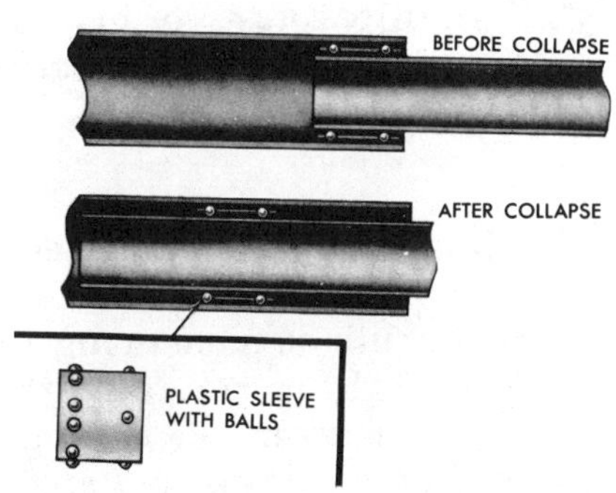

FIGURE V-5 Detail view of ball bearing type collapsible steering column. (Courtesy of Chevrolet Motor Division)

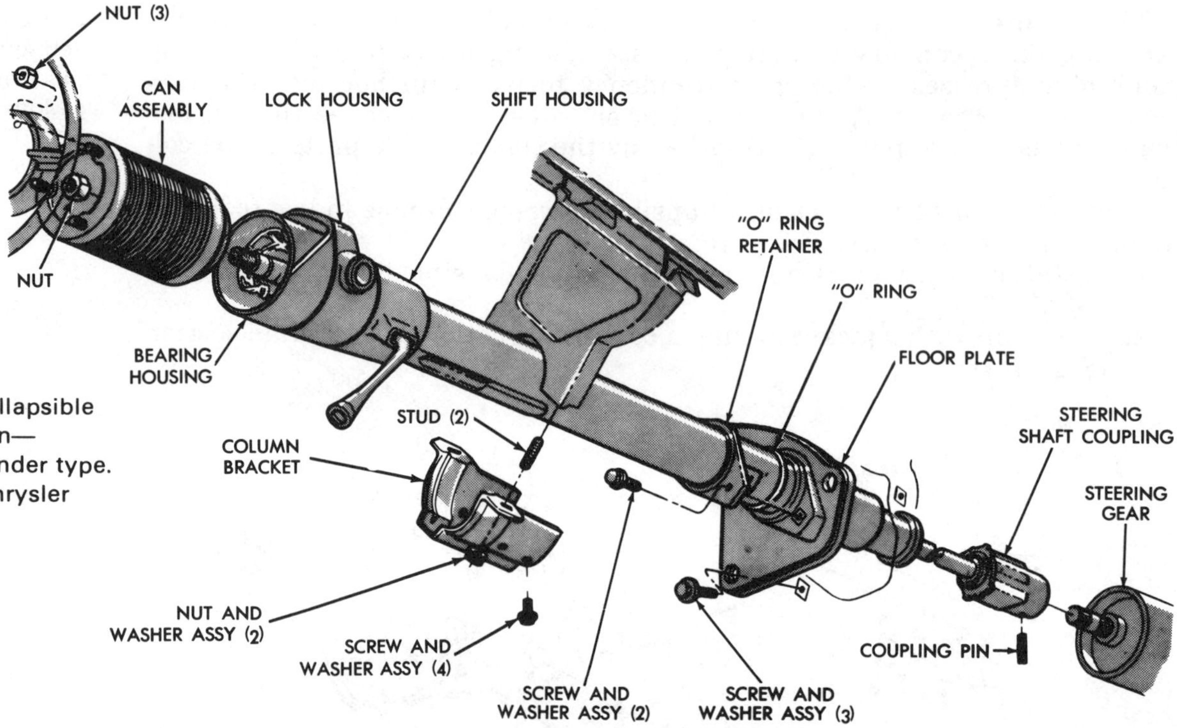

FIGURE V-6 Collapsible steering column—corragated cylinder type. (Courtesy of Chrysler Motors Corp.)

OBJECTIVE V-A1 (manipulative)

Working alone, you will test and report in writing to your instructor all functions and measurements of one steering column with at least five of the six column features listed below:

1. Collapsible steering column inspection and measurement only
2. Movable steering wheel: tilt, telescoping, or lateral movement
3. Ignition switch and lock
4. Hazard warning light switch
5. Turn signal switch
6. Horn switch

Time recommended: 15 minutes

Preparation:

1. Review pages 372 through 379.
2. Study pages in manufacturer's or general repair manuals, where necessary.
3. Carefully observe instructor's demonstration of measurement and function testing of a steering column.

Materials and Tools Needed:

1. Steel rule or measuring tape
2. Flashlight or trouble light
3. Manufacturer's or general repair manual with service procedure and wiring diagrams included.

Procedure:

1. With specification of steering-jacket or shaft length, follow the manufacturer's recommendations for inspection and measurement. Record this measurement and other test results on Steering Column Job Report Sheet V-A1.
2. Move steering wheel through all positions where applicable. If the wheel cannot be turned, the lock function is OK. After unlocking column, check for binding between positions and as the wheel is turned.
3. With the key in the lock position and the transmission selector in "Park" or "Reverse," turn on the ignition switch. Turn the key further to the "start" position and start the engine.
4. Place the shift lever in "Park" or "neutral" with the parking brake on.
5. Turn on the hazard warning light switch and check the front and rear of the vehicle to see if its lamps are flashing.
6. Move the turn signal switch either right or left and check the lamp flashers at both front and rear. Turn the steering wheel one complete turn in the same direction as the indicated turn. Turn the steering wheel in the opposite direction to see if the turn signal canceling mechanism is operational.
7. Make the same tests as step 6, except move the switch and steering wheel in opposite directions to test the other side of the switch, its lamp operation, and the canceling mechanism.
8. With the ignition switch on, quickly press all parts of the horn ring or button, to test horn functions. Report all test results on Steering Column Job Report Sheet V-A1.

COLLAPSIBLE COLUMN SERVICE

Service work on collapsible steering columns should begin with a visual inspection. Look for evidence of front end collision, then examine the lower end of the column itself. Inspect the mesh and/or bellows section, looking for bulges or bends. The steering gear should have its flexible coupling intact and aligned. Check the column-support brackets in the engine and passenger compartments. Collapsing movement is indicated by separation of the mounting capsules from the bracket.

If the column shows signs of collapse, there may or may not be problems with the other column functions. Thoroughly check all column functions by test driving the car.

The only positive method of determining steering column collapse is by measurement of a specific distance on the column. Figs. V-7 and V-8 show examples of jacket and shaft measurement. Since there are no general methods, refer to the manufacturer's service manual or general repair manuals for procedures and specifications. Some columns when collapsed can be renewed with minor replacement parts; however, it is standard trade practice to replace the entire collapsible section of the column or column assembly.

Caution:

1. Do not bump, drop, hammer, or apply side pressure to the unit.
2. Use only correct replacement parts and tools designed for that column.
3. Tighten all fasteners to recommended torque specifications.

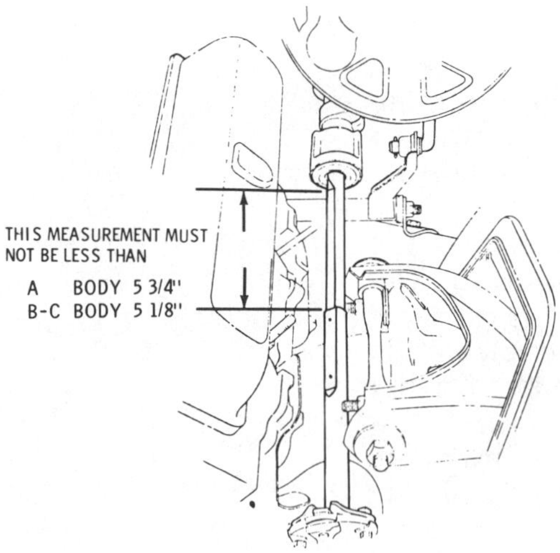

FIGURE V-7 Measurement points of shaft of collapsible steering column. (Courtesy of Oldsmobile Division)

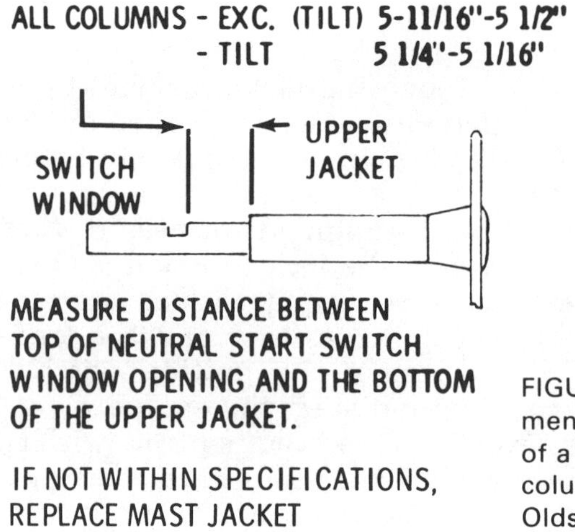

FIGURE V-8 Measurement points on the jacket of a collapsible steering column. (Courtesy of Oldsmobile Division)

MOVABLE STEERING COLUMNS

Steering columns may have additional features which allow the position of the steering wheel to be adjusted to suit the individual needs of drivers. These conveniences allow tilting and/or telescoping of the steering wheel. Some columns may have either or both capabilities. A column with both features is shown in Fig. V-9.

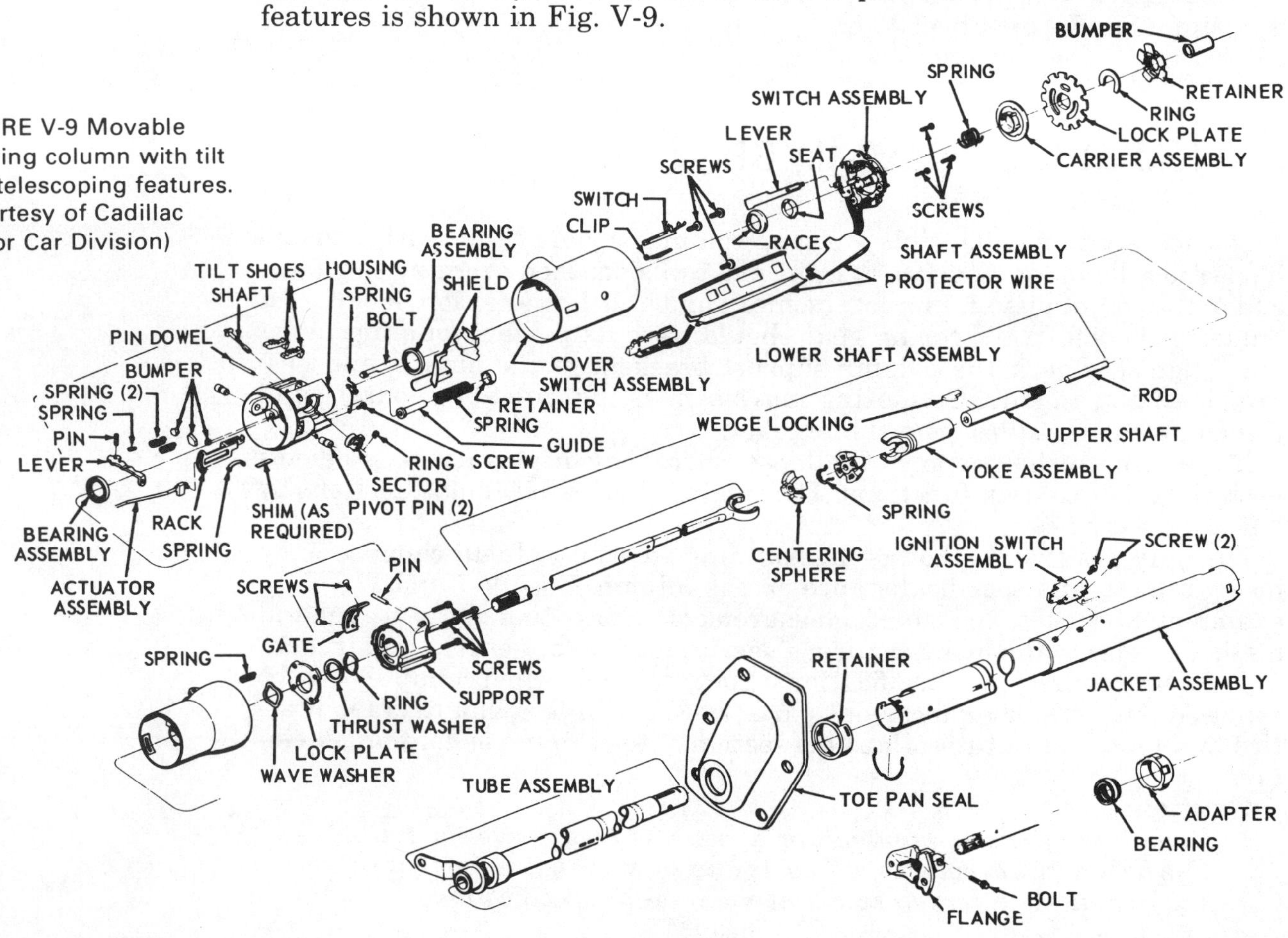

FIGURE V-9 Movable steering column with tilt and telescoping features. (Courtesy of Cadillac Motor Car Division)

The average tilting column has 15 degrees of upward-downward movement. The movement is normally achieved by pulling upward (toward the driver) on a small spring-loaded lever located on the left side of the column just below the turn-signal level (Fig. V-10. This allows the steering wheel to be positioned at the desired angle. When the lever is released, the wheel should lock into position.

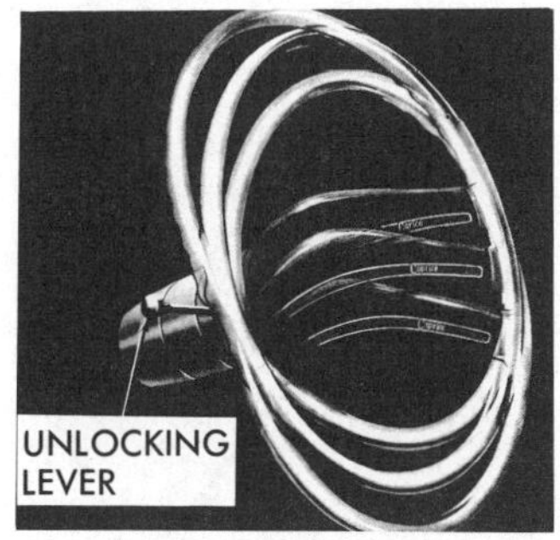

FIGURE V-10 Tilt steering wheel in different positions and location of lever. (Courtesy of Chevrolet Motor Division)

Telescoping columns usually permit adjustment of the distance between the driver and the wheel. The column can be secured in any position within an approximate range of 2.5 inches (Fig. V-11). The release and lock for this telescoping action is normally located on the steering wheel itself (Fig. V-12).

Diagnosis of problems in movable columns involves making a visual inspection and manipulative test. Look for too much or too little space between steering wheel and column jacket. Make a test of all positions of tilt and telescoping movement.

If either visual or manipulative tests are negative, the steering wheel will have to be removed to service the column.

To remove a steering wheel, the following general directions will apply. *(If trouble is encountered in the removal, review the procedures in the manufacturer's service manual or general-repair manual.)*

1. Remove the positive battery cable or disconnect the directional signal connector at the lower end of the column inside the passenger compartment.
2. Remove the horn-button cover. Small horn buttons are usually held in place by friction, (Fig. V-13) while larger buttons with padded surfaces can usually be removed by loosening the screws that pass through the wheel spokes from the back side (Fig. V-14).

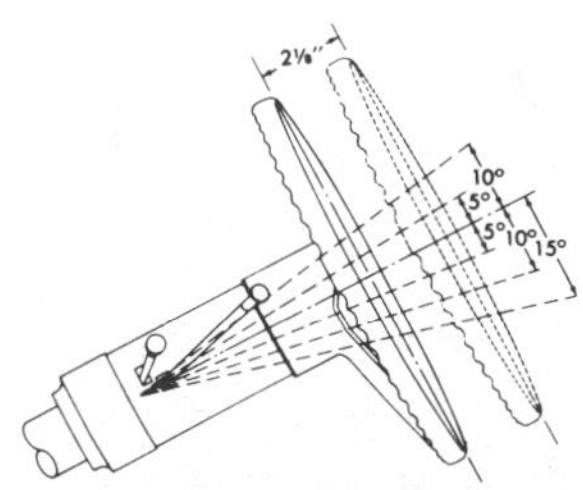

FIGURE V-11 Typical tilt-telescope steering wheel with range of angles and distance. (Courtesy of Cadillac Motor Car Division)

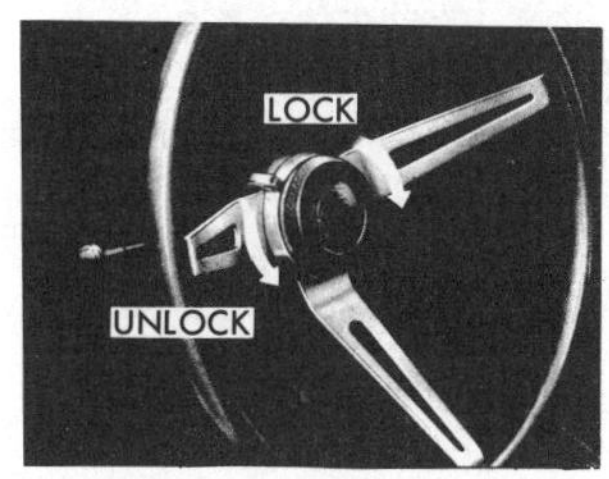

FIGURE V-12 Release-and-lock mechanism of typical telescoping steering wheel. (Courtesy of Chevrolet Motor Division)

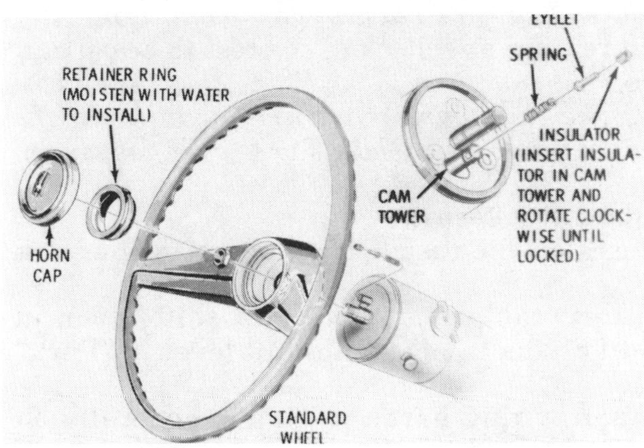

FIGURE V-13 Horn button assembly with friction retainer ring. (Courtesy of Oldsmobile Division)

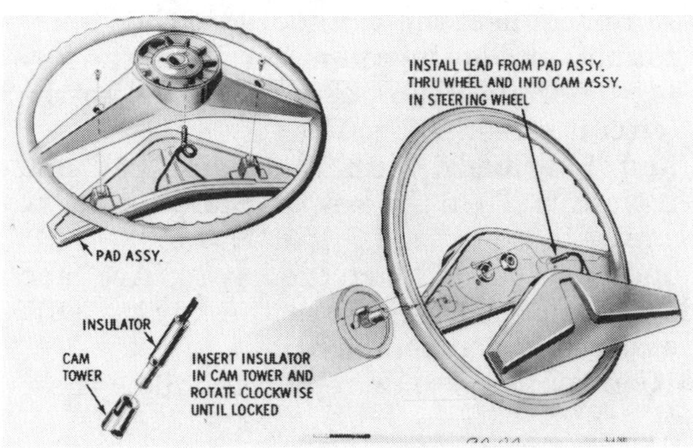

FIGURE V-14 Horn button assembly with screws passing through from backside of steering wheel spokes. (Courtesy of Oldsmobile Division)

FIGURE V-15 Removing steering wheel with puller. (Courtesy of Cadillac Motor Car Division)

3. Mark the steering shaft, nut, and the wheel with a center punch so the wheel can be reassembled in the same position.
4. Remove the horn switch parts, the steering shaft nut, and the washer.
5. Select the proper length and size of bolts that will attach the steering-wheel puller to the metal frame of the wheel.
6. Place the steering wheel puller in position with the proper puller bolts. Use an adapter to protect the steering shaft threads (Fig. V-15).
7. Remove the steering wheel by turning the puller center bolt in a clockwise direction.

Further service work will require special tools and specifications. Refer to the manufacturer's service manual.

COLUMN IGNITION SWITCHES AND LOCKS

FIGURE V-16 Typical fuse panel. Check for discoloration or separated metal link inside glass tube, indicating blown fuse. (Courtesy of Chevrolet Motor Division)

In this section we will discuss the operation and minor service of ignition switch/lock devices, and turn signal switches.

The ignition switch controls the current flow in the primary ignition circuit. In the modern steering column, this switch is designed to act as an antitheft device by controlling the ignition circuit and locking the steering wheel and shift lever.

To troubleshoot the ignition switch, you must have an ohmmeter, two wire-piercing awls, two 30-inch test leads, and a wiring diagram for the car being tested.

All test specifications listed below indicate a switch in good condition. Any deviations from the reading would require a recheck of the test to be certain no operator error has taken place.

Note: *During all tests, wiggle the key and wiring to check for intermittent open or closed circuits. Also, meter readings other than 0 or ∞ are reasons for switch replacement.*

Follow the sequence below for best test results:

1. Check for a blown fuse or fuses on the fuse block or on a fusible link under the hood (Fig. V-16).
2. Disconnect the positive battery cable and place the shift lever in "Neutral" or "Park." (For Ford products, place the shift lever in "Park" only.)
3. Disconnect the steering column wiring harness. This is generally on the lower end of the column, but may be in the engine compartment.
4. After viewing the wiring diagram, connect the ohmmeter to the "Batt" lead and "Ign" 1 lead on the column end of the connector. With the key on "Off," the ohmmeter reads ∞. (circuit open), (Fig. V-17). With the key on "On," meter reads 0 (circuit closed) (Fig. V-18).

FIGURE V-17 Ohmmeter indication showing an open circuit.

5. Connect the meter to "Batt" and "Sol" leads. With the key on "Off" and "On," the meter reads ∞ (open circuit). Turn the key to "Start" position: the meter reads 0 (closed circuit).
6. Connect the meter to "Batt" and "Acc" leads. Turn the key to "Acc" and "Run" positions: meter reads 0 (closed circuit). Turn the key to "Off" and "Start" positions: the meter reads ∞ (open circuit).

When diagnosis reveals the ignition switch is to be replaced, follow these general directions:

1. Disconnect the positive battery cable.
2. Lower the steering column or remove the steering wheel and turn the signal switch.
3. Place switch in the "On" or "Lock" or "Accessory" position, depending on the make of car.
4. Remove or depress the clip and remove the switch.

The lock mechanism within the switch may or may not require replacement at this time. Refer to service manuals for specific procedures not covered above.

FIGURE V-18 Ohmmeter indication showing a closed circuit.

HORN SWITCH

When the horn does not function, the switch on the steering wheel is usually not the problem.

The horn switch merely serves as a grounding device to complete the activating circuit. When this ground occurs, the horn-relay contacts close, completing a circuit from the battery to the horn and causing it to vibrate and make sound.

When making electrical checks of a possible malfunctioning switch, it is usually best to give the entire circuit a brief check. Proceed as follows:

1. Disconnect the horn wire at the horn.
2. Run a test lead from a 12-volt source to the horn. The horn should blow if its ground is good.
3. Attach a 12-volt test light to the wire disconnected from the horn. Ground the other end of the test light.
4. After turning on the ignition switch, push the horn button on the steering wheel. The lamp should light. If the lamp does not light:
   a. Remove the horn button on the steering wheel and check for bent metal contacts or a disconnected wire.
   b. Remove the lower steering column connector. Look on the wiring diagram and find the horn switch lead. Ground this horn-switch lead on the connector that goes to the horn relay. If it is operating properly, the relay will click.
5. If the horn relay worked when connected directly to the battery (Step 1) but did not work when the lead was grounded (Step 4b), replace the horn relay.
6. If the horn relay did work (Step 4b), the problem lies in the column or horn switch. Repair or replace as necessary. Refer to "Steering Wheel Removal" and "Turn Signal," "Hazard Warning Switch" sections.

HAZARD WARNING SWITCH

Hazard warning lights and blown fuses can affect the function of the turn signal switch, so a check of these fuses should be given top priority when either fails.

In situations where electrical failure of the turn signal and hazard warning switch has occurred, a specific procedure for diagnosis should be followed. Refer to the manufacturer's or general-repair manuals.

TURN SIGNAL SWITCH

Turn signal switches have two common types of malfunctions. One is mechanical and the other is electrical. In both situations, the nature of the customer complaint will normally point to the problem area.

The mechanical problem is often one where the canceling mechanism breaks, causing the lights to continue flashing after a turn is completed. In this instance, the service procedure calls for steering wheel removal and replacement of the turn signal switch assembly (Fig. V-19). This assembly is commonly made of tough plastic and permanently sealed to make it an unrepairable unit.

To replace a turn signal switch, follow these general directions:

1. Disconnect the positive battery cable.
2. Remove the steering wheel (see "Steering Wheel Removal," previous section).
3. Remove the lock plate or other pieces to expose the turn signal switch.
4. Disconnect the wiring harness at the lower column.
5. In most cases the connector cannot be fed up through the jacket of the column. It is therefore necessary to remove all wires from the connector by inserting a small steel strip to lift the locking tabs of the wire terminals.
6. Tie and wrap a piece of mechanic's wire to the lower end of the harness.
7. Remove the screws holding the switch in position and pull the switch upward, being certain the mechanic's wire is attached to the bottom end of the harness.
8. When the switch and harness are above the steering shaft, untie the mechanic's wire.
9. Wrap the mechanic's wire around the lower end of the new harness wires and switch assembly.
10. At the lower end of the column, pull the mechanic's wire while feeding the new wires into position.
11. Place the new switch into position with screws.
12. Reconnect the wires into their correct positions in the connector and plug the connector ends together.
13. Before replacing the steering wheel, connect the battery and test the electrical functions of the new switch.
14. If your results are positive, reassemble the steering wheel and other parts.

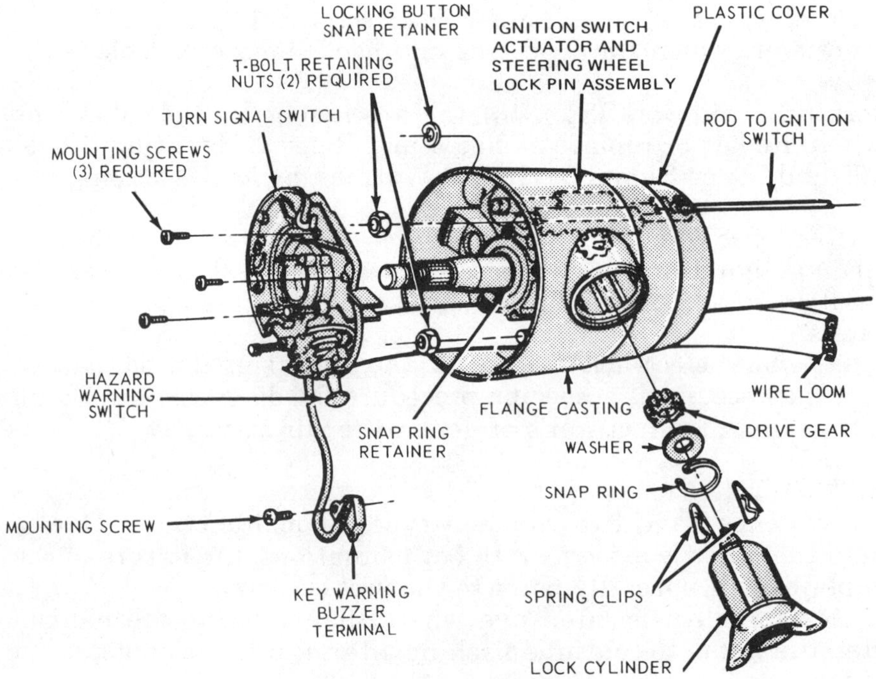

FIGURE V-19 Typical turn signal switch assembly. (Courtesy of Ford Customer Service Division)

STEERING COLUMN CHECK

Student name________________________________Date____________________

Make of car______________Model______________Year____________________

Time started______________Time finished______________Total time______________

Flat-rate time ____

1. Length of column______________measured from______________

   to______________

2. Is it within specifications? ________________________________

   Lock function OK:______________Not OK:______________

3. Ignition switch function OK: ______________Not OK:______________

   If defects encountered, list: ________________________________

   ____

   ____

4. Hazard warning switch function OK:______________ Not OK:______________

   All lamps flash?________________________________

5. Turn signal switch:

Left side	Right side
All lamps flash:______________	All lamps flash:______________
Canceling operation:______________	Canceling operation: ______________

6. Horn operation OK:______________Not OK:______________

7. Other defects encountered or recommendations for repairs: ______________

   ____

   ____

   ____

DATE COMPLETED______________INSTRUCTOR'S SIGNATURE ______________

TEST V-A

Name______________________________Date ________________________

Directions: Select the best answer to each question and circle the correct letter at the left.

1. The collapsible steering was designed to:
   I. Reduce cost of collision repairs
   II. Reduce injury to driver during collisions
   A. I only
   B. II only
   C. Both I and II
   D. Neither I nor II
2. Collapsible steering columns can best be inspected for damage by:
   I. Visual inspection
   II. Measurement of specific distance on the column
   A. I only
   B. II only
   C. Both I and II
   D. Neither I nor II
3. Acceptable method(s) of visual diagnosis of movable steering columns include:
   I. Too much space between steering wheel and jacket
   II. Binding steering wheel at upper-travel limits
   A. I only
   B. II only
   C. Both I and II
   D. Neither I nor II
4. Caution(s) to be observed when servicing collapsible steering columns include:
   I. Not applying side pressure to column
   II. Using only minimum effort to stretch column
   A. I only
   B. II only
   C. Both I and II
   D. Neither I nor II
5. Manipulative testing(s) of movable columns include:
   I. Applying heavy hand pressure on wheel
   II. Pulling upward with heavy pressure while trying to turn the steering wheel
   A. I only
   B. II only
   C. Both I and II
   D. Neither I nor II
6. Removing a cable from the battery before removing a steering column will prevent:
   I. The horn from blowing
   II. Accidental fires
   A. I only
   B. II only
   C. Both I and II
   D. Neither I nor II

7. Whenever the steering wheel is to be removed:
   I. Raise the front wheels off the ground
   II. Turn steering wheel until it is in the straight-ahead position
      A. I only
      B. II only
      C. Both I and II
      D. Neither I nor II
8. If a steering wheel puller is not readily available:
   I. Use a hammer to strike the shaft, thereby loosening the wheel
   II. Do not proceed until puller can be obtained
      A. I only
      B. II only
      C. Both I and II
      D. Neither I nor II
9. Fuses are often used to control excessive current in:
   I. Ignition circuits
   II. Hazard warning and turn-signal circuits
      A. I only
      B. II only
      C. Both I and II
      D. Neither I nor II
10. When removing wiring harness from modern steering columns in the car:
   I. It should be cut at the upper end and pulled through at the bottom
   II. Tie a piece of wire around the lower connector and pull upward
      A. I only
      B. II only
      C. Both I and II
      D. Neither I nor II
11. The ignition switch has:
   I. A locking function only
   II. An electrical function only
      A. I only
      B. II only
      C. Both I and II
      D. Neither I nor II
12. The ignition switch on steering columns, after 1968, controls:
   I. Turn signal-switch power source
   II. Movement of telescoping column
      A. I only
      B. II only
      C. Both I and II
      D. Neither I nor II
13. When testing for possible ignition switch malfunctions, among the first checks should be for:
   I. Blown fuses
   II. Loose column connector
      A. I only
      B. II only
      C. Both I and II
      D. Neither I nor II
14. When using the ohmmeter to test ignition switches:
   I. The battery should be disconnected
   II. The meter will read 0 when the circuit is open
      A. I only
      B. II only
      C. Both I and II
      D. Neither I nor II

15. Turn signal switches are usually:
    I. Unrepairable
    II. Located beneath the steering wheel
        A. I only
        B. II only
        C. Both I and II
        D. Neither I nor II
16. Hazard warning lights should flash:
    I. At all four corners of the car
    II. Only with the ignition switch in the "On" position
        A. I only
        B. II only
        C. Both I and II
        D. Neither I nor II
17. Horn buttons are fixed to the steering wheel by:
    I. Friction
    II. Screws
        A. I only
        B. II only
        C. Both I and II
        D. Neither I nor II
18. The horn switch on the column is:
    I. Connected directly to the horn
    II. A grounding switch
        A. I only
        B. II only
        C. Both I and II
        D. Neither I nor II
19. To test the function of the horn alone:
    I. Connect a jumper wire from the positive battery post to horn terminal
    II. Connect a jumper from the negative post to the horn button
        A. I only
        B. II only
        C. Both I and II
        D. Neither I nor II
20. To test the function of the horn button and relay:
    I. Connect a test light to the wire removed from the horn and push horn button
    II. Connect a jumper to the horn terminal and to the negative battery post
        A. I only
        B. II only
        C. Both I and II
        D. Neither I nor II

DATE COMPLETED ______________________ INSTRUCTOR'S SIGNATURE ______________________

GLOSSARY OF SUSPENSION AND STEERING TERMS

Ackerman principle: The design of steering parts that allows for a different turning radius between the inside and outside front wheel in a turn. Turning radius or toe-out on turns.
AIC: Automotive Information Council.
Alignment: An adjustment to align; to bring into proper adjustment the front wheels of an automobile.
AMA: Automobile Manufacturers Association.
Antisway bar: *See* **Stabilizer bar.**
Apex: The point of a cone.
API: American Petroleum Insititute.
ASIA: Automotive Service Industry Association.
ASME: American Society of Mechanical Engineers.
Atmospheric pressure: Pressure of the atmosphere or air due to its weight. Average value is 14.7 pounds per square inch at sea level and reduces to 12.2 pounds per square inch at 10,000 feet above sea level.
Axial: Having the same direction or being parallel to the axis of rotation.
Axle: A crossbar supporting a vehicle on which wheels turn. Can be of I-beam design.
Back lash: The clearance or "play" between two parts, such as meshed gears.
Balance: A round object such as a tire that has its weight equally distributed around its circumference.
Ball bearing: An antifriction bearing consisting of a hardened inner cone and an outer cup, with hardened steel balls interposed between the cup and the cone.
Ball joint: A hardened steel stud threaded on one end, tapered in the middle, and formed into a ball on the other end. They are commonly used as suspension and steering-linkage connectors on independent front suspensions.
Ball joint suspension: Suspension system on late model automobiles that has a metal ball stud housed in a bearing socket. It replaces the upper and lower outer-control-arm bushings and the kingpins.
Ball-and-nut steering: Steering gear, either power or manual, that uses a worm with recirculating ball bearings between the worm and the nut. The nut drives the pitman or sector shaft.
Base circle: The part of the cam with the smallest diameter, such as the smallest circle of an eccentric cam.
Bell crank: A moving arm that pivots near the middle. It is used to change the direction of motion.
Broach: To finish the surface of metal by pushing or pulling a multiple edge cutting tool over or through it.
Bypass valve: A valve that provides an alternate path for a flowing substance like power-steering fluid.
Camber: The inward or outward tilt of the front wheels at the top.
Cam washer: A washer that has the hole for a bolt drilled off center. Used on eccentric-type alignment controls.
Case harden: To harden the surface of steel.
Casing: The tire casing, made of nylon, rayon, fiberglass, or steel cords to which rubber is vulcanized.
Castellete: Formed to resemble a castle battlement, as in a castellated nut. Slots for cotter pin or tie-wire placement.
Caster: The forward or backward tilt of the steering knuckle at the top.
Center of gravity: An imaginary point about which the vehicle's weight is balanced.
Centrifugal: A force in a direction away from the turning center of an object moving in a curved path.
Chassis: The vehicle frame, suspension, and running gear. Everything except the car body.
Clockwise rotation: Rotation of any part in the same direction as the movement of the hands on a clock.
Coil spring: A spring-steel rod that is formed into the shape of a coil.
Compression: *See* **Jounce.**
Concentric: Two or more circles having a common center such as a washer that has the hole drilled in dead center.
Control arms: The horizontal arms of the front suspension by which the front wheels or rear axles are attached to the frame; used to support the front end or stabilize the differential housing of the vehicle.
Cornering force: The side-slippage force generated between the tire and pavement when turning.
Cornering wear: The abrasive tire wear that results from cornering too fast.

Counterclockwise rotation: Rotation of any part in a direction opposite to the movement of the hands on a clock.
Crocus cloth: An extremely fine abrasive cloth that is used for polishing.
cid; Cubic inch displacement.
cu. in.: Cubic inch. A cube that has six equal square sides, each one being 1 inch square.
Curb weight: The net weight of a vehicle including a full supply of fuel, oil, and water, but excluding all passengers.
Degree: Abbreviated *deg.* or indicated by a small ° placed alongside a figure. May be used to designate temperature or sections of a circle, one degree being 1/360 of a circle.
Diagnosis: Refers to the use of any instrument and/or procedure that will allow the technician to determine the problem and the cause of any malfunction.
Dial indicator: A gauge that uses a dial and needle to register measurements in thousandths of an inch.
Direct-acting shock absorber: A shock absorber that shortens or lengthens in action; also called a *telescoping* shock absorber.
Distortion: A warpage or change in form from the original shape.
Dynamic balance: To balance a tire so that it will rotate smoothly on its axis without any bouncing or shimmy.
Eccentric: Two or more circles, one surrounding the other and each having a different center. An eccentric bushing would have a different center for the outside than it would for the hole through the bushing.
Extreme pressure lubricants: EP is used to designate a lubricant to which an ingredient has been added to increase the lubricant's ability to withstand high pressure, as between gear teeth.
Fatigue: A breakdown of material through a large amount of flexing or bending. The first signs are cracks followed shortly by breakage.
Foot-pound: (Or ft-Lb.) A measure of the amount of energy or work required to lift or pull 1 pound 1 foot.
Forged: A part made by forging.
Forging: A process of hammering or compressing a red-hot metal part to form it into a desired shape, such as a forged steering knuckle.
Frame: The assembly of metal structural parts and channel sections that support the engine and the body and is supported by the wheels and suspension system.
Frame gauges: Gauges that hang from the frame to check for frame alignment.
Free play: Looseness in a linkage between the start of application and the actual movement of the device, such as the movement in the steering wheel before the wheels start to turn.
Frequency: The number of oscillations, vibrations, cycles, or changes in direction in a unit of time.
Friction: The resistance to slipping or skidding.
Front suspension: A series of linkages by which the front wheels are attached to the frame; used to support the front end of the vehicle and keep the front wheels in proper alignment.
Fulcrum: The hinge point of a lever.
Gear ratio: The number of revolutions made by the driven gear as compared to the number of revolutions made by a drive gear of different size. For example, if the drive gear makes three revolutions while the driven gear makes one revolution, the gear ratio would be 3 to 1.
Handling: An abstract term to describe the steering control quality of a vehicle.
Hot shrinkage: A method of straightening metal panels or tubular sections by heating to a cherry red and then cooling rapidly.
Hydraulics: A study of fluids.
id: Inside diameter.
Idler Arm: A lever or arm used to support the passenger or left side of the steering linkage.
in.: inch.
Integral: The whole made up of parts. An integral power steering unit is one in which the gearbox and the power assist are in one unit.
Jounce: The compression of a suspension spring.
Keyed: A part that is prevented from rotating within or around another part by a small metal device called a *key*.
Kingpin: A steel pin that attaches the steering knuckle to the vertical support or axle, around which the knuckle can pivot.
Kingpin inclination: (Steering-axis inclination) The amount in degrees the upper end of the kingpin or knuckle-support pin is tilted inward toward the center of the car.

Knuckle: (Steering knuckle) A movable joint of which the front-wheel spindle is a part, which can be turned to left or right to steer the car on its desired course. (Most often found between the upper and lower ball joint)
Knuckle pin: *See* **Kingpin.**
Knuckle support: The part that connects the outer ends of the upper and lower suspension support arms, and to which the steering knuckle is attached.
Knurl: A roughened surface caused by a sharp wheel that displaces metal outward as its sharp edges push into the metal surface.
lb.: Abbreviation for *pound,* a unit of weight.
Lever: (Arm) A rigid bar or beam capable of turning about one point called a fulcrum.
Leverage: The mechanical advantage gained by the use of a lever.
Linkage: A system of links and levers connected together to transmit motion or force.
Liter: A metric measure equal to 2.11 pints.
Micrometer: An instrument used for measuring accuracy to one-thousandth of an inch.
Millimeter: The metric equivalent of .039370 of an inch. One inch is the equivalent of 25.4 mm.
NSPA: National Standard Parts Association.
od: Outside diameter.
Ohmmeter: A battery powered device used to measure electrical resistance or circuit continuity.
Oscillation: A vibration back and forth or up and down.
Oversteer: The tendency of the rear of a vehicle to lose traction and swing around into a spin while making a turn. Commonly called fishtailing.
Parallel: Straight lines that are the same distance apart from end to end.
Pitman arm: An arm that is connected between the steering gear and the center link of the steering linkage.
Pitman shaft: A shaft that transmits turning motion from the worm gear to the pitman arm. The sector half of the worm-and-sector gear is on the pitman shaft.
Planetary: A type of gear train having pinion gears rotating around a central gear.
Plies: the layers of fabric in a tire casing. Each layer is considered a ply.
Preload: The amount of load imposed on a bearing before actual operating loads are imposed. It is accomplished with an adjusting nut, such as the spindle nut on front wheel bearings.
Preventative maintenance: Service procedures that are done before a malfunction occurs to prevent failure.
psi: Pounds per square inch.
Pull: (Lead) The tendency of a vehicle to pull or lead to the right or left of the highway.
Quench: Cool to the temperature at which the flame will go out.
Race: The part of a bearing on which the rollers or balls roll.
Radial: Moving straight out from the center.
Radius: The distance from the center of rotation to the arc or circumference of the circle made by the rotation.
Rear axle: The drive between the differential side gear and the wheel.
Rebound: An expansion of a suspension spring after it has been compressed as the result of jounce.
Retainer: A device to hold parts together.
Ride: The characteristic feel as one rides in a vehicle.
Road shock: (Wheel fight or steering kick) A shock, movement, or kick transmitted to the steering wheel through the steering linkage, caused by irregularities in the road surface or a faulty steering system.
Roll: Movement around the center of gravity.
Roll steer: The steering effect as a result of body lean during a turn.
Run out (lateral): A wheel that does not run true, but moves in and out on the side wall as it rotates.
Run out (radial): A wheel that is out of round.
SAE: Society of Automotive Engineers.
Scrub radius: The distance between the intersection of an imaginary center line of the steering-axis inclination and the center line of the tire on the road.
Scuffing: Sliding a tire on the road surface.
Sector: One section of a gear. Specifically, the gear sector on the pitman shaft in many steering gears.
Servo action: A large action that results from a small input action.
Shackle: Swinging support by which one end of a leaf spring is attached to car frame.
Shim: A spacer to adjust and maintain the distance between two parts.
Shimmy: (Wobble) A rapid series of oscillations of the steering knuckle and wheel and tire assembly about the steering knuckle pin.

Shock absorber: A hydraulic device to dampen or stabilize the up-and-down movement of the car frame by controlling the compression and rebound of the springs.
Side sway: a force causing a lean one way then the other.
Skid: A tire sliding on the road surface.
Slip angle: the angle between the tire heading and the actual direction of movement.
Spindle: A shaft or pin around which another part rotates.
Splined: Sized grooves on a shaft and in a hole that match to prevent torque slippage.
Spring: An elastic device that yields under stress or pressure, but returns to its original state or position when the stress or pressure is removed.
Spring Rate: The unit of measurement of the amount of spring deflection (compression).
Sprung weight: That part of the car which is supported on springs (the frame and body, for example).
Stabilizer bar: A torsional bar and linkage used to eliminate sway and decrease the side-rolling tendency of the car frame and body, often called an antisway bar.
Static balance: Is balance at rest, or still balance. It is the equal distribution of weight of the wheel and tire around the axis of rotation, in such a manner that the wheel assembly has no tendency to rotate by itself regardless of its position.
Steering gear: A device consisting of a worm-and-gear sector or cross shaft and roller that transmits the driver's effort with increased leverage to the steering linkage to guide the car on its course.
Steering linkage: That system of links, rods or tubes, and arms or levers connected together to transmit the force or motion of the steering gear to the front wheels.
Steering system: The complete steering mechanism consisting of the steering wheel, steering gear, and steering linkage that is provided as a means for the car driver to turn the front wheels to guide the car on its course.
Strut rod: A metal piece used to control movement of front end lower control arms.
Support arms: *See* **Control Arms.**
Tandem: One directly in front of the other and working together.
Thrust washers: Washers with bearing surfaces that are designed to retain end thrust.
Tie rods: In the steering system, the rods that link the center link to the steering arms.
Tire: The casing and tube assembled on a car wheel to provide pneumatically cushioned contact and traction with road.
Toe-in: the turning in of the front wheels. Wheels are closer together at the front than at the back of the wheels.
Torque: Turning or twisting effort, usually measured in foot-pounds or inch-pounds.
Torque wrench: A special wrench that indicates the amount of torque being applied to a nut or bolt.
Torsion-bar spring: A long, straight bar, fastened to the frame at one end and to a control arm at the other. Spring action is produced by the twisting of the bar.
Tracking: the following of the rear wheels directly behind, or in the tracks of, the front wheels.
Tramp: the action of a wheel to jump or hop up and down.
Troubleshooting: The detective work necessary to run down the cause of a trouble. Also implies the correction of the trouble by elimination of cause.
Turning radius: Difference in angles between the two front wheels and the car frame during turns. Inner wheel in a turn turns out or toes out more.
Understeer: The tendency of a moving vehicle to maintain a straight ahead line while the front wheels are being turned in either direction. Commonly known as *pushing* or *plowing*.
Unsprung weight: the weight of that part of a vehicle that is not supported by the springs.
Vertical wheel support: The vertical shaft or beam connecting the upper and lower support arms to which the steering knuckle is attached.
Wander: (Weave) To ramble or move about without control from a fixed course.
Wheel alignment: The mechanics of properly adjusting all the factors of the front and rear wheels so as to cause the car to steer a true course with the least effort and to reduce tire wear to a minimum.
Wheel spindle: The shaft extending from the front steering knuckle, and about which the front wheel rotates.
Wheel tramp: Tendency for the wheel to move up and down so it repeatedly bears hard, or "tramps," on the pavement. Sometimes called *high-speed shimmy*.
Wire, Mechanics: Noninsulated soft steel wire used as locking wire or as a wire pull.
Worm: Type of gear on the lower end of the steering shaft.

Index

Page numbers in **bold type** refer to illustrations.